*Statistics
with
Applications
to
the Biological
and
Health Sciences*

RICHARD D. REMINGTON

Dean, School of Public Health
The University of Michigan

M. ANTHONY SCHORK

Department of Biostatistics
The University of Michigan

PRENTICE-HALL, INC., ENGLEWOOD CLIFFS, NEW JERSEY

Statistics
with
Applications
to
the Biological
and
Health Sciences

PRENTICE-HALL INTERNATIONAL, INC. *London*
PRENTICE-HALL OF AUSTRALIA, PTY. LTD. *Sydney*
PRENTICE-HALL OF CANADA, LTD. *Toronto*
PRENTICE-HALL OF INDIA PRIVATE LTD. *New Delhi*
PRENTICE-HALL OF JAPAN, INC. *Tokyo*

STATISTICS WITH APPLICATIONS TO
THE BIOLOGICAL AND HEALTH SCIENCES

RICHARD D. REMINGTON and M. ANTHONY SCHORK

Library of Congress Catalog Card Number 71–100588

Printed in the United States of America
13-846188-0

Current Printing (last digit):
17   16

To BETTY and HELY

Preface

An important question in the preparation of a text in biostatistics is the relation of such a text to existing teaching and learning materials in this and allied fields. Other books on biostatistics have tended to take one of two courses: reliance on verbal explanations of concepts with little in the way of general mathematical formulation or rather spare verbal exposition with concentration on formulas and numerical illustrations.

The approach we take is quite different. We hold strongly the view that the fundamental concepts of statistics are reasonable and can be made to appear reasonable if sufficient space is allowed for appeals to intuition and common sense. Thus, for example, we devote considerable attention to the establishment and reinforcement of the concept of sampling distributions. In our experience, this notion is only infrequently locked into the mind of the beginning statistics student. It is so fundamental that no real understanding of the processes of statistical inference can occur without it.

In teaching statistics to students from the many segments of the health sciences we also find that examples chosen from outside biology are of little help either in establishing the ideas of statistics or in building interest and enthusiasm for acquiring such ideas. We believe this to be a fundamental rather than an incidental difficulty in using general teaching materials with such students. The notions of statistics are brand new; the task of transmitting these notions becomes unnecessarily, perhaps impossibly, greater. The examples we use either come from real data or are suggested by real and, we hope, reasonable scientific situations in biology.

We do not avoid mathematical formulas and expressions in this book. However, we attempt to make each such formula seem natural and reasonable by developing it upon a foundation of intuition and common sense.

The book should be useful to students of medicine, public health, dentistry, veterinary medicine, nursing, pharmacy, and the other health sciences. Students from the basic areas of biology should also find it useful. Research workers and teachers in the biological and health sciences who wish to develop a basic understanding of statistical methods useful in their work may also find the material appropriate to their needs.

The mathematical preparation needed for an understanding of the book consists of a single course in algebra, perhaps only dimly recalled. We have attempted to work at a fairly slow but steady mathematical pace at least in the early chapters to give the students an opportunity to reestablish some basic skills. Some material from elementary mathematics is developed within the text itself, particularly when we are unable to assume that such material would have been a part of an elementary algebra course.

We are indebted to many persons who have assisted in preparing this book. A number of typists have given careful attention to the manuscript at various stages of its preparation. These include Mrs. P. M. A. Blair, Miss E. K. Atherton, Mrs. Cheryl Symons, Mrs. Rebecka Andress, and Miss Marjorie Forster. Mrs. Rosalie Karunas provided much editorial and proofreading assistance. To all those individuals and to our wives and families for their patience and encouragement during the tedious hours of preparation of the work go our heartfelt thanks.

RICHARD D. REMINGTON
M. ANTHONY SCHORK

Houston, Texas
Ann Arbor, Michigan

Contents

4 PROBABILITY, 59

5 POPULATIONS, SAMPLES, AND INFERENCE, 87

6 SOME IMPORTANT DISTRIBUTIONS, 112

7 ESTIMATION, 148

8 HYPOTHESIS TESTING, 192

9 CHI-SQUARE TESTS FOR FREQUENCY DATA, 229

10 REGRESSION AND CORRELATION, 253

*Statistics
with
Applications
to
the Biological
and
Health Sciences*

1

General Introduction to Biostatistics

Biostatistics is the scientific discipline concerned with application of statistical methods to problems in biology. It is an applied discipline rather than a basic or fundamental study. Its roots lie in mathematics, but its branches touch all areas of biology, extending, for example, from basic cellular physiology to community surveys designed to study the frequency and the cause of home accidents. Although biology has long been an important area for application of statistical methods, only in recent years has a specific field known as biostatistics emerged. With the accumulation of knowledge in both the biological and mathematical sciences, it is inevitable that statistical practice in biological fields demands training and interest in both areas. Thus, the field of biostatistics has evolved as a medium of statistical expression to parallel this increase in knowledge.

1-1 The Bases of Biostatistics

Mathematics is the fundamental study on which biostatistics rests, and mathematical skills and attitudes are an important part of the training of a biostatistician. The development of new statistical methods involves mathematical proof and deductive reasoning. Probability theory is the specific portion of mathematics most directly applicable to statistics; however, since probability theory rests on the basic mathematical areas of analysis, algebra and geometry, statistics also depends upon these areas.

Pure mathematics is not directly concerned with the description of physical or biological phenomena, although much of the important work in mathematics has been stimulated by observations of such phenomena. Mathematics is a logical system involving the process of deduction from prescribed sets of axioms through the use of specific rules of inference and combination. The biostatistician, on the other hand, while relying heavily on mathematical knowledge, must always be concerned with the gap between this knowledge and the physical and natural universe.

In applying mathematical results to natural phenomena we necessarily approximate, simplify, and idealize. However, if we regard the growth of scientific knowledge as a continuing process with no definite beginning or end, then the study of such approximations and simplifications can be seen in its proper perspective and understood to be fundamental to the process through which scientific knowledge is accumulated.

1-2 Definition of Statistics

Although no single definition of statistics is satisfactory for all purposes, the following statement will be useful:

> Statistics is the study of methods and procedures for collecting, classifying, summarizing, and analyzing data and for making scientific inferences from such data.

The breadth and scope of this definition are obvious. The statement covers much of the activity of the scientist. However, note that the direct concern of statistics is with the data and scientific observations themselves, rather than with the biological material involved in the study. It is not possible and probably not desirable to make rigid lines of demarcation between biology, biostatistics, statistics, and mathematics. Although the organization of man's knowledge demands classification, the attempt at rigid adherence to and observation of boundaries between these fields must inevitably lead to frustration.

Statistics, according to the definition given, breaks naturally into 2 reasonably distinct subcategories—descriptive statistics and inferential statistics.

1-2-1 DESCRIPTIVE STATISTICS

The area of descriptive statistics involves, through the use of graphic, tabular, or numerical devices, the abstraction of various properties of sets of observations. Such properties include the frequency with which various values occur, the notion of a typical or usual value, the amount of variability in a set of observations, and the measurement of relationships between 2 or more variables.

Notice that the field of descriptive statistics is not concerned with the implications or conclusions that can be drawn from sets of data. Descriptive statistics serve as devices for organizing data and bringing into focus their essential characteristics for the purpose of reaching conclusions at a later stage. Naturally, however, the conclusions reached will depend strongly upon the selection of descriptive and summarizing techniques. Indeed, the failure to choose appropriate descriptive statistics has often been responsible for faulty scientific inferences. An example of this is the process of selecting certain portions of sets of data and eliminating other parts that do not conform to the investigator's preconceived opinions about the true state of nature.

1-2-2 INFERENTIAL STATISTICS

A basic characteristic of experimental science is the necessity for reaching conclusions on the basis of *incomplete information*. For example, Mendel, in studying the way pea plants differed from one another in height, color of seeds, color of pods, and color of flowers necessarily had to make his conclusions on the basis of a relatively small group of plants compared with the entire population of pea plants of a particular type. In making a statement about, say, color of flowers, Mendel's conclusions were dependent upon the particular sample of plants available for study. He certainly did not expect color variations found in the group of plants he studied to be identical with those in the universe of all possible pea plants of that type. However, he was able to reach general, genetically accurate conclusions about the inheritance of color in the universe of pea plants.

Alternatively, suppose a geneticist wonders whether heritability of color trait in a newly discovered flower is the same as that noted in Mendel's pea plants. He knows the familiar $1:2:1$ pattern developed by Mendel and wishes to see if the newly observed plants are the same. Again, only a limited number of plants of the new type will be available for study, and the geneticist is thus subject to the possibility of making an error whether he concludes that the new type is the same as or different from the old.

Although this example is chosen from the field of botany, it illustrates the situation confronting experimental biologists in any field of interest. In testing the safety and efficacy of a new hypotensive drug, for example, the internist will have only a limited number of hypertensive patients with whom to work. It is unlikely that a second internist will be interested in the particular group of patients studied in the experiment but rather in those patients who present themselves to his clinic. Thus, the second man is interested in the extent to which generalizations to his patients can be made from the published experimental results in a different group. Of course, this type of generalization is necessarily inductive rather than deductive, in that we attempt to reach conclusions concerning a large group on the basis of studying only a small subset.

In statistical terminology this inductive procedure involves making inferences about an appropriate *population* or universe in light of a single subset or *sample*. The *population* is the full set of individuals to whom we limit any discussion or inferences, while the *sample* is a subset of that population. We will refine and expand these concepts later in the book.

Statistical inference is concerned with the procedures whereby such generalizations or inductions can be made. This concern is most often limited to the quantitative aspects of the generalization, but more and more often in practice the biostatistician is asked to contribute to the process of reaching substantive conclusions as well. In many respects the biostatistician in this role functions as an applied philosopher of science. It should be pointed out clearly, however, that the statistician is not necessarily in as good a position to make substantive inferences as is the experimenter. In fact, often the reverse is true. However, study of the quantitative aspects of the inferential process provides a solid basis on which the more general substantive process of inference can be founded. Furthermore, the biostatistician will frequently be in a position to make contributions to a substantive inference by virtue of experience with, and abstraction from, other similar problems.

It should be obvious from this discussion that the process of scientific inference involves the highest degree of cooperation and association between the experimental investigator and the biostatistician. This degree of cooperation and understanding can probably be realized at its highest level only if the experimenter knows a reasonable amount of statistics and if the biostatistician knows a reasonable amount from the area of investigation. In this time of explosive growth of applied research, we hear much about interdisciplinary or cooperative investigation. The work of the biostatistician requires an interdisciplinary orientation, since only a high level of interplay between statistician and investigator can successfully avoid all the misinterpretations and misapplications with which scientific literature abounds.

1-3   Study Material versus Study Observations

The investigator studies mice or human beings or blood or pea plants or insects, whereas the statistician's concern is with *numbers*. It is important to emphasize here the distinction between the material of the investigation and the numerical abstraction of that material with which the statistician must work. The process of assigning numbers to biological or physical observations is an important, indeed, crucial step in scientific research. Both descriptive and inferential statistical methods deal primarily with numerical data, whereas the investigator's real interest is with the biological material of the study.

In studying blood pressure, for example, we are concerned with the amount of work done by the heart and by the muscular coats of arteries all over the body. In order to get at this, we place an inflatable rubber pouch wrapped in cloth around the upper arm of a subject, and inflate the pouch, thereby squashing the skin, blood, extracellular fluid, and muscle under the pouch or cuff. We place a stethoscope over the brachial artery at a point where it comes relatively near the surface at the antecubital fossa (hollow in front of the elbow joint) and listen while simultaneously reducing the pressure in the rubber pouch. Soon we begin to hear rhythmic sounds coordinated with the cardiac systole (beating of the heart). The sounds go through a number of stages of varying intensity and timbre and eventually disappear. We record the pressure in the cuff at certain key points in this series of sounds—frequently the point at which sounds first appear and the point at which sounds disappear.

We then may act as if we know a great deal about the status of the subject's peripheral circulation when, in fact, our knowledge depends on the extent to which the true pressure inside the brachial artery is a good indicator of the pressure in other arteries, the extent to which we can estimate the pressure inside the brachial artery by listening to sounds produced over the artery, the extent to which auditory acuity is sufficient to permit accurate definition of the critical sounds heard in the stethoscope, and the extent to which the subject's blood pressure at the moment is indicative of his typical or usual blood pressure. Yet, after all this, what does the investigator record and the statistician manipulate? Two numbers, e.g., 120 mm Hg systolic blood pressure, and 80 mm Hg diastolic pressure.

Now surely in working with such figures we are guilty of oversimplification at many stages. If our interest is in the subject's actual blood pressure, even in the brachial artery, then we have passed through a many-linked chain in going from that actual pressure to the numbers we write down for statistical analysis.

The example of blood pressure is probably a reasonable one, since many of the observations that we must work with in biology are less accurately determined than blood pressure, and, on the other hand, many observations and measurements are more accurately determined. For example, consider the problem of assessing a human subject's personality characteristics. The scaling and measurement problems for this variable must certainly be much more difficult than they are for a blood pressure determination. At the other end of the scale, consider the determination of a person's ABO blood group, an attribute of a subject which is stable throughout his life and which is not subject to a great deal of observer interpretation and measurement error, at least compared to blood pressure. The categorization of a person's personality type or blood group, of course, is scaled in numerical form in a different way from blood pressure—blood pressure being recordable on an ordered numerical scale.

The process of assigning numerical scales to physical, biological, or other natural phenomena has been extensively studied by psychologists. We can appreciate the fact that these problems become acute in the social sciences, and that it is only natural that social scientists should have contributed much of our knowledge in this area. Festinger and Katz (14, Chap. 11) present an elementary discussion of scaling and provide further insight and references into the nature of the problem.

The foregoing discussion should not lead us to despair completely of being able to obtain useful information by studying numerical observations. On the contrary, if we fail to record our observations in numerical form, we frequently have a strong feeling that our knowledge of the phenomena we study is incomplete and faulty. With all its limitations, numerical scaling is still the best way we know of getting at the natural phenomena in which we are interested. The purpose of this discussion is to make the reader conscious of the limitations of this approach and of the fact that we are indeed oversimplifying when we work with numerical data, since our basic interest is in the study material itself.

Throughout this book we shall tend to take measurements for granted. The statistical techniques discussed will assume that we are given a set of observations in which we have sufficient confidence to justify statistical treatment. This assumption in many cases is unwarranted. Careful investigation and research dictate, for every set of data, thorough consideration of whether the numerical observations actually provide sufficient information concerning the natural population to justify further manipulation or statistical treatment. In this area, however, there are no fixed rules. The individual who expects to be told, "Here is a formula whereby you can decide whether these data are amenable to statistical treatment," will be frustrated and disappointed. No such formula exists. There is no substitute in this area for experience and intellectual maturity. One cannot set on a printed page rules for attaining experience and intellectual maturity. Only time, study, thought, and hard work can provide these. It is even incorrect to assume that 2 experienced and mature statisticians or investigators would come to the same conclusions concerning adequacy or suitability of a given set of data for further statistical treatment. Statistics is both an art and a science, and at no point does its artistic content become more evident than here. The question, "Does this set of numbers provide important information concerning the natural phenomena of interest?" has no objective, black-and-white answer. The most we can hope for is the realization on the part of the investigator that the act of assigning a number involves an abstraction and a departure, however small, from the intent of his investigation.

1-4 Some Popular Concepts and Misuses of Statistics

Many people think of statistics as beginning and ending with lengthy tabulations of figures and numerical data. Although numerical data form the grist for the statistician's mill, consideration of problems of statistical inference immediately reveals that this is but a small part of the field of statistics.

Nearly everyone has heard the statement originally made by Disraeli, "There are three kinds of lies: lies, damned lies, and statistics," or the statement "Figures don't lie, but liars figure," or "Statistics can prove anything." Certainly the statistical method is a 2-edged sword and can be incorrectly used. The fields of marketing and advertising provide many examples of misuses of all sorts. While the use of pseudo-statistics and pseudo-experiments in modern advertising for the purpose of motivating consumers is well understood by many educated persons, there seems to be a less general awareness that similar methods can be unwittingly used by a well-meaning investigator, and that equally erroneous conclusions may often be reached.

In this connection the reader is urged to study Huff (27). Many of the standard techniques for abusing data are presented in this reference. While the emphasis of pedagogically acceptable presentation should be on correct methods of data handling and inference, some appreciation for incorrect methods is essential to an understanding of the statistical method.

The misapplication of statistical method in scientific research is rarely due to a conscious or subconscious effort to mislead; rather, it is often the result of insufficient deliberation and study of the particular experimental problem. Unfortunately, the hustle and bustle of modern America have penetrated the research laboratory, where the pressure to publish or to obtain more and larger research grants has frequently denuded the experimental method of what is perhaps its most essential ingredient—careful deliberation and consideration of the meaning, significance, and logical structure of the research. Certainly the statistician holds no special qualifications in the area of careful consideration and reflection into experimental inference, but all too frequently today the investigator abdicates his responsibility in these areas to the statistician to the point that modern training in biostatistics must include consideration of these aspects of scientific research.

It is probably safe to say that a majority of the statistician's problems and difficulties in handling experimental data are caused in one way or another by haste and consequent superficiality if not downright errors on the part of the investigator.

The worker who hurries into an experiment, hurries the data collection, and rushes the publication of his results runs the risk of wasting his entire research effort while saving a few moments. Careful work requires careful and time-consuming thought. The gray-flannel-suit approach to research may eventually rob science of the dignity and respect that it has historically both demanded and earned.

2 Graphs and Tables: General Principles

An important part of many scientific publications is the presentation by graphs or tables of various research results. Frequently a reader who has insufficient time to read the entire paper reads the introduction and conclusion and looks over the tables and graphs. For this reason the tables and graphs may reach a wider audience than the textual portions of the report. On such a premise, the importance of well-constructed, comprehensive graphic and tabular presentations becomes obvious.

There are a number of principles concerning the construction of graphs and tables which hold quite generally and which should be presented at the outset.

2-1 General Principles Concerning the Construction of Graphs

1. Graphs should be fully self-explanatory. Under the premise that many readers will not read the detailed text of a report, the contents of a graph should be as complete as possible. It is disconcerting to a busy reader to be forced to thumb through a paper looking for some essential item needed to clarify the contents of a graph. If the title fails to clarify the frame of reference, then the reader may be at a loss to interpret the contents of the graph itself. The title should include information concerning who or what the subjects or experimental material are, what observations are abstracted from those subjects or material, and what restrictions of time and place apply to the graph. For example, a presentation of birth rates in the state of Michigan should never be headed merely "Birth Rates," but might well be modified to say "Birth Rates per 1,000 Population, White Race, Michigan, 1920–1960." If the length of title becomes a problem, then

additional essential material can frequently be included in a footnote. Whatever the mechanics are, the essential principle is that the title of a graph should provide as complete information about its contents as is possible. In fact, the graph should be as self-contained as possible, requiring as little outside information for clear interpretation as is feasible.

2. Vertical and horizontal scales should be clearly labeled and units should be identified. Most graphs present numerical information in scaled form. The scales must be labeled in order to describe fully the variable presented on the scale, and for measurement variables the units of measurement should be identified. For example, a graph showing change of weight of young guinea pigs with time might be labeled: "weight (grams)" and "age (days)."

3. Do not try to include too much information in a single graph. Many graphic presentations fail because of an attempt to present too much information on a single graph. It is better to include several graphs than to compress information too much. A device frequently used for the presentation of many curves or trend lines is the presentation of a series of small graphs, each presenting one such line. A safe rule of thumb is to avoid graphs containing more than 3 curves. Of course, this does not imply that all graphs containing 2 curves are good and that all graphs containing 4 curves are bad.

4. Graphs are intended to give an overview rather than a highly detailed picture of a set of data. Do not include too much detail in a graph. Detailed presentations should be reserved for tables. The purpose of a graph is to present a general picture of the information contained in a set of data. Graphs condense detail to permit us to see the forest rather than the trees. If your main interest is in the trees, use a table. The inclusion of too much detail in a graph will tend to obscure the essential points.

5. In general, avoid the inclusion of numbers within the body of a graph. This statement applies to graphs in which numerical scales—vertical, horizontal, or both—are used to present information. In such graphs, other numerical entries will be confusing. In bar charts we frequently see numbers at the head of each bar showing its height. It is better to include a vertical scale at the left side of the graph, since numbers at the head of each bar will tend to obscure the natural comparisons of relative magnitudes of bar heights.

2-2 *General Principles Concerning*
the Construction of Tables

1. Tables should be fully self-explanatory. Just as for graphs, enough information should be provided in the title and in column and row headings

of the table to permit the reader easily to identify its contents. The goal should be a table that can stand by itself, providing information to the reader as rapidly, concisely, and completely as possible.

Since the title will generally be the first thing read in detail, it should provide all essential information concerning the table's contents, and should specify time, place, experimental or study material, and essential relationships presented in the table. Although titles in general should be short, this shortness is entirely a secondary requisite to completeness. However, footnotes to tables may in some cases be used to provide essential information that cannot be included in the title for reasons of length.

2. Units should be stated for each numerical variable.

3. The function of ruling is to provide clarity of interpretation; unnecessary ruling should be avoided. Items that would tend to be confused by the reader should be separated by ruling, either in the form of horizontal or vertical lines. Very frequently, spacing can provide the same effect as ruling, and whenever this is the case, spacing is preferred.

In reading a table ruled lines tend to stop the eye of the reader. For this reason, ruling should be included only where it materially improves our ability to read the table; for example, to set off the title of the table, to divide major row and column headings, and to close the table at the bottom.

The use of ruling is sometimes carried to the point that every single numerical entry is contained in its own little box. Such extensive ruling is rarely necessary. Horizontal lines in particular should be kept to a minimum, whereas it is frequently helpful to separate vertical columns of figures by vertical lines. Even in this case, however, spacing can often adequately replace ruling.

4. Do not try to include too much information in a single table. The author of a table is almost inevitably more familiar with the contents of the table than the reader. For this reason, the principle of extreme simplicity in tabular presentation, with reduction of the contents to the minimum essential for organic unity, is recommended. Multiple tables, each containing a simple, concise, and understandable presentation, are much preferable to a single table that attempts to present all the information at once.

5. Numerical entries of zero should be explicitly written rather than indicated by a dash or a dotted line. These latter notations should be reserved for data that are missing or unobserved. Zero is a number, and numerical observations of zero should be explicitly presented as such. For example, if a survey shows no cases of poliomyelitis in a particular county in a particular year, then the entry should indicate this fact. If the information from that particular county was incomplete or otherwise unavailable, a dash or a dotted line should be used. Thus, the dash or dotted line represents uncertainty of an actual value, or may even represent a logical impossibility, e.g., the number of calves born to bulls.

6. A numerical entry should not begin with a decimal point. The reader runs some risk of interpreting a leading decimal point as a flyspeck or other foreign object. This misinterpretation can be avoided quite simply merely by showing a leading zero immediately to the left of the decimal point. For example, write 0.5 instead of .5.

7. Numbers indicating values of the same characteristic should be reported to the same number of decimal places. In a tabular presentation showing average weight by age, if the decision is made to indicate average weight to the nearest hundredth of a pound, then this decision should be reflected in the tabulated average weight in every age group. Not only is the appearance of a table improved by following this practice, but also tabulated entries can be used more readily in subsequent computations.

2-3 Misleading Graphs

Figure 2-1 shows the change in heart disease mortality rate for the state of Michigan from 1900 to 1960. The graph illustrates that the rates have increased.

However, if it is our purpose really to impress readers with the magnitude of this increase, a simple device will make the trend appear much more

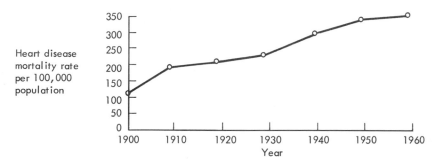

Figure 2-1   Heart disease mortality rates, Michigan, 1900–1960.
Source: **Michigan Health Statistics 1961, Michigan Department of Health (34).**

striking. If we compress the horizontal scale and stretch the vertical scale, a much more striking view of this trend is presented. Figure 2-2 shows the results.

The thing to keep in mind is that the same basic data have been used in both presentations. Yet, the impression given to the reader is vastly different in the 2 cases. When one is preparing a graph, it is important to realize that the reader will be given an overall impression as well as the facts themselves. As responsible investigators, we must be aware of the impression our presentations will convey. As consumers of research presentations, or, for that matter, of any presentation in graphic form, we should be conscious of our own impression formed from the graph and ask whether this impression is the product of the data or of the person preparing the graph.

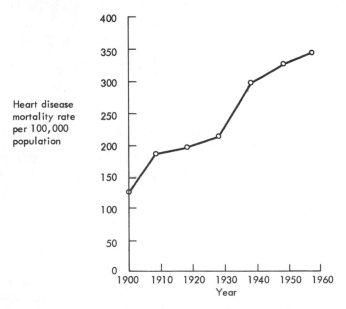

Figure 2-2   Heart disease mortality rates, Michigan, 1900–1960. *Source:* **Michigan Health Statistics 1961, Michigan Department of Health (34).**

It would be difficult to say that there was anything overtly incorrect in presenting Fig. 2-2 instead of Fig. 2-1. However, the individual preparing a graph has an obligation that goes beyond merely presenting the facts. He must make a conscious effort to understand the impression his graph will give and to avoid letting his own unsupported biases influence this impression. We have done our job if we have shown you that graphs can be manipulated, thereby planting in you some seeds of wariness.

2-3-1   THE TRUNCATED BAR CHART

Suppose 2 drugs, A and B, are studied in a clinical trial against a particular disease. Suppose that of 100 patients tested with drug A, 52 patients are cured, whereas drug B cures 56 patients out of 100 tested. Figure 2-3 shows the results in the form of a bar chart.

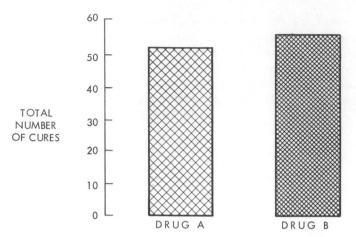

Figure 2-3   Total number of cures from each of two drugs.

Suppose, however, we wish to present the results in a slightly different form, possibly somewhat more favorable to drug B. Figure 2-4 shows one way of making such a presentation.

Here, the vertical scale has been truncated or broken so that it begins at 50 rather than at 0, and of course our eye tells us that drug B is far superior

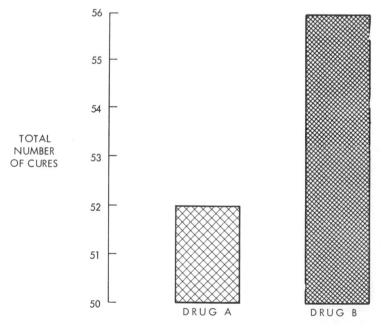

Figure 2-4   Total number of cures from each of two drugs.

to drug A. Again, both figures present the same data. The reader is left to draw his own conclusions.

Of course, the device at work here is that of breaking the vertical scale. In this situation, there is no direct way to compare the vertical lengths of the 2 bars. Perhaps the most negative aspect of Fig. 2-4 is that the reader is

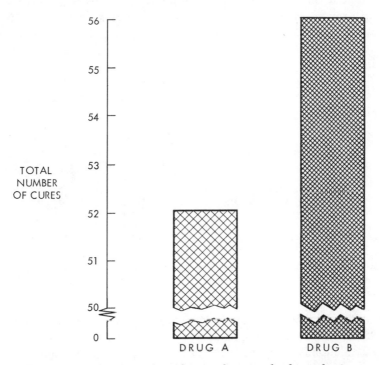

Figure 2-5   Total number of cures from each of two drugs.

given no direct indication that a scale break has been used. Such breaks should be used only in cases of necessity, e.g., when the bars required if no break were used would be excessively tall. In any case, a break should be clearly identified for the reader, probably both by breaking the vertical scale on the left-hand side of the graph and by breaking the bars themselves, as shown in Fig. 2-5.

2-3-2   THE INFLATED PICTOGRAPH

The pictograph is a means of graphic presentation frequently used by economists but also of some importance in other fields of application. A pictographic presentation compares magnitudes by comparing objects relevant to the subject matter. For example, a pictographic unit for presenting wheat production might be a bag of wheat; for representing steel production, a small Bessemer converter; or for representing birth rate, a baby.

Unfortunately, like many other graphic techniques, pictographic presentations can be misleading. Suppose we want to present the fact that the birth rate in Michigan in 1950 is twice as large as it was in 1920. This fact can be illustrated quite nicely by drawing a picture of a baby to show the birth rate in 1920 and 2 babies to show the birth rate in 1950, as shown in Fig. 2-6.

However, if we really want to impress our readers with the magnitude of this increase, we might use the device of Fig. 2-7. Here the same baby is

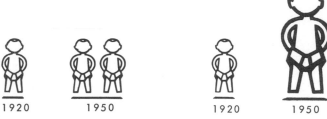

Figure 2-6   Total number of live births, Michigan, 1920 and 1950.

Figure 2-7   Total number of live births, Michigan, 1920 and 1950.

used to illustrate the birth rate in 1920, but in order to show twice the rate for 1950, a baby twice as high and twice as wide has been drawn. Naturally this presentation is more impressive, since it emphasizes the increase in birth rate more strongly. The difficulty is that when we observe a pictograph, we tend to be influenced by area or perhaps even by volume. The 1950 baby in Fig. 2-7 has not twice but 4 times the area of the 1920 baby. As such, it becomes obvious to us why the trend is emphasized more in Fig. 2-7 than in Fig. 2-6.

To summarize, pictographs should compare relative magnitudes by an increase or decrease in the number of basic objects shown rather than by an increase in the size of the basic object.

This discussion should serve to warn the reader to be cautious in interpreting graphic presentations. In everyday reading we may be confronted by graphs even more misleading than those discussed here. It is a difficult matter to attempt to separate numerical facts from impressions obtained from a graph. However, only by knowing that misleading practices exist can we hope to avoid them both as consumers and producers of graphic material. One way to avoid constructing misleading graphs is never to construct graphs, always presenting tables instead. This is too extreme a position. Good graphs can enhance a written presentation, providing variety and relief to the reader. Furthermore, there are many situations in which an overview is preferable to full detail. Just be sure to check yourself and your graphs for hidden biases. Huff (27) presents many more examples of misleading graphic presentation. While the examples are from a much broader area of application than biology, the points are well made and nearly all have implications for biological applications.

3   *Descriptive Statistics*

3-1 *Introduction*

As mentioned earlier in this text, the field of statistics may logically be divided into 2 major categories—descriptive statistics and statistical inference. In this chapter we investigate the former area. An investigator faced with a large or even a moderately large set of observations is usually interested in certain characteristics of the entire group. A clinician investigating a particular treatment for a certain disease seeks to describe the patients who have been under his treatment in recent months. A veterinarian studying a disease outbreak in a particular state is interested in characterizing herds of animals in which the disease has occurred at a high frequency. A dentist considering the possible relationship between fluoride ion concentration and dental caries is interested in describing the caries activity in children within a particular age range, living in a community in which the water supply contains an adequate concentration of naturally occurring fluoride ion. A surgeon investigating a new procedure for treatment of cerebral vascular disease is interested in describing the occurrence of operative and postoperative complications in his patients. A physical educator is interested in the percentage of middle-aged men who adhere to each of 2 different fitness-improving regimes, one involving primarily running and walking, the other involving sports activities.

The single characteristic that all the above examples have in common is that the investigator wishes to describe groups of observations with regard to some general level of a particular characteristic of interest, with regard to the degree of variability of that characteristic, or with regard to the frequency with which various values of the characteristic occur. The goal in these instances is description of the data rather than inference from the data.

However, before we consider specific descriptive techniques, we will discuss a convenient notational device.

3-2 The Summation Sign and Its Properties

In statistics the operation of addition is used so frequently that a notational shortcut is needed to indicate summation. The Greek capital sigma ($\sum$) is most often used as a summation sign. Appearing just to the left of an algebraic symbol, this sign is read "the sum of," e.g., $\sum x$, "the sum of x."

However, by itself the symbol $\sum$ is of limited usefulness and can lead to ambiguities. For example, suppose that we want to sum only part of the x's or suppose that there are several different kinds of x's. Suppose, for instance, that our x's are heights of a group of college students and we want the sum of the heights of the freshmen only.

This leads us to introduce first some sort of tagging or labeling device. Subscripts are frequently used for this purpose. Suppose we have measured the heights of n students. We might let x_1 be the height of the first student, x_2 the height of the second student, etc., letting x_n be the height of the nth student. The letter i is frequently used as a roving index to help us keep track of the observations; i.e., we let x_i be the height of student i. It is important to realize that i identifies only the individual in whom we are interested and not his height.

Now we are in a position to introduce a precise notation for the sum of the heights of the n students. One way of writing this is $x_1 + x_2 + x_3 + \cdots + x_n$. A shorter way, using our shorthand, is $\sum_{i=1}^{n} x_i$. This symbol, which is now complete, is read "the sum of x_i, where i ranges from 1 to n." The $i = 1$ written below the $\sum$ tells us that this is the first value taken by i (we start with the first student). The n written above the $\sum$ tells us that this is the last value taken by i (we finish with the last or nth student). Now suppose the freshmen in our sample are numbered from 1 through m inclusively. We can indicate the sum of their heights by $\sum_{i=1}^{m} x_i$.

The $\sum$ is merely an operator which tells us what to do with the x_i's— namely, sum them. The i is an index or tag. The entries above and below the $\sum$ tell us the range of the index over which we are to sum.

3-2-1 PROPERTIES OF $\sum$

The summation operator $\sum$ has several important properties that it inherits from the operation of addition. To present these, let us introduce another variable, y, to denote, say, weight. That is, let y_i be the weight of the ith student.

Property 1

$$\sum_{i=1}^{n} (x_i + y_i) = \sum_{i=1}^{n} x_i + \sum_{i=1}^{n} y_i .$$

In words this property says that if, for each student, we add height and weight together and then sum across all n students, we get the same number as if we added the sum of all the heights to the sum of all the weights. Intuitively, this is clear, but its truth can be verified rigorously, and doing so will help us to understand $\sum$ a little better. All we will use is the definition of $\sum$ and the knowledge that when we add numbers together, the order of addition is unimportant. Each equal sign is justified by a phrase written above it.

definition of Σ

$$\sum_{i=1}^{n} (x_i + y_i) = (x_1 + y_1) + (x_2 + y_2) + \cdots + (x_n + y_n)$$

rearranging order   definition of Σ

$$= x_1 + x_2 + \cdots + x_n + y_1 + y_2 + \cdots + y_n = \sum_{i=1}^{n} x_i + \sum_{i=1}^{n} y_i .$$

Property 1a

$$\sum_{i=1}^{n} (x_i - y_i) = \sum_{i=1}^{n} x_i - \sum_{i=1}^{n} y_i .$$

Demonstration of this relation follows the above pattern and is left as an exercise.

Property 2

If c is a constant (has the same value for each student), then $\sum_{i=1}^{n} c = nc$.

This just says that if you add the same number to itself n times, the result is nc. To demonstrate:

$$\sum_{i=1}^{n} c = \underbrace{c + c + c + \cdots + c}_{n \text{ terms}} = nc.$$

Property 3

$$\sum_{i=1}^{n} (x_i + c) = \left(\sum_{i=1}^{n} x_i \right) + nc.$$

This is a consequence of properties 1 and 2 as follows:

$$\sum_{i=1}^{n} (x_i + c) = \overset{\text{property 1}}{\underset{\downarrow}{\sum_{i=1}^{n} x_i}} + \overset{\text{property 2}}{\underset{\downarrow}{\sum_{i=1}^{n} c}} = \left(\sum_{i=1}^{n} x_i\right) + nc.$$

Property 4

$$\sum_{i=1}^{n} cx_i = c \sum_{i=1}^{n} x_i.$$

This says that multiplying each height by a constant and adding the resulting numbers together gives the same result as adding the heights and then multiplying the sum by the constant. This property follows by factoring:

$$\overset{\text{definition of } \Sigma}{\underset{\downarrow}{\sum_{i=1}^{n} cx_i}} = cx_1 + cx_2 + cx_3 + \cdots + cx_n$$

$$\overset{\text{factoring out } c}{\underset{\downarrow}{=}} c(x_1 + x_2 + x_3 + \cdots + x_n) = \overset{\text{definition of } \Sigma}{\underset{\downarrow}{c \sum_{i=1}^{n} x_i.}}$$

3-2-2 DOUBLE SUMS

To introduce this idea, we shall consider an actual set of data. These are 30 determinations of systolic blood pressure taken from a subject participating in a program designed to study sources and strengths of variation of blood pressure readings. This subject had his blood pressure read by 6 physicians at each of 5 visits.

Table 3-1 shows the observations classified in 2 ways: according to the visit at which the reading was made, and according to the physician who made the reading.

Table 3-1

SYSTOLIC BLOOD PRESSURE READINGS ON ONE SUBJECT TAKEN
AT 5 VISITS BY 6 OBSERVERS

Visit *Number*	*Physician Number*					
	1	2	3	4	5	6
1	118	112	116	118	116	122
2	120	116	112	112	120	112
3	114	120	112	117	108	120
4	118	116	118	116	112	118
5	118	108	122	116	110	120

In order to straighten out these 2 classifications, a system of double subscripts is used. That is, we use one index for visit number and one for physician number. The letters i and j are frequently used in such situations. Thus, in place of the table of actual data we will suppose that the classification scheme has r rows and c columns. In the above example $r = 5$ and $c = 6$. We let x_{ij} denote the reading in the ith row and jth column. Specifically, in the above set of data, $x_{12} = 112$, $x_{34} = 117$, $x_{22} = 116$, etc. Table 3-2 shows the general situation.

Table 3-2

GENERAL ROW BY COLUMN
CLASSIFICATION OF DATA

$i =$	$j = 1$	2	3	$\ldots$	c
1	x_{11}	x_{12}	x_{13}	$\ldots$	x_{1c}
2	x_{21}	x_{22}	x_{23}	$\ldots$	x_{2c}
3	x_{31}	x_{32}	x_{33}	$\ldots$	x_{3c}
.	.	.	.	$\ldots$	.
.	.	.	.	$\ldots$	.
.	.	.	.	$\ldots$	.
r	x_{r1}	x_{r2}	x_{r3}	$\ldots$	x_{rc}

We now consider several types of sums, remembering to keep the indices i and j straight. For example, the sum of the entries in the second row is $\sum_{j=1}^{c} x_{2j}$ (the row and hence its subscript is fixed; only the column changes). The sum of the entries in the third column is $\sum_{i=1}^{r} x_{i3}$.

We now ask the question: How can we denote the sum of all the readings in the table? One way to get this sum would be to find the sum of the entries in each row separately and then add these sums together. The row sums are

$$\sum_{j=1}^{c} x_{1j}, \sum_{j=1}^{c} x_{2j}, \sum_{j=1}^{c} x_{3j}, \ldots, \sum_{j=1}^{c} x_{rj}.$$

Adding these together, we have

$$\sum_{j=1}^{c} x_{1j} + \sum_{j=1}^{c} x_{2j} + \sum_{j=1}^{c} x_{3j} + \cdots + \sum_{j=1}^{c} x_{rj}.$$

But this is lengthy and can be written more compactly if we note that it is a sum in which the individual terms are sums. As we go from term to term the only thing that changes is the first or row subscript which varies from

1 to r. We have already used i to index rows, so it is a very natural extension to write the final sum as

$$\sum_{i=1}^{r} \sum_{j=1}^{c} x_{ij}.$$

This is called a double sum and it very much simplifies notation.

For practice with these ideas, find the individual column sums first and then add them together over all the columns. Of course, you will again get the grand total of the whole table, but this time it comes out in the form

$$\sum_{j=1}^{c} \sum_{i=1}^{r} x_{ij}.$$

This shows that in a double sum the order of the summation signs may be interchanged without affecting the result.

The notation of multiple summations such as the double sum may be extended as far as we like. We can imagine, for example, a third classification. In the blood pressure data above, we might take similar measurements on several different subjects. This gives us the 3 classification variables: visit, physician, and subject. We can think of this as several tables like Table 3-2 stacked on top of each other in banks. Let us suppose that there are b of these banks. We must then introduce a new indexing subscript, say k, so that x_{ijk} will represent the observation in the ith row, jth column, and kth bank. In this case, then, by extending the reasoning we used to arrive at the double summation, we can see that the grand total of all the observations may be indicated by

$$\sum_{i=1}^{r} \sum_{j=1}^{c} \sum_{k=1}^{b} x_{ijk}, \quad \text{a triple summation.}$$

Such sums as this are used frequently in statistics.

3-2-3 DOT NOTATION

The frequency of use of various kinds of summations in statistics has led to a further compression of notation involving the use of dots in place of summation signs. In this system, whenever a subscript is replaced by a dot, we know that the missing subscript has been "summed out."

We illustrate this point by referring to Table 3-2. Instead of the sigma notation for the row totals:

$$\sum_{j=1}^{c} x_{1j}, \sum_{j=1}^{c} x_{2j}, \sum_{j=1}^{c} x_{3j}, \ldots, \sum_{j=1}^{c} x_{rj}$$

the dot notation is

$$x_1., x_2., x_3., \ldots, x_r..$$

The dot notation for the column totals is

$$x._1, x._2, x._3, \ldots, x._c.$$

The grand total would then become $x..$.

3-3 Measures of Central Tendency

In describing groups of observations, we frequently wish to describe the group by a single number. In describing the heights of a number of adult males, for example, a typical or usual height can be of use in characterizing the group. For this purpose, of course, we would not use the largest or the smallest value as the single representative, since these values represent extremes rather than typical values. We would be more likely to seek a central value. Measures descriptive of a typical or representative value in a group of observations are frequently called measures of central tendency. In another sense we are attempting to locate the entire group with respect to the variable under study, and these measures are also called measures of location.

In our discussion of measures of central tendency it is important to realize that such measures apply to groups rather than to individuals. You may have heard that there is no average man in the United States, since the average man might have 2.1 children, 0.8 automobile, 0.1 umbrella, and so forth. There is widespread confusion concerning the meaning of averages or measures of central tendency. Many people think of an average as an attribute of an individual, but as we have seen an average is a group, not an individual, characteristic.

3-3-1 ARITHMETIC MEAN

The simplest and in many respects the most obvious measure of central tendency we might choose is the simple average of the observations in the group, i.e., the value obtained by adding the observations together and dividing this sum by the number of observations in the group. The term "average," however, is used today with varying precision. In fact there are many kinds of averages and we can never be certain that when an author uses the word, "average," he intends to specify the quantity we have just defined. Because of this confusion statisticians have quite uniformly decided to use the terminology "arithmetic mean" to denote the sum of a group of observations divided by their number. When the term "mean" is used alone, we will always be referring to the arithmetic mean.

EXAMPLE

The following 5 observations are radiologic counts, each count taken from the same source material counted for 0.1 minute:

$$4, 5, 9, 1, 2.$$

The arithmetic mean of these observations is $\frac{21}{5}$ or 4.2. Note that the value 4.2 is not attained by any of the individual counts but refers to the group as a whole. Generally, using the summation notation developed earlier, we can denote the arithmetic mean of the group of observations $x_1, x_2, \ldots, x_n$ by

$$\frac{\sum_{i=1}^{n} x_i}{n}.$$

This expression is often reduced still further to the single term $\bar{x}$. Each of these expressions is merely a notational device to describe the arithmetic mean.

3-3-2 MEDIAN

Another very frequently used measure of central tendency is the median, the middle value in a set of observations ordered by size. For example, to compute the median of the five radiologic counts given in Sec. 3-3-1 we must first arrange the observations in order of size, obtaining

$$1, 2, 4, 5, 9.$$

The median is then the observation in the middle position in the set of ordered observations, that is, the number 4. The initial operation of arranging the observations in order of magnitude is very important. Had this not been done, and if the order given originally for the set of observations had been used, the middle value would have been 9—a ridiculous result for a measure of central tendency. Errors in the computation of the median are frequently due to the omission of this initial ordering operation.

The process just described is useful for data consisting of an odd number of observations, but obviously cannot be applied directly if we have an even number. For example, if another count of 2 had been observed in the previous experiment the resulting ordered set of observations would have been 1, 2, 2, 4, 5, 9. In this case there is no middle observation, but rather there are 2 middle observations, 2 and 4. The median is taken to be the value halfway between the 2 middle observations, 3 in this instance. This value is simply the arithmetic mean of the 2 middle observations, i.e., $(2 + 4)/2 = 3$.

3-3-3 MODE

Another common measure of central tendency is the mode. The mode is the most frequently occurring value in a set of observations. For example, in the series of radiologic counts 1, 2, 2, 4, 5, 9, the mode would be 2.

3-3-4 COMPARISON OF THE MEAN, MEDIAN, AND MODE— ADVANTAGES AND DISADVANTAGES

Relative to the 5 radiologic counts 4, 5, 9, 1, 2, we observe that the arithmetic mean 4.2 is larger than the median 4.

This distinction points out a useful contrast between the mean and the median—the mean is sensitive to extremes, whereas the median is not. The extreme value 9 in this instance has increased the arithmetic mean, while the median, looking only at the middle value in the ordered set of observations, is not affected. Further insight to this distinction is provided by a physical characterization of the arithmetic mean. Consider a weightless bar marked off in units and imagine each observation in the group as contributing a unit of weight that can be hung from its appropriate point on the bar. For the above set of observations, weights would hang from the points labeled 1, 2, 4, 5, 9. The point at which the bar would balance is the arithmetic mean. Thus the mean is a kind of fulcrum or center of gravity of a set of observations. In this instance the bar would balance around the point 4.2. From this representation it is clear just why the arithmetic mean is sensitive to extreme values. If the largest value were increased beyond 9 say to 15, the arithmetic mean would follow that increase to maintain balance. However, as noted earlier, the median would be unaffected.

For some purposes we may wish to take into account extreme values. In these cases the arithmetic mean is useful. However, in other instances it can give a very misleading picture of a set of observations. For example, if the set of counts had been 1, 1, 1, 1, 17, the arithmetic mean, which is again 4.2, would be a rather poor representative of the set, since no observation is close to this value. On the other hand, the median, 1 in this case, seems to characterize the group quite adequately. However, the median appears in many cases to ignore useful information. Although it takes all the observations into account, at least by virtue of noting their relative positions, the only values directly influencing the computation of the median are the 2 middle observations in the case of an even set or the single middle observation in the case of an odd set. In some cases this can result in loss of useful information. The mode has one possible advantage relative to the mean and median—it does describe an individual, in fact, several individuals. By definition the mode must be attained by more than one observation in the sample. However, this advantage is of dubious value, since our purpose is to characterize groups rather than individuals. Furthermore, the same set of observations might well have more than one mode. If our set of 6 radiologic counts had been 1, 2, 2, 4, 4, 9, then both the values 2 and 4 would have been modes. It should be clear from the definition that any set of observations has only one arithmetic mean and one median. Sets of observations having more than one mode are said to be multimodal. Sets with 2 modes, for example, are said to be bimodal. An even worse complication can arise. For instance, the radiologic counts 4, 5, 9, 1, 2 have no mode, since no observation occurs most frequently.

3-3-5   OTHER MEASURES OF CENTRAL TENDENCY

3-3-5-1   Midrange.   The midrange is the value midway between the smallest and largest values in the sample, i.e., the arithmetic mean of the largest and the smallest values. For example, in the set of radiologic counts 4, 5, 9, 1, 2 the midrange is $(9 + 1)/2$ or 5. It is clear that the midrange will be influenced by extreme values.

3-3-5-2   Geometric Mean.   The geometric mean of a set of observations is the nth root of their product. The computation of the geometric mean requires that all observations be positive, i.e., greater than zero. For example, the geometric mean of the 2 radiologic counts 4 and 9 is $\sqrt{4 \cdot 9} = \sqrt{36}$ or 6. A general formula for computing the geometric mean of the set of observations $x_1, x_2, \ldots, x_n$ is $\sqrt[n]{x_1 \cdot x_2 \cdots \cdots x_n}$. The geometric mean is sometimes denoted by $\bar{x}_g$. An interesting property is that the logarithm of the geometric mean is the arithmetic mean of the logarithms of the individual observations. This result is expressed by the formula

$$\log \bar{x}_g = \frac{\sum\limits_{i=1}^{n} \log x_i}{n} .$$

The geometric mean is used in microbiology for computing average dilution titers, but has a variety of other uses as well.

3-3-5-3 Harmonic Mean. The harmonic mean of a set of observations is the reciprocal[1] of the arithmetic mean of the reciprocals of the observations. That is, if the observations are $x_1, x_2, \ldots, x_n$, then the harmonic mean is

$$\frac{n}{\sum\limits_{i=1}^{n} \dfrac{1}{x_i}}.$$

The harmonic mean is often denoted by $\bar{x}_h$. Freund (19) gives an interesting example of the usefulness of the harmonic mean. He suggests the problem of determining the average velocity of a car that has traveled the first 10 miles of a trip at 30 miles per hour and the second 10 miles at 60 miles per hour. At first glance the average velocity would seem to be the simple average of 30 and 60, i.e., 45 miles per hour. However, this kind of average is usually defined to be total distance divided by total time. Here the total distance is 20 miles, whereas the total time is $\frac{1}{3}$ hour plus $\frac{1}{6}$ hour or $\frac{1}{2}$ hour, producing an average velocity of 40 miles per hour rather than 45 miles per hour. It is interesting to note that this average would be available as the harmonic mean of the velocities 30 and 60. That is, 2 divided by $\frac{1}{30} + \frac{1}{60} = 40$.

3-3-5-4 Weighted Mean. In certain circumstances not all observations have equal weight. For example, certain observations might be measured more precisely than others. Consider the following illustration: 3 well-standardized laboratories in the same area are investigating throat cultures for presence of beta hemolytic streptococci. Laboratory A investigates 50 cultures of which 25 are positive, so the positivity rate is 0.50 or 50 per cent. Laboratory B examines 80 cultures, finding 60 positive for a positivity rate of 0.75 or 75 per cent. Laboratory C examines 120 cultures, finding 30 positive for a rate of 0.25 or 25 per cent. In describing the average rate of positivity for the 3 laboratories we might be tempted to take the arithmetic mean of 50 per cent, 75 per cent and 25 per cent, obtaining an average of 50 per cent. A little reflection would show this procedure to be a poor one. It fails to consider that the 3 laboratories have examined different numbers of cultures. It seems plausible to argue that laboratory C's rate should be given most weight, since the greatest number of cultures were examined there. Similarly, laboratory A's rate should be given least weight. A better procedure would be to divide the total number of positive tests, 115, by the total number of tests, 250, which yields an overall rate of $\frac{115}{250}$ or 46 per cent. This is a weighted average of the individual percentages, with weights equal to

---

[1] The reciprocal of a number b is $1/b$.

numbers of cultures investigated in the individual laboratories. Thus

$$46\% = \frac{50(50\%) + 80(75\%) + 120(25\%)}{50 + 80 + 120}$$

$$= \frac{11,500\%}{250}.$$

In general, if we have observations $x_1, x_2, \ldots, x_n$ with weights $w_1, w_2, \ldots, w_n$, the weighted average of the x's is defined to be

$$x_w = \frac{\sum_{i=1}^{n} w_i x_i}{\sum_{i=1}^{n} w_i}.$$

To see how this formula fits the situation involving the laboratory tests, note that in this example $n = 3$, $x_1 = 50$ per cent, $x_2 = 75$ per cent, $x_3 = 25$ per cent, $w_1 = 50$, $w_2 = 80$, $w_3 = 120$.

Another example of a weighted mean, familiar to college students, is the grade point average. In this instance letter grades are assigned numerical values, typically 4 for A, 3 for B, 2 for C, etc. A student's grades in a series of courses are averaged under this system, each grade being weighted by the number of credit hours for the particular course it represents.

3-4 Measures of Variability

Again considering radiologic counts, we examine the 2 sets: 1, 4, 4, 4, 7 and 4, 4, 4, 4, 4. These sets have much in common. They have the same arithmetic mean, median, mode, and midrange. In other words, they are similar in central tendency. However, the first set of observations is more variable than the second. Variability and its measurement are of fundamental importance in the biological sciences.

3-4-1 GENERAL PROPERTIES OF MEASURES OF VARIABILITY

In examining these 2 sets of radiologic counts notice that the first set varies from a low value of 1 to a high value of 7, whereas the second set has 4 as both its lowest and highest value. The first set shows some variability, whereas the second set shows none. In attempting to define a measure of variability it seems reasonable to require that it take the value zero if and only if the observations show no variability and that it take some positive value for observations showing some variability. Furthermore, the greater the degree of variability in a set of observations, the greater the value of the measure of variability. Using these standards, we can proceed now to judge several proposed measures of variability.

3-4-2　RANGE

A reasonable measure of variability, the range, might be obtained by subtracting the lowest value in a set of observations from the highest value. For the first set of radiologic counts this is $7 - 1 = 6$ and for the second set $4 - 4 = 0$. Considering the criteria we have established for measuring variability, we note that the range fills the bill, at least in our example. It is positive for the first set of observations, which shows variability, and zero for the second set. A common mistake in using the range is to assign 2 numbers instead of one, i.e., we are tempted to say that the range of the first set is from 1 to 7. This is incorrect statistical usage, since the range and indeed any reasonable measure of variability should be a single number, not 2 numbers.

The range has several advantages. It is easy to compute, and its units are the same as the units of the variable being measured. In a study of heights, for example, the range would be obtained by taking the largest height minus the smallest height. If heights were measured in inches, the difference would also be measured in inches. On the other hand, the range also has disadvantages. First, if we had taken a set of 100 radiologic counts instead of 5, we might well have observed a few counts perhaps as large as 10 or 12, increasing the range. The range does not take into account the number of observations in the sample, but only the largest observation and the smallest observation, wherever they may be. Since we expect large samples to include occasional extreme values, we expect a larger range. A measure of variability should not depend on the number of observations collected. A second disadvantage of the range is that it makes no direct use of many of the observations in the sample. Observations between the smallest and largest in a set are used only to determine which observations are the smallest and the largest. Some use of the actual values of intervening observations seems desirable. The range also suffers from its dependence upon extreme observations. An extremely large count, say a count of 25, would severely affect the range of the radiologic counts above.

3-4-3　A TRIAL BALLOON

In attempting to overcome some of the undesirable properties of the range and in particular to make use of each observation in the set, consider the following procedure. Compute the arithmetic mean of the set of observations; then determine the difference between each observation and the arithmetic mean. Add these differences across the entire set of observations and then divide by the number of observations in the set to obtain an average difference. A measure such as this one has several potential advantages. It uses every observation and corrects for variation in the number of observations by the final division. We will compute this measure for the 2 sets of radiologic counts, 1, 4, 4, 4, 7 and 4, 4, 4, 4, 4. The arithmetic mean of the first set is 4. The first observation is 1. Subtracting the arithmetic mean, we obtain -3, i.e., $1 - 4 = -3$. Continuing, we have $4 - 4 = 0$, $4 - 4 = 0$, $4 - 4 = 0$, $7 - 4 = 3$. Adding together these differences, we obtain $(-3) + 0 + 0 + 0 + (+3)$ or 0, and dividing by 5, we still obtain 0. In the second set of observations the arithmetic mean is also 4. If we carry through the above steps, we again obtain a value of zero. Of course, we want a value of zero for the second set, which showed no variability. The zero obtained for the first set is disturbing. This set clearly does have some variability. Something is wrong with the proposed measure.

In fact, this measure takes the value zero for any set of numbers that can be written down, and is therefore useless as a measure of variability, or indeed as a measure of anything else. The difficulty is that we are not asking how far each observation is from the arithmetic mean, but also in what direction it lies. The signs of the differences have a canceling effect and exactly eradicate the overall contribution of the observations. There seems to be the germ of an idea here, however, and perhaps we can make use of it. In particular, if we ignore the sign and treat each difference in terms only of its magnitude, we may have something.

3-4-4 MEAN DEVIATION

The mean deviation is defined exactly as is the measure proposed in the preceding section, except that the signs of the differences are ignored and each difference is taken to be positive. That is, in the first set of radiologic counts with arithmetic mean 4 and first observation 1, we take $1 - 4 = -3$, but now ignore the sign and merely record 3 (understood to be $+3$). Proceeding in this manner, we obtain deviations of 0, 0, 0, and 3. Adding these deviations together, we obtain 6. Dividing the sum by 5, we obtain 1.2 as our measure of variability. In the second set of observations the application of this procedure still produces zero, as it should. In mathematics the operation of ignoring the sign of a quantity occurs very frequently, and a special notation is used to represent it. If we wish to show that the quantity c is to have its sign ignored, we enclose it in vertical lines and write $|c|$. This symbol is read, "the absolute value of c." If the first observation in a set is denoted by x_1 and if the arithmetic mean is denoted by $\bar{x}$, then the difference is $x_1 - \bar{x}$, and the absolute value of the difference is $|x_1 - \bar{x}|$. If this operation is repeated for each observation, if the deviations are added together, and if the result is divided by the number of observations n, we obtain a formula for the mean deviation, denoted by M.D.

$$\text{M.D.} = \frac{\sum\limits_{i=1}^{n} |x_i - \bar{x}|}{n}.$$

The mean deviation has the properties we called desirable for measures of variation. It is scaled in the same physical units as the observations themselves, is relatively easy to compute, takes the value zero for sets with no variability, and is positive otherwise. Its major difficulty is technical—mathematical analysis with absolute values is a bit complicated.

3-4-5 VARIANCE

Another mechanism is available for solving the problem of the canceling effect of positive and negative differences. If each difference is squared before summing, the square cannot be negative. In fact, the square of no number, positive or negative, is negative. Applying this procedure to the first set of radiologic counts, we obtain

$$(1 - 4)^2 + (4 - 4)^2 + (4 - 4)^2 + (4 - 4)^2 + (7 - 4)^2$$

$$= (-3)^2 + (0)^2 + (0)^2 + (0)^2 + (3)^2 = 9 + 0 + 0 + 0 + 9 = 18.$$

It would seem reasonable to divide 18 by 5, the sample size. However, at this point most statisticians prefer to divide not by the sample size, but by one less than the sample size for reasons to be clarified later in this book. This produces a value of $\frac{18}{4} = 4.5$. This quantity is called the variance. It is generally denoted by s^2, and its defining formula is

$$s^2 = \frac{\sum\limits_{i=1}^{n} (x_i - \bar{x})^2}{n - 1}.$$

A simplified formula is available for computing the variance.

$$s^2 = \frac{n \sum\limits_{i=1}^{n} x_i^2 - \left(\sum\limits_{i=1}^{n} x_i \right)^2}{n(n - 1)}.$$

This formula is mathematically equivalent to the defining formula but is easier to compute and more accurate.[2] Applying the computing formula to the first set of radiologic counts, we see that

$$\sum_{i=1}^{n} x_i^2 = 1^2 + 4^2 + 4^2 + 4^2 + 7^2 = 98$$

and that

$$\sum_{i=1}^{n} x_i = 1 + 4 + 4 + 4 + 7 = 20$$

whence

$$n \sum_{i=1}^{n} x_i^2 - \left(\sum_{i=1}^{n} x_i \right)^2 = 5(98) - (20)^2 = 490 - 400 = 90$$

and finally

$$s^2 = \frac{90}{5(4)} = 4.5.$$

You may think that we have taken leave of our senses in claiming that the second formula is easier for computational purposes than the defining formula. All we can say is that life is rarely as simple outside the covers of a textbook as inside. This radiologic count example, although useful for illustrative purposes, is simpler than most problems encountered in the cold, cruel world. For example, here there is no problem of rounding error, the sample size is small, and the observations themselves are small, well-behaved whole numbers. Unfortunately, life is rarely like this, and you will become convinced even by exercises included in this book that the computing formula is superior for computation. The quantity

$$n \sum_{i=1}^{n} x_i^2 - \left(\sum_{i=1}^{n} x_i \right)^2$$

occurring as the numerator of the computing formula for the variance is called the large variance. It arises frequently in statistics and is very useful.

A little reflection will convince you that the variance as defined here satisfies the desirable properties for measures of variability. It does have one disadvantage, however—its physical units are not the same as the physical units of the observations themselves. In fact, the variance is measured in squared units. For example, if we observe heights and if x_i is the height of the ith individual measured in inches, then $x_i - \bar{x}$ is again measured in inches, but $(x_i - \bar{x})^2$ is now in square units, square inches in this case.

---

[2] This statement applies to desk calculator computations, but it requires modification when one is discussing high speed electronic computers.

Summing such quantities does not alter their physical units, nor does the division by $n - 1$, and thus the quantity s^2 is measured in square physical units. This difficulty can be corrected quite simply by taking the square root of the quantity s^2.

3-4-6 Standard Deviation

The standard deviation s is defined to be the positive square root of s^2, the variance. Its advantage over the variance is that its physical units are the same as the physical units of the individual observations. We never take s to be negative. Its defining formula is

$$s = \sqrt{\frac{\sum_{i=1}^{n}(x_i - \bar{x})^2}{n - 1}}.$$

Its computing formula is

$$\sqrt{\frac{n\sum_{i=1}^{n}x_i^2 - \left(\sum_{i=1}^{n}x_i\right)^2}{n(n - 1)}}.$$

For the first set of radiologic counts we have

$$s = \sqrt{4.5} = 2.12.$$

The second set of radiologic counts, of course, has a standard deviation of zero.

3-5 The Effect on the Mean, Variance, and Standard Deviation of Uniform Changes in the Observations

3-5-1 Effect of Adding a Constant

If a constant is added to each observation in a set, there is a predictable effect upon the mean, variance, and standard deviation. Intuitively, if the same value is added to every observation, it seems clear that the average should be increased by the same amount. The addition of a constant shifts the entire set of observations by a certain number of units and should not affect their variability. Thus, we expect on intuitive grounds that the variance and standard deviation will be unaffected by adding a constant to every observation. We illustrate this by the set of radiologic counts 1, 4, 4, 4, 7 considered earlier. Suppose the constant value 2 is added to each observation. The resulting set of observations is then 3, 6, 6, 6, 9, and the arithmetic mean becomes $\frac{30}{5} = 6$. Note that 6 is 2 units larger than 4, the arithmetic mean of the original set. The variance becomes

$$s^2 = \frac{5(3^2 + 6^2 + 6^2 + 6^2 + 9^2) - (30)^2}{5(4)}$$

$$= \frac{5(198) - 900}{20} = \frac{90}{20} = 4.5,$$

exactly the value in the original set. Of course, the standard deviation is also unchanged at 2.12. This supports our intuition at least with regard to this numerical example. We can prove that these results hold for any set of observations $x_1, x_2, \ldots, x_n$. If a constant c is added to each observation, the new observations become $x_1 + c, x_2 + c, \ldots, x_n + c$. The arithmetic mean of this set is

$$\frac{\sum_{i=1}^{n}(x_i + c)}{n} = \frac{\sum_{i=1}^{n} x_i + \sum_{i=1}^{n} c}{n} = \frac{\sum_{i=1}^{n} x_i + nc}{n} = \frac{\sum_{i=1}^{n} x_i}{n} + c = \bar{x} + c,$$

where $\bar{x}$ is the mean of the original set. This establishes the fact that increasing a set of observations by a constant increases the arithmetic mean by the same constant. Since that constant might itself be negative, the same argument establishes the fact that decreasing a set of observations by a constant decreases the arithmetic mean of that set by the same constant. With respect to the variance of the new set of observations we have, using the defining formula,

$$\frac{\sum_{i=1}^{n}[(x_i + c) - (\bar{x} + c)]^2}{n - 1} = \frac{\sum_{i=1}^{n}(x_i + c - \bar{x} - c)^2}{n - 1} = \frac{\sum_{i=1}^{n}(x_i - \bar{x})^2}{n - 1} = s^2,$$

where s^2 is the variance of the original set. This shows that the addition of a constant to a set of observations does not alter the variance. Of course, as a consequence the standard deviation will also be unaltered.

3-5-2 EFFECT OF MULTIPLYING BY A CONSTANT

Just as predictable effects are found when a constant is added to each observation, definite effects are produced when each observation is multiplied by a constant.

Intuition indicates that if each observation is multiplied by the same amount, the arithmetic mean is also changed by that constant. The variability measures, on the other hand, cause more of an intuitive problem. The process of multiplying each observation by a constant squeezes the observations together if the constant is less than one or spreads them apart if the constant is greater than one. Thus, variability will be affected.

Let us examine the radiologic counts, 1, 4, 4, 4, 7. Suppose each observation is multiplied by 3. The resulting set is 3, 12, 12, 12, 21. The arithmetic mean is now $\frac{60}{5} = 12$, or 3 times the previous mean of 4. The variance becomes

$$s^2 = \frac{5(3^2 + 12^2 + 12^2 + 12^2 + 21^2) - (60)^2}{5(4)}$$

$$= \frac{5(882) - 3600}{20} = \frac{810}{20} = 40.5.$$

Recall that the variance of the original set was 4.5 and notice that the new variance is $3^2 \times 4.5 = 9 \times 4.5 = 40.5$. The new standard deviation is $\sqrt{40.5} = 6.36 = 3 \times 2.12$, where the previous standard deviation was 2.12. In other words, the variance has been multiplied by the square of the multiplicative constant, while the standard deviation has been multiplied by the constant itself. We shall prove these results in general for any set of observations $x_1, x_2, \ldots, x_n$. If each observation is multiplied by a constant c, the resulting values are $cx_1, cx_2, \ldots, cx_n$. The arithmetic mean of this set is

$$\frac{\sum_{i=1}^{n} cx_i}{n} = \frac{c \sum_{i=1}^{n} x_i}{n} = c\bar{x},$$

where $\bar{x}$ is the mean of the original set. This establishes the fact that multiplying each observation in a set by a constant multiplies the mean of that set by that same constant. Note that in this argument c may be either greater than or less than one. The variance of the new set is

$$\frac{\sum_{i=1}^{n} (cx_i - c\bar{x})^2}{n - 1}.$$

Now,

$$\frac{\sum\limits_{i=1}^{n} (cx_i - c\bar{x})^2}{n-1} = \frac{\sum\limits_{i=1}^{n} [c(x_i - \bar{x})]^2}{n-1}$$

$$= \frac{\sum\limits_{i=1}^{n} c^2(x_i - \bar{x})^2}{n-1} = \frac{c^2 \sum\limits_{i=1}^{n} (x_i - \bar{x})^2}{n-1} = c^2 s^2,$$

where s^2 is the variance of the original set. This proves that multiplication of a set by a constant multiplies the variance by that constant squared. The new standard deviation should be $\sqrt{c^2 s^2} = cs$, i.e., the constant times the old standard deviation, but notice a small problem here. If c is negative, cs will be negative. The difficulty comes from taking the square root of $c^2 s^2$. We have said that we will always use the positive square root when computing the standard deviation from the variance. This will produce $|c|s$ as the new standard deviation and keep the standard deviation from becoming negative— an unhealthy state of affairs for any self-respecting standard deviation.

Summarizing the information in Secs. 3-5-1 and 3-5-2, we have x_i, the ith observation in a set with mean $\bar{x}$, variance s^2, and standard deviation s. The new sets have observations $x_i + c$ or cx_i, where c is a constant. The descriptive statistics of these new sets are given in Table 3-3.

Table 3-3

THE EFFECT OF ADDITIVE AND MULTIPLICATIVE
CONSTANTS ON THE DESCRIPTIVE STATISTICS
OF A SET OF OBSERVATIONS

| | *Set of Observations* | |
Descriptive Statistic	$x_i + c$	cx_i
Mean	$\bar{x} + c$	$c\bar{x}$
Variance	s^2	$c^2 s^2$
Standard deviation	s	$|c|s$

3-6 Discrete and Continuous Variables

A categorization of data that has proved useful in organizing statistical procedures is the distinction between discrete and continuous variables. In simple terms a discrete variable is one which *inherently* contains gaps between successive observable values. More formally, we will define a discrete variable to be a variable such that between any 2 (potentially) observable values there lies at least one (potentially) unobservable value. For example, a count of the number of bacterial colonies growing on the surface of an agar plate is a discrete variable. Whereas counts of 3 and 4 are potentially observable, one of $3\frac{1}{2}$ is not. Any variable in the form of a count will be discrete, although not all discrete variables will be of this form.

In contrast, a continuous variable has the property that between any 2 (potentially) observable values lies another (potentially) observable value. A continuous variable takes values along a continuum, i.e., along a whole interval of values. Lengths and weights are examples of continuous variables. A man's height might be either 69 inches or 69.01 inches, but it could also potentially assume any intermediate value such as 69.005 inches. An essential attribute of a continuous variable is that, unlike a discrete variable, it can never be measured exactly. With a continuous variable, there must inevitably be some measurement error. A continuous variable could be represented exactly only by an infinite decimal, and no one has yet written down all the digits in such a number. (We recommend against your attempting this. You can undoubtedly find a more profitable life's work or at least one offering a greater prospect of fulfillment.) This implies that when we write down a value for a continuous variable we are only approximating its actual value by a number that reflects the precision of the measuring instrument used. If, for instance, we are measuring heights to the nearest half inch, our recorded value will be a member of the sequence . . . , 68.5, 69.0, 69.5, 70.0, 70.5, . . . , where the three dots signify continuation of values in a similar pattern. If on the other hand we measure to the nearest tenth of an inch, the recorded value will be a member of the sequence . . . , 68.9, 69.0, 69.1, 69.2, ... Thus, an important principle concerning continuous variables is the following: continuous variables are always *recorded* in discrete form, the size of the gap between adjacent recordable values being determined by the precision of measurement.

3-7   Grouped Data

As sets of data become larger, it becomes progressively more difficult to see the forest instead of the trees. That is, large arrays of individual observations are extremely difficult to inspect directly in order to get a notion of their central tendency, variability, or other general characteristics. A device that is often used to overcome this difficulty is grouping. If observations taking similar values are collected into groups or classes and if we retain information concerning the range of values contained within each class as well as the number of observations falling into the class, we can often overcome the difficulties attendant to too much detail in the basic set of data.

3-7-1   FREQUENCY TABLE

A common device for organizing and representing grouped data is the frequency table. This table shows the classes and the frequency with which observations fall into each class.

A class or group of observations will be defined by the smallest and largest recorded values (potentially) falling into the class. Of course, all intermediate values will also fall into the same class. These 2 extreme values are called the lower and upper class limits. It is important, particularly for continuous variables, to note that the class limits refer to recorded values rather than actual values and that furthermore these limit values need not occur within the particular set of data at hand. By actual value of a continuous variable we mean that infinite decimal that represents the true value of the characteristic. The recorded value includes the error inherent in the measuring device and might be called the measured value. In dealing with systolic blood pressure the lower and upper class limits of a class might be 120 mm Hg and 129 mm Hg, respectively, even though the value 129 may not occur within the set of observations under study.

Once a class has been defined by giving its lower and upper class limits, we count the number of observations at or between the limits. This number is called the class frequency.

Although the same set of data can be organized into many different equally valid frequency tables, there are a number of general guide lines to follow in choosing the classes.

1. The first class interval should be constructed to contain the smallest observation; the last interval, the largest observation.

2. The larger the number of observations, the larger the number of classes.

3. Class intervals should be mutually exclusive, i.e., nonoverlapping. This implies that the upper class limit of one class will be smaller by one unit of measurement precision, e.g., $\frac{1}{10}$ inch, than the lower class limit of the next higher class.

4. Class intervals should be exhaustive of the data i.e., every observation should fall into some class. Points 3 and 4 guarantee that every observation will be uniquely classifiable into one and only one class.

5. If possible, classes should be so constructed that frequently occurring values fall near the middle of a class. This is because the detailed values of individual observations are lost after classification in the frequency table. Thus, an intermediate, representative value will be taken to stand for all the observations in the class and a smaller error due to grouping will occur if this point is observed.

6. If possible, classes should be of equal length. In other words, the quantity—upper class limit minus lower class limit—for a particular class should be the same across all classes. Although this property cannot always be attained, many headaches, some requiring major analgesia, will be avoided when it is. In fact, point 6 might come under the heading, "Preventive Medicine."

7. Be sensible! Choose intervals which relate to the phenomenon under study. In dealing with weights recorded to the nearest pound, for example, classes like 150–156, 157–163, 164–170, 171–177, . . . may be acceptable mathematically, but it is hard to imagine an applied situation in which they would be as useful as, say, 150–159, 160–169, 170–179, ... Chosen classes should reflect as closely as possible a reasonable and realistic representation of the actual variable under study and above all should be natural instead of artificial.

Following all the guide points listed, one can probably group the same set of data into several equally acceptable frequency tables. Thus, there will not necessarily be any single best frequency table. On the other hand, it

Table 3-4

WEIGHTS IN GRAMS OF BOTH KIDNEYS
OF 50 NORMAL MEN AGED 40–49

374	363	252	305	323
309	358	332	387	329
323	355	403	349	327
288	361	277	303	311
301	265	208	293	256
345	311	322	356	310
358	388	307	350	342
340	240	379	470	247
329	260	319	362	329
309	288	369	288	358

may be necessary for good cause to violate one or more of these points. If the cause is both known and good, then forge ahead. These principles are intended to be helpful rather than restrictive.

We shall now turn to a detailed example of the construction of a frequency table. The observations in Table 3-4 represent the weight in grams of both kidneys in 50 presumably normal men aged 40 to 49 years.

Notice the difficulty in obtaining any detailed sense of central tendency or variation in the array of individual observations. A little searching determines that the smallest observation is 208 gm and the largest is 470 gm, producing a range of 262 gm to be divided into classes. Pondering, consulting the oracle, observing the phase of the moon, or even referring to the list of points given earlier in this section might inspire you to think that 20 gm divisions would be a reasonable choice and that the first such class might have limits 200 and 219. Notice that 20 individual potential weights to the nearest gram are located within this interval including the 2 limits, while the only observed value from the data falling into the interval is 208. If equal-length intervals are taken, then the frequency table is determined by fixing

Table 3-5

FREQUENCY TABLE OF WEIGHTS IN GRAMS OF
BOTH KIDNEYS OF 50 NORMAL MEN AGED
40–49

Class Limits	Frequencies
200–219	1
220–239	0
240–259	4
260–279	3
280–299	4
300–319	10
320–339	8
340–359	10
360–379	6
380–399	2
400–419	1
420–439	0
440–459	0
460–479	1

the first interval, finding the succeeding intervals, and counting the number of individual observations falling into each class. The resulting table in final form is represented as Table 3-5.

From this tabulation it appears that a reasonable central value would probably lie somewhere within the interval 320–339, that the observations are reasonably well balanced or symmetric about this interval, and that the large majority (all but 2) of the observations lie between 240 and 419.

An apparently innocent choice of class intervals can result in induced irregularities in a frequency table. For example, in recording heart rate in the usual manner by palpating the radial artery pulse, a common practice is to count the pulse for 15 seconds and then multiply by 4 to obtain a reading in counts per minute. In dealing with a set of one-minute pulse counts derived in this manner, an apparently reasonable and logical choice of classes is 50–59, 60–69, 70–79, 80–89, ... However, any pulse count obtained in this manner must be a multiple of 4, and thus within the interval 50–59 the only values that can occur are 52 and 56. In the next interval 60–69, however, the 3 values 60, 64, 68 can occur. Thus, the number of potential values per interval will alternate in successive intervals in the pattern 2, 3, 2, 3, ... This can only introduce artificial irregularities and a kind of sawtooth appearance in the tabulated frequencies. If, for example, we used the intervals 50–61, 62–73, 74–85, . . . or the intervals 50–57, 58–65, 66–73, . . . , such irregularities due to grouping would be avoided, since each interval would contain 3 potential values in the former case and 2 in the latter. At first, such intervals seem unnatural, but in terms of the actual measurement process they are more defensible. If heart rates were counted for a full minute, the original set of limits would probably be preferred.

Since class limits are the most extreme recorded values that can occur within a given class, it is natural to ask for the extremes of the actual values of a continuous variable that will produce a recorded value within the given class. These extreme values will be called the *class boundaries*. In the case of the total kidney weights we have the continuous variable, weight, measured to the nearest gram. Thus, any actual weight from barely over 321.5 gm to barely under 322.5 gm will produce a recorded value of 322 gm.

An apparently reasonable question would be—what about a weight of exactly 321.5 gm? Remember that we assumed a precision of measurement of one gm. If we have a better measuring device or wish more precision, we will perhaps be recording to the nearest tenth of a gram, in which case a recorded value of 322.0 gm would include all actual values from barely over 321.95 gm to barely under 322.05 gm. Thus, these extreme values are inevitably outside the precision of measurement. On the other hand, again considering measurement to the nearest gram, we might have a great deal of difficulty deciding in a particular case whether the nearest whole number of grams is 321 or 322 for a value that apparently falls "exactly" halfway between these values. In this case remember that the odds against the actual value being exactly 321.5, i.e., 321.5000. . . with infinitely many zeros after the 5 are absolutely and positively overwhelming. There are simply too many (depressingly many too many) different possible actual values near 321.5 for the value in this case to lie right on that value. In other words, try a little harder and you should be able to make a decision as to whether 321 or 322 is closer. If it really seems to be a toss-up, then go ahead and toss a coin, and throw the cantankerous observation one way for a head, the other for a tail.

Using the classes given earlier for the kidney weights, we can then determine class boundaries. They will be 199.5–219.5, 219.5–239.5, 239.5–259.5, ... Notice that the upper class boundary of one class is identical to the lower class boundary of the next higher class. This is in distinction to the class limits, which show gaps. The limits are recorded values and are thus discrete, i.e., gappy, whereas the boundaries are actual values of a continuous variable and thus must not show any gaps.

It is not always true that class boundaries are exactly midway between the upper and lower class limits of 2 adjacent classes as they are in this example. In the case of measurement data involving lengths or weights or dial readings they will almost always have this property, but in certain other cases, such as determination of age, they may not.

In many countries it is common practice to record age in complete years, or age at last birthday; i.e., we say that a person is 23 years of age from his 23rd birthday until his 24th birthday. Of course, age, which is really the length of time since birth, is a continuous variable. In recording age in completed years for a particular study we might be using classes with the following limits: 20–24, 25–29, 30–34, ... To get at the corresponding class boundaries, say for the class 20–24, we must ask for the extremes of actual age that will produce recorded ages in this interval. A little reflection shows that these extremes are 20 and 25, since any actual age from 20 to barely less than 21 produces a recorded age of 20, and similarly any actual age from 24 to barely less than 25 produces a recorded age of 24. Thus, the class boundaries corresponding to the given class limits are 20–25, 25–30, 30–35, . . . for age recorded *in completed years*.

If, however, we are dealing with age *to the nearest year* the same class limits, 20–24, 25–29, 30–34, . . . , will produce boundaries 19.5–24.5, 24.5–29.5, 29.5–34.5, ...

To carry this example one stage further, in some countries it is conventional to assign an individual the recorded age of one at the moment of birth, the age of 2 one year later, etc., a practice which makes at least approximate embryologic sense. In this case, a person with recorded age of 20 might have an actual age—actual time since birth—of from 19 to barely less than 20. This produces for our given class limits, class boundaries of 19–24, 24–29, 29–34, ...

In summary, class limits take into account measurement and recording practice, whereas class boundaries refer to the actual, underlying, continuous scale.

The class midpoint—sometimes called the class mark—for a continuous variable is the point midway between the 2 class boundaries of a given class, or in other words the arithmetic mean of the boundaries. It is used as a representative value for all the observations falling within the class, which of course have lost their individual identities after being grouped into the frequency table.

To summarize these concepts, we present in Table 3-6 an expanded version of the frequency table for kidney weights.

Notice that the class midpoints may, as in this case, reflect a degree of precision not present in the individual observations. This is not surprising, since they are computed from the class boundaries, which may themselves show this increase in precision, referring as they do to actual rather than measured values.

It is rare that a frequency table presented in a report of an investigation will contain all the columns shown in Table 3-6. Ordinarily, columns for

Table 3-6

EXPANDED FREQUENCY TABLE OF WEIGHTS IN GRAMS OF BOTH KIDNEYS
OF 50 NORMAL MEN AGED 40–49

Class Boundaries	Class Limits	Class Midpoints	Class Frequencies
199.5–219.5	200–219	209.5	1
219.5–239.5	220–239	229.5	0
239.5–259.5	240–259	249.5	4
259.5–279.5	260–279	269.5	3
279.5–299.5	280–299	289.5	4
299.5–319.5	300–319	309.5	10
319.5–339.5	320–339	329.5	8
339.5–359.5	340–359	349.5	10
359·5–379.5	360–379	369.5	6
379.5–399.5	380–399	389.5	2
399.5–419.5	400–419	409.5	1
419.5–439.5	420–439	429.5	0
439.5–459.5	440–459	449.5	0
459.5–479.5	460–479	469.5	1

class limits and class frequency will suffice. However, for calculation of descriptive statistics the additional information is important, as will be seen shortly.

Discrete data present no special problems of grouping distinct from the continuous case. However, in the discrete case, there is ordinarily no difference between recorded values and actual values, and class boundaries become in a sense an artificial construct. If class boundaries are needed for graphic or other purposes, it is conventional to assign the point halfway between the class limits of adjacent classes as the "boundary point." This results in a value that may have no meaning in the underlying scale, e.g., a half-unit for count data, but should be regarded as an attempt to approximate a discrete scale by an imputed continuous scale. If boundary values are needed only to compute class midpoints, you will notice that this system of constructing boundaries yields the same midpoints as the procedure of taking the point midway between the class limits, or equivalently taking the mean of these limits. Again, this may produce midpoint values that could not be attained by an individual, but since the midpoint is intended to represent the entire group of observations in the class rather than a single observation, this is not surprising. Recall that a similar situation holds for measures of central tendency and that this point was discussed in the introductory paragraphs of Sec. 3-3.

An example of this situation would be the classification of radiologic counts into intervals with the following class limits: 0–1, 2–3, 4–5, 6–7, 8–9, 10–11, 12–13. Here the midpoint values will be 0.5, 2.5, 4.5, 6.5, 8.5, 10.5, 12.5, none of which is attainable by an *individual observation*. However, when we consider choosing a value to represent the *group* of observations in the interval 2–3 in which both 2's and 3's can occur, the value 2.5 seems quite reasonable.

3-7-2 DESCRIPTIVE STATISTICS FOR GROUPED DATA

For notational purposes, suppose that n observations have been classified into c classes and that these classes have midpoints $x_1, x_2, \ldots, x_c$ with corresponding class frequencies $f_1, f_2, \ldots, f_c$.

As usual, we wish to find the sum of the observations divided by their number, but in this case x_1 is the representative of all the values in the first class, x_2 the representative of all those in the second class, etc. Thus we might say that x_1 occurs f_1 times, x_2 occurs f_2 times, etc. and that the sum of the observations in the first class is $f_1 x_1$, the sum of those in the second class $f_2 x_2$, etc. The sum of all observations is

$$f_1 x_1 + f_2 x_2 + \cdots + f_c x_c = \sum_{i=1}^{c} f_i x_i.$$

The number of observations is $n = \sum_{i=1}^{c} f_i$. As a result, the arithmetic mean for grouped data is

$$\bar{x} = \frac{\sum_{i=1}^{c} f_i x_i}{n}$$

In the case of the kidney weights presented in Table 3-6, we have

$$\bar{x} = \frac{(1)(209.5) + (0)(229.5) + (4)(249.5) + \cdots + (1)(469.5)}{50}$$

$$= \frac{16,275.0}{50} = 325.5 \text{ gm.}$$

This will, of course, give a different result from that obtained by computing $\bar{x}$ on the ungrouped data as they appear in Table 3-4, since actual observations are used in place of class midpoints. The ungrouped sample mean for these data is 324.6 gm, a fairly close level of agreement, which can be expected for data with a reasonably large number of observations grouped into a reasonably large number of classes.

In Sec. 3-4-5 the variance was defined for ungrouped data and a simplified computing formula was given. This can be modified directly for grouped data as follows:

Defining formula for variance for grouped data

$$s^2 = \frac{\sum_{i=1}^{c} f_i(x_i - \bar{x})^2}{n - 1}.$$

Computing formula for variance for grouped data

$$s^2 = \frac{n \sum_{i=1}^{c} f_i x_i^2 - \left(\sum_{i=1}^{c} f_i x_i\right)^2}{n(n - 1)}.$$

As in the case of the mean for grouped data, the factors f_i are inserted because each midpoint value x_i occurs f_i times, as does each $(x_i - \bar{x})^2$ in the defining formula and each x_i^2 in the computing formula. Note further that the upper limits of all summations are changed from n to c. This is due to the fact that there are only c different x_i values now, one for each class. The $\bar{x}$ in the defining formula here, of course, refers to the grouped mean.

The grouped standard deviation presents no difficulty, since it is the positive square root of the grouped variance.

For the kidney weights the grouped variance is $s^2 = 2187.8$ gm^2, and the grouped standard deviation is $s = 46.77$ gm.

For comparison, if we compute the variance and standard deviation of the ungrouped data we obtain 2223.1 and 47.15, respectively.

Notice that this example is somewhat artificial in that we would not ordinarily be able to calculate both grouped and ungrouped means, variances, and standard deviations from the same set of data. If it is possible to compute them, the ungrouped measures are always to be preferred. However, data are sometimes available only in grouped form, necessitating grouped measures. Grouped computations were formerly made for reasons of computational simplicity even when the ungrouped data were available. With the increasing availability of high-speed electronic computing equipment such reasons no longer carry much weight, however. Means, variances, and standard deviations for grouped data from discrete variables are computed in the same manner as for continuous variables.

For some purposes it is useful to define, corresponding to each class in a frequency table, a cumulative frequency. The cumulative frequency corresponding to a given class in a frequency table is defined to be the total number of observations less than the upper class boundary of the class. The cumulative frequency thus is a count not only of the observations in the class but also of all the observations in classes of lower numerical value on the scale of measurement.

Table 3-7 shows the cumulative frequencies for the kidney weights.

Table 3-7

CUMULATIVE FREQUENCIES FOR KIDNEY WEIGHTS
IN GRAMS OF 50 NORMAL MEN AGED 40–49

Class Boundaries	Cumulative Frequencies
199.5–219.5	1
219.5–239.5	1
239.5–259.5	5
259.5–279.5	8
279.5–299.5	12
299.5–319.5	22
319.5–339.5	30
339.5–359.5	40
359.5–379.5	46
379.5–399.5	48
399.5–419.5	49
419.5–439.5	49
439.5–459.5	49
459.5–479.5	50

The median can also be calculated from grouped data. Here we seek the point on the scale of measurement that divides the n observations into 2 equal-sized groups, each containing $n/2$ observations. If n is odd, $n/2$ will not be a whole number. Here, we think of a given observation as occupying a certain distance along a scale of measurement, and thus half an observation would occupy half that distance.

To clarify this point consider the kidney weight data. Here we wish to estimate the point along the weight scale below and above which $n/2 = 25$ of the observations occur. From Table 3-7 we see that 22 of the observed weights are below 319.5 gm and 30 are below 339.5 gm. Therefore, the median value we seek must lie between 319.5 and 339.5. Within this interval lie 8 observations, each of which is considered to occupy $\frac{1}{8}$ of the length of the interval, i.e., $(\frac{1}{8})(20)$ or 2.5 gm. Since at 319.5 we have accumulated only 22 of the required 25 observations, we need to proceed through the interval an additional distance corresponding to the 3 units needed to reach 25. But each such unit occupies 2.5 gm, and therefore we require an additional $3(2.5 \text{ gm}) = 7.5$ gm to reach the median point. The median of these data is $319.5 + 7.5 = 327$ gm. For comparison notice that the ungrouped median for the same data is 325 gm.

In applying a general formula for the grouped median, we must first locate the class containing the median. Suppose that the median is located within the ith class. Let L_i be the lower class boundary of that class, and let f_i be the class frequency, while U_i denotes the upper class boundary. Finally, let r be the number of observations remaining after L_i to reach the median point: $r = 3$ in the kidney example. Then

$$\text{Grouped median} = L_i + \frac{r}{f_i}(U_i - L_i).$$

If n is odd, r will be expressed to a half-unit, e.g., $1\frac{1}{2}$, $5\frac{1}{2}$, etc. This causes no difficulty in applying the formula. To see how the formula works on the kidneys, note that

$$L_i = 319.5, \qquad f_i = 8, \qquad U_i = 339.5, \qquad r = 3,$$

$$\text{Grouped median} = 319.5 + \tfrac{3}{8}(339.5 - 319.5)$$

$$= 319.5 + \tfrac{3}{8}(20) = 319.5 + 7.5 = 327 \text{ gm.}$$

A similar procedure can be applied to discrete variables; note that class boundaries must be constructed in the manner described in Sec. 3-7-1.

3-7-3 GRAPHIC DEVICES FOR GROUPED DATA

After data have been grouped into a frequency table, it is a simple matter to construct a histogram as a graphic representation of the grouping. In fact, many workers initially construct a histogram in place of a frequency table. The histogram has as its horizontal axis a scale of values of the measured variable. Ordinarily class boundaries are explicitly marked on this scale. As vertical axis the histogram has a frequency scale. Above the interval on the horizontal scale corresponding to a particular class, a rectangular bar of height equal to the corresponding class frequency is erected. The result for the kidney weights is shown in Fig. 3-1.

It is important to note that each observation is assigned one unit of area. When the eye observes such a graph, relative frequencies of occurrence of various values are determined by comparing relative areas.

It is not always possible in classifying data to maintain equal-length intervals as has been done in this example, although point 6 of Sec. 3-7-1 indicates the desirability of doing so. It may, in the kidney weights, for example, have been necessary to construct a single interval of length 40 gm between the boundaries 379.5 and 419.5 in place of the pair of 20 gm intervals

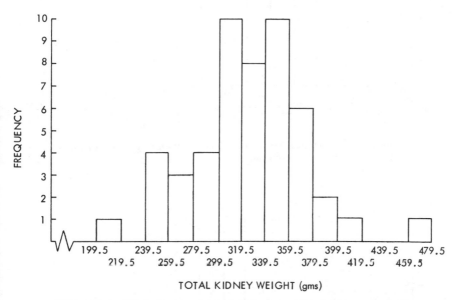

TOTAL KIDNEY WEIGHT (gms)

Figure 3-1 Histogram of total kidney weight in grams of 50 normal men aged 40–49.

used here, because of such things as aberrations in the measuring instrument, loss of detail in the data, or clerical error. This would produce a total frequency of 3 in the longer class. Now consider the effect of plotting a single histogram bar of height 3 above the double-length interval 379.5–419.5. This would have the visual effect of assigning twice the area to these 3 observations that had been assigned any other 3 observations, clearly a distortion of the true frequencies. Figure 3-2 illustrates this effect.

To preserve the proper relative areas when the interval length is doubled as in this case, the plotted height of the histogram bar must be cut in half in compensation. The proper result is shown in Fig. 3-3.

This histogram bears a closer resemblance to that of Fig. 3-1, as it should.

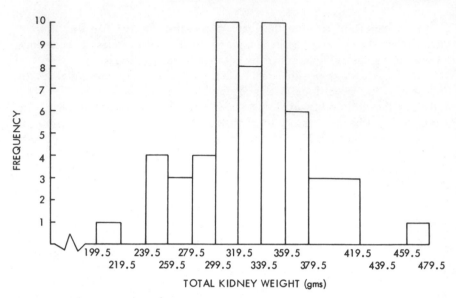

Figure 3-2   Distortion of histogram of kidney weights due to failure to allow for unequal length intervals.

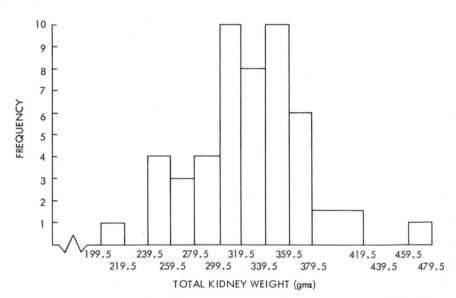

Figure 3-3   Histogram of kidney weights making proper allowances for unequal length intervals.

In general, if an interval is t times the basic interval length, the frequency f_i in the interval should be plotted on the histogram at a height of f_i/t. On the other hand, if an interval is only $1/t$ times as long as the basic interval length, the frequency should be plotted at a height of tf_i. This procedure has the effect of maintaining the basic unit of area and avoiding distortion.

Another useful graphic device for portraying grouped data is the *frequency polygon*. Here a frequency scale like that of the histogram is used, but the horizontal axis is marked off at class midpoints instead of boundaries.

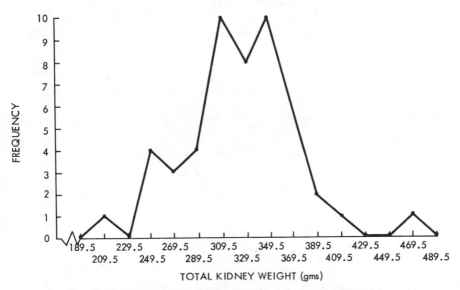

Figure 3-4 Frequency polygon of kidney weights of men aged 40–49.

A single point is located above the midpoint at the proper height, and adjacent points are connected. It is customary and convenient to add an interval of zero frequency at either end of the range used in the data in order to tie the polygon down to the horizontal axis at either end. The result for the kidney weights is shown in Fig. 3-4.

Again, a change in basic interval length results in a distorted appearance unless the same alteration used for the histogram is made in plotting the vertical position above the affected intervals. In the frequency polygon the basic unit is also a unit of area. You might enjoy dusting off a little plane geometry to show that the total area under a frequency polygon plotted as in Fig. 3-4 is identical to the total area under the corresponding histogram (Fig. 3-1).

In Figs. 3-1 and 3-4 the use of lower class boundaries and class midpoints, respectively, as scale positions for the horizontal scale has produced rather peculiar-appearing values, each ending in 9.5 gm. It would seem more natural to replace these values by multiples of 10 gm, and thus to shift the entire scale by 0.5 gm. This can be done without serious effect and may, in fact, improve the general impression given by the curve. A shift of $\frac{1}{2}$ unit is ordinarily (as in this case) imperceptible visually. Remember, though, that in the case of computation of numerical descriptive statistics for grouped data the proper values of class boundaries and class midpoints should be used as called for by the appropriate formulas.

The *cumulative frequency polygon* is a broken-line graph like the frequency polygon, but here the resemblance between the two stops. The

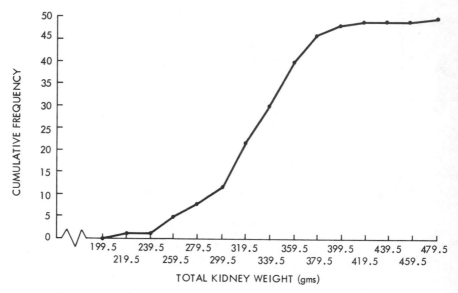

Figure 3-5 Cumulative frequency polygon of kidney weights of men aged 40–49.

vertical scale in this device is a scale of cumulative frequency, and the horizontal axis is marked off at class boundaries. Above each boundary point the cumulative frequency, that is, the total number of observations with values less than that boundary value, is plotted. The result for the kidney weights is shown in Fig. 3-5.

The cumulative frequency polygon as we pass from left to right along the horizontal axis is never-decreasing; i.e., the polygon itself either rises away from the horizontal axis or stays level, but never dips down. This is a consequence of the definition of cumulative frequency and of the procedure for constructing the polygon.

Unlike the histogram and frequency polygon, the basic unit in the cumulative frequency polygon is a unit of length rather than area. The height of the polygon above the horizontal axis is the critical feature. As a consequence, class intervals of unequal length produce no plotting difficulty. The cumulative frequency is computed for each of the boundary points regardless of spacing, and the corresponding plotting position on the vertical scale is unaffected.

As in the case of the histogram and frequency polygon, a $\frac{1}{2}$ unit shift in horizontal scale position to produce more even scale points is often made. If the cumulative frequency polygon is to serve only a graphic function, this is again a reasonable or possibly a superior practice. If, however, the curve is used to determine percentiles, as discussed in Sec. 3-8, the original plotting points at the exact class boundary values should be retained in order to avoid a systematic error.

No special problems arise in the graphic treatment of grouped data from discrete variables. Any of the devices presented here may be used. It is, of course, necessary to construct class boundaries, but the methods and discussion of Sec. 3-7-1 apply.

3-8   Percentiles, Quantiles, and Percentile Ranks

The *percentiles* of a set of observations divide the total frequency into hundredths. That is, the 30th percentile is that value of the variable below which 30 per cent of the observations lie; below the 90th percentile lie 90 per cent of the observations, etc. If the vertical scale of a cumulative frequency polygon is changed to include percentage of total frequency, then the percentiles can be determined directly by finding the appropriate percentage on the vertical scale, reading across until the cumulative polygon is intersected, and then reading the percentile value as the value on the horizontal axis just below the point of intersection. For example, from Fig. 3-5 (here the conversion from frequency to percentage can be made merely by doubling the vertical scale values, since $n = 50$) we find that the 80th percentile is 359.5 gm, the 20th percentile approximately 290 gm, etc. From this discussion, it follows that the median is the 50th percentile. For the kidney weights in Fig. 3-5 we find that the 50th percentile is approximately 326 gm. Recall that the calculated value is 327 gm, the disagreement being due to slight imprecisions in drawing and reading from a graph. Formulas for calculating other percentiles from grouped data can be constructed; they follow a similar pattern to that for the median.

In addition to percentiles, which divide the total frequency into hundredths, there is often a need for quantities dividing total frequency into larger equal parts, e.g., thirds, fourths, fifths, or tenths. The division points for these various partitions are called *tertiles*, *quartiles*, *quintiles*, and *deciles*, respectively. Thus, the first quartile is the same as the 25th percentile, the second quintile is the 40th percentile, the seventh decile is the 70th percentile, etc. The second quartile and the fifth decile are simply other terms for the median.

The term *quantile* is a generic term for a division point relative to any partition. That is, percentiles, tertiles, quartiles, quintiles, and deciles are all examples of quantiles.

The *percentile rank* of a particular numerical value of the observed variable is a percentage P such that the specified value of the variable is the Pth percentile of the set of observations. Thus, the percentile rank of the median is 50. In the kidney weights we may read approximate percentile ranks from Fig. 3-5 as follows: the percentile rank of 325 gm is 48, the percentile rank of 370 gm is 86, etc., the vertical scale again being adjusted to show percentage not frequency.

Be sure to note the fact that a percentile, or any quantile, for that matter, is a value of the observed variable, whereas a percentile rank is a percentage.

3-9 General Comments and Conclusions Concerning Descriptive Statistics

Statistical description and the abstraction of information from sets of observations are essential and basic parts of all statistical theory and practice. This topic is often given rather short shrift, particularly by statisticians primarily concerned with inference. To be sure, the problems of a purely mathematical nature associated with induction and inference are often more intricate and more sophisticated than those arising from data description. However, sound description and summarization are the basis of such problems. For workers interested in reaching sound scientific conclusions from their data, adequate and thoughtful data reduction and description are essential. An excellent inferential technique, though solidly based in mathematical theory, may fail to reach the subject matter itself and may be founded upon a weird distortion or caricature of the data. The biostatistician must be sure that his statistical gangplank is anchored in solid mathematics at one end and in solid subject matter at the other. Sound statistical description goes a long way toward accomplishing this goal.

EXERCISES

3-1 Let $x_1 = 3$, $x_2 = 5$, $x_3 = 8$, $x_4 = 10$, $x_5 = 9$, $x_6 = 7$, $x_7 = 4$, $x_8 = 1$, $x_9 = 7$, $x_{10} = 6$. Find:

a. $\sum_{i=1}^{10} x_i$.

b. $\sum_{i=5}^{9} x_i$.

c. $\sum_{i=1}^{10} x_i^2$.

d. $\sum_{i=2}^{6} x_i^3$.

e. $\sum_{i=1}^{10} cx_i$, where $c = 3$ and where $c = \frac{1}{2}$.

f. $10 \sum_{i=1}^{10} x_i^2 - \left(\sum_{i=1}^{10} x_i \right)^2$.

3-2 Show that the following relationships are true:

a. $\sum_{i=1}^{n} (a + b) = na + nb$.

b. $\sum_{i=1}^{n} (ax_i + b) = a \sum_{i=1}^{n} x_i + nb$.

c. $\sum_{i=1}^{n} na = n^2 a$.

d. $\sum_{i=1}^{n} \frac{a}{n} = a$.

e. $\sum_{i=1}^{n} \frac{x_i}{c} = \frac{1}{c} \sum_{i=1}^{n} x_i$.

Hint for f and g: To show 2 expressions are unequal, it is only necessary to show a particular set of values of the x_i for which they are not equal.

f. $\sum_{i=1}^{n} x_i^2 \neq \left(\sum_{i=1}^{n} x_i \right)^2$.

g. $\sum_{i=1}^{n} x_i y_i \neq \sum_{i=1}^{n} x_i \cdot \sum_{i=1}^{n} y_i$.

3-3 In this exercise let $\bar{x} = \dfrac{\sum\limits_{i=1}^{n} x_i}{n}$. Show:

a. $\sum\limits_{i=1}^{n} (x_i - \bar{x}) = 0.$

b. $\dfrac{\sum\limits_{i=1}^{n} (x_i + a)}{n} = \bar{x} + a.$

c. $\dfrac{\sum\limits_{i=1}^{n} ax_i}{n} = a\bar{x}.$

3-4 Using the data of Sec. 3-2-2 on systolic blood pressure readings by doctors and visits, find:

a. $\sum\limits_{i=1}^{5} \sum\limits_{j=1}^{6} x_{ij}$

b. $x_{1.}, x_{2.}, x_{3.}, x_{4.}, x_{5.}$

c. $\sum\limits_{i=1}^{5} x_i.$

d. $\sum\limits_{i=2}^{4} \sum\limits_{j=3}^{5} x_{ij}$

3-5 Using a method similar to that developed in Sec. 3-7-2 for computing the median from grouped data, find a generalized formula for constructing the Pth percentile from grouped data.

3-6 The following data are the number of patients obtaining contact lenses at a local eye specialist on 3 consecutive days: 3, 7, 2. Compute the following descriptive statistics for this set of values: (a) the arithmetic mean, (b) the geometric mean, (c) the harmonic mean, (d) the median, (e) the mode, (f) the midrange, (g) the range, (h) the variance, and (i) the standard deviation.

3-7 Suppose that the following represent the number of children for 10 physicians on a particular hospital staff: 3, 2, 0, 1, 4, 7, 3, 2, 4, 2. Find the following descriptive measures: (a) the arithmetic mean, (b) the median, (c) the mode, (d) the midrange, (e) the range, (f) the variance, and (g) the standard deviation.

3-8 Thirteen sheep were fed pingue (a toxin-producing weed of the southwest United States) as part of an experiment by Aanes (1) and died as a result. The time of death in hours after the administering of pingue for each sheep was as follows: 44, 27, 24, 24, 36, 36, 44, 44, 120, 29, 36, 36, 36. For these data compute: (a) the arithmetic mean, (b) the median, (c) the mode, (d) the range, (e) the variance, (f) the standard deviation.

3-9 Using the data of Table 3-4, find: (a) the arithmetic mean, (b) the median, (c) the mode, (d) the midrange, (e) the range, (f) the variance, (g) the standard deviation.

3-10 The following sample of serum cholinesterase indices in normal individuals are taken from Kaufman (30).

2.29	1.95	1.52	1.93	1.59	1.92	2.59	1.92	1.15	1.70
2.67	1.75	1.67	1.65	1.09	2.55	1.35	1.48	1.04	2.27
3.09	1.92	1.40	2.23	2.02	1.61	1.84	1.35	1.18	2.14
2.65	1.92	2.13	1.32	1.16	1.72	1.65	1.21	1.08	1.44
2.52	1.46	1.23	1.75	2.26	1.71	1.68	1.37	1.39	1.89
1.75	1.15	1.83	1.16	2.60	1.65	1.22	1.30	1.44	1.55
2.12	1.70	1.91	1.24	2.76	1.57	1.27	1.24	1.26	2.17
1.54	1.86	1.78	1.32	3.27	1.86	1.25	1.04	1.43	1.88
1.94	1.04	2.10	1.03	2.54	1.15	1.59	1.03	1.40	1.42
1.82	1.06	1.92	1.51	2.14	1.69	1.70	1.13	1.52	2.15

Compute: (a) the arithmetic mean, (b) the median, (c) the mode, (d) the midrange, (e) the range, (f) the variance, (g) the standard deviation.

3-11 Classify the following variables as discrete or continuous: (a) cranial circumference of a newborn infant, (b) systolic blood pressure, (c) heart rate, (d) red blood cell count, (e) number of DMF teeth, (f) pH of an unknown liquid, (g) age of an experimental rabbit, (h) length of incubation for a chicken egg, (i) presence or absence of atheromatous plaques in an adult, (j) weight of an overactive pituitary gland.

3-12 Suppose that by use of a special antitoxin each sheep's life in Exercise 3-8 was prolonged by: (a) 10 hours, (b) 3 times the original time period. Find the 6 descriptive statistics requested in 3-8 for these situations. (c) Is there a simple way of getting the mean, variance, and standard deviation in these cases? (d) How about the median, mode, and range?

3-13 Group the data of Table 3-4 into a frequency table in a fashion different from that presented in the text. Then find for your grouped data: (a) the mean, (b) the median, (c) the variance, (d) the standard deviation.

3-14 Using the grouping developed in Exercise 3-13, construct: (a) a histogram, (b) a frequency polygon, (c) a cumulative frequency polygon.

3-15 Using the cumulative frequency polygon of Fig. 3-5 and the one developed in Exercise 3-14, find: (a) the 30th percentile, (b) the 70th percentile, (c) the percentile rank of kidneys weighing 300 gm. (d) Compare your answers to those found using Fig. 3-5.

3-16 Using the data of Exercise 3-10, construct a frequency table and find for the grouped values: (a) the mean, (b) the median, (c) the variance, (d) the standard deviation.

3-17 Using the frequency table developed in Exercise 3-16, construct a frequency polygon and a cumulative frequency polygon.

3-18 Using Fig. 3-5, find: (a) the quartiles, (b) the quintiles, and (c) the deciles of the kidney weights.

3-19 Using the cumulative frequency polygon developed in Exercise 3-17, find: (a) the 50th percentile, (b) the 95th percentile, (c) the percentile ranks of a serum cholinesterase index of 1.99 and of 2.52.

3-20 Frequently when one is confronted with data having large numerical values, computation of the various descriptive measures is simpler when coded values are used. (a) For example, the following are weights of 5 adult males in pounds: 183, 187, 175, 170, 182. Find the mean and variance of these data as follows: first, subtract 180 from each value; second, find the mean and variance of the coded data (original value minus 180), and third, convert these descriptive measures back to the original units (use the results of Sec. 3-5). (b) In a similar fashion find the mean and variance, using the data of Exercise 3-8 with the following coded or transformed observations:

(1) $x - \bar{x}$.

(2) $\dfrac{x - \bar{x}}{s}$.

(3) $5 + \dfrac{x - \bar{x}}{s}$.

(4) $50 + 10 \cdot \left(\dfrac{x - \bar{x}}{s} \right)$ [*Note:* the coded measures of (3) and (4) are frequently used in educational and psychological statistics and are called standard scores].

3-21 Code the grouped data of Table 3-6 by subtracting 329.5 from each midpoint and dividing the differences by 20. Then find the mean and variance of the coded data. Use the coded mean and variance and the results of Sec. 3-5 to find the mean and variance of the original observations.

4   Probability

The basis of statistical inference is a branch of mathematics called probability theory, but do not let this lofty title confuse you. The origins of the subject matter are prosaic and reflect man's desire to achieve one-upmanship and solvency at the gambling table. We shall not attempt to describe in detail the historical development of this subject. The classic work by Todhunter (50) and the recent work by David (8) provide adequate background for further reading.

A brief word about some of the more interesting characters in this history is perhaps justified, however, in a work dealing with biostatistics. Girolamo Cardano (1501–76), whose name is often anglicized to Jerome Cardan, was a well-known physician, mathematician, and gambler, and, according to David (8, p. 60), "... was the first mathematician to calculate a theoretical probability correctly ...." Cardano is a controversial figure who led an exciting, varied, and often tragic life. A biography by Ore (39) provides additional background.

The real development of probability came much later with the exchange of correspondence between the mathematicians Pascal and Fermat in the mid-17th century on the solution of the "problem of points," concerned with the equitable division of stakes in a gambling game interrupted before its completion. The problem was apparently posed to Pascal by the Chevalier de Méré, a dedicated gambler. The later work of James Bernoulli, De Moivre, and Laplace served to establish probability as a field for valid intellectual inquiry. In spite of its somewhat questionable origin, the subject has by now made a contribution to virtually every field of scientific investigation.

In the interest of preserving historical context and because of their naturalness, we shall use cards and dice quite freely to provide examples. In an attempt to justify such usage in a work on biological applications of statistics, we can only say that modern dice have their origin in that rather basic bone, the talus, valued since ancient times as a gambling implement because of its hardness and relative symmetry. Furthermore, any description of the behavior of the highest species of vertebrates should include what takes place around a card table.

4-1 Definition of Probability

Definitions of probability adequate to provide a basis for full and rigorous development of its mathematical theory have been surrounded with controversy. We shall use a definition adequate for the immediate purpose, but must note that a more sophisticated definition would be necessary to treat the subject at a more advanced level.

Definition

If a procedure can result in n equally likely outcomes, n_A of which have attribute A, then we say that attribute A has *probability* n_A/n and we write $P(A) = n_A/n$.

From this statement we can deduce several important properties of probability:

1. $P(A) \geq 0$, for any attribute A.[1] Since n_A is a count, it is at least zero and thus $n_A/n = P(A)$ is at least zero. (The case $n = 0$ results from highly uninteresting procedures, those with no outcomes, and will not be considered here.)
2. $P(A) \leq 1$, for any attribute A. From the definition, the n_A outcomes are among the n total outcomes and thus $n_A \leq n$ and $n_A/n = P(A) \leq 1$.
3. If $\bar{A}$ denotes the absence of attribute A, i.e., the occurrence of "not A," then $P(A) + P(\bar{A}) = 1$ or $P(\bar{A}) = 1 - P(A)$. This follows from the fact that $n - n_A$ outcomes fail to show the attribute A and thus $P(\bar{A}) = (n - n_A)/n = 1 - (n_A/n) = 1 - P(A)$. The attributes A and $\bar{A}$ are said to be *complementary*.

EXAMPLE 1

Consider a true die (singular of dice) tossed fairly. The probability of an even-numbered outcome (2, 4, or 6) is $\frac{1}{2}$. Here 3 of the 6 equally

---

[1] The following symbols represent inequalities:

$$>\quad \text{``is greater than.''}$$
$$<\quad \text{``is less than.''}$$
$$\geq\quad \text{``is greater than or equal to.''}$$
$$\leq\quad \text{``is less than or equal to.''}$$

likely outcomes have the required attribute of evenness: $n_A = 3$, $n = 6$, and $P(A) = \frac{3}{6} = \frac{1}{2}$.

EXAMPLE 2

Consider drawing a single card from a well shuffled bridge deck containing 52 cards of 4 suits (clubs, diamonds, hearts, spades) and 13 denominations (deuce through ace). The probability of drawing a heart is $\frac{13}{52} = \frac{1}{4}$.

Notice that the concept "equally likely" is at the basis of the definition of probability. This supposes that the fundamental procedure or experiment of interest can be broken up into a set of basic outcomes, no one of which is more likely to occur than any other. In gambling procedures, great care is taken to insure that this will be true or nearly true. Dice produced from uniform material are carefully machined and are often thrown from a cup against a barrier; cards are carefully printed, are of uniform size and composition, and are shuffled carefully and thoroughly; roulette wheels are carefully balanced and vigorously spun. The care expended to insure equally likely outcomes seems to be proportional to the stakes involved in the game, but the end in view is the production of procedures that will satisfy the assumptions of our definition. An example will show how we go astray if equally likely outcomes are not assured.

EXAMPLE 3

Suppose that a penny and a nickel are tossed together. What is the probability of getting 2 heads?

Incorrect solution: There are 3 outcomes—2 heads, 1 head, and zero heads; one of these has the desired attribute; therefore, the required probability is $\frac{1}{3}$.

Correct solution: The solution just given fails because 2, 1, and zero heads are not equally likely. In fact, exactly 1 head is twice as likely as either exactly 2 heads or exactly zero heads. The following diagram shows the proper breakdown into equally likely outcomes, represented by the 4 columns.

Equally Likely Outcomes

	1	2	3	4
Penny:	H	T	H	T
Nickel:	H	H	T	T

Outcomes 2 and 3 both consist of exactly 1 head. Only outcome 1 consists of exactly 2 heads, and the correct probability is thus $\frac{1}{4}$. If you are unconvinced, take 2 coins and toss them awhile, making a record of the proportion of tosses resulting in exactly 2 heads. If you still do not believe it, come around sometime. We shall be happy to give you only 2 to 1 odds (corresponding to probability $\frac{1}{3}$) against the occurrence of 2 heads instead of the appropriate 3 to 1. We shall, however, insist upon a neutral coin supplier and tosser.

4-2   A Priori Probabilities

Such probabilities as those we have been discussing are sometimes called a priori probabilities, since they can be computed before the actual tossing of a die, drawing of a card, or tossing of 2 coins. Any a priori probability has the property that if a long sequence of trials of the procedure is carried out, the proportion of outcomes with the attribute will get closer and closer to the a priori value as the sequence of trials increases in length. If the true die of Example 1 is tossed repeatedly, the proportion of even outcomes will approach $\frac{1}{2}$.

In general, if we perform the procedure n' times and observe n'_A outcomes with the attribute A, then n'_A/n' approaches n_A/n, the a priori probability of A, as n' increases.

This property of probability is useful in dealing with some situations in which the basic procedure cannot conveniently be broken down into equally likely outcomes. For example, we may be playing a simple (in more ways than one) dice game like that of Example 1 with a Mississippi riverboat gambler whose outstanding features include a long waxy mustache, an embroidered brocade vest, and a large diamond stickpin in a wide bouffant silk cravat. After (hopefully) a short time we may perceive that the outcomes odd and even do not seem equally likely. A flicker of suspicion that the die may be loaded crosses our mind. Now what? How do we compute the correct probability? Of course we cannot compute it exactly, but if we begin to keep track of the running proportion of even outcomes, we know that this proportion will approach the true unknown probability. If that proportion seems to be approaching $\frac{1}{4}$ instead of $\frac{1}{2}$, the next move is up to us. This is not a textbook on coping with the criminal mind, although we shall have more to say on the subject in Chap. 8.

Suppose, as a second example, that we wish to estimate the probability that a live birth in Michigan produces a male child. We know that there are 2 outcomes, ignoring the small number of true hermaphrodites, but that these outcomes are not quite equally likely. If we record the sex in a long sequence of such births, we may estimate the correct probability.

The estimation of a probability using an observed relative frequency depends upon the assumption that it is meaningful to speak of such a probability. This assumes an underlying stability of the phenomenon under investigation. Gross instability may make it impossible for us to observe a relatively undisturbed sequence of events or may make our observed relative frequency applicable only to the unique, precise set of circumstances of observation. If this latter situation holds, then the calculated result cannot be used except in a historical descriptive sense and is no help at predicting the occurrence of future or related events.

4-3 Personal or Subjective Probability

Mathematicians and statisticians have lately renewed their interest in a kind of probability very different from the a priori and relative observed frequency probabilities we have been discussing. There are attributes or events which do not fit the formulation of our definition and which do not allow for the stable repetition of the relative frequency estimation procedure exemplified by the Mississippi gambler and Michigan male births. Yet in many such situations it seems both possible and reasonable to assign a numerical value between zero and one to our degree of belief in the statement expressing the attribute or event. In fact, we often use conventional a priori probabilities in this way. For example, our degree of belief that the true die will show an even number is adequately measured by $\frac{1}{2}$, the point midway between zero and one, the full range of variation of probability.

Examples of statements to which the a priori definition or relative frequency estimation of probability fail to apply are

1. There is intelligent life on Mars.
2. There is intelligent life on Earth.
3. The rain in Spain stays mainly in the plain.
4. A cure for cancer will be discovered within the next 5 years.
5. Coronary atherosclerosis is produced by a virus.

A common feature of these statements is that an element of personal subjectivity enters the numerical assessment of the likelihood that each is true. It is, moreover, impossible to conceive of a long sequence of repeated trials to get at relative frequencies in these cases. We might expect quite general agreement on statements 1 and 2 (although there is often ample reason to measure the truth of statement 2 with a number considerably short of 1). Our assessment of statement 3 might either reflect direct knowledge of Spanish meteorology and geography, confidence in Alan Jay Lerner, or pure guesswork. Quantitation for statements 4 and 5 will surely reflect personal experience, knowledge, and prejudice.

The probabilities we shall study will not be of the subjective or personal variety. This is perhaps more because of the lack of availability of comprehensive methodology and useful application of such techniques to biological problems to date than because of their lack of appropriateness. In the future the work of the so-called neo-Bayesian school of statisticians who are actively investigating the subjective basis of probability and statistics may have progressed sufficiently to allow routine use of techniques based upon this kind of probability. In our judgment, however, that day is still to come.

4-4   Principles of Enumeration

Our definition of probability depends upon the quantities n_A and n. These must be determined by considering the procedure under study and the particular attribute in which we are interested. In other words, we must learn to count.

4-4-1   A MULTIPLICATION PRINCIPLE

If an event can occur in a ways and if a second event can occur in b ways independent of the way in which the first event has occurred, then the 2 events together can occur in $a \cdot b$ ways.

EXAMPLE

Suppose a clinical laboratory has available 4 methods for determining a patient's serum cholesterol level and 2 methods for determining his blood glucose level; then the patient's serum cholesterol and blood glucose levels can be determined in $4 \cdot 2 = 8$ different ways. The following scheme illustrates the situation.

Method of determining serum cholesterol level: I, II, III, IV.

Method of determining blood glucose level: 1, 2.

Method of determining cholesterol and glucose levels: I1, I2, II1, II2, III1, III2, IV1, IV2.

For each choice of a method for cholesterol there are 2 choices of a method for glucose, giving 8 choices in all. Notice that the principle requires that the choice of the way in which the second event occurs must not depend upon the particular choice made for the first event. The reason for this is illustrated by the following situation.

EXAMPLE

Suppose again that there are 4 methods for serum cholesterol and 2 for blood glucose, but suppose further that if method IV is chosen for cholesterol, then method 2 must be used for glucose. This is presumably for reasons of economy or efficiency. Then the only possible choices for the 2 procedures are I1, I2, II1, II2, III1, III2, IV2. The muliplication principle has failed because of violation of the assumption of independent choice.

The multiplication principle can be extended to more than 2 events. For instance, if the first event can occur in a ways, the second in b ways independent of the choice for the first, and if a third event can occur in c ways independent of the choices for the first 2 events, then the 3 events together can occur in $a \cdot b \cdot c$ ways. Extension to larger numbers of events is made in a similar manner.

4-4-2 AN ADDITION PRINCIPLE

If one event can occur in a ways and if a second event can occur in b ways, then one or the other of the events can occur in $a + b$ ways, provided the 2 events cannot occur together.

EXAMPLE

A student nurse is assigned 2 male patients and 5 female patients to bathe as part of her morning duties. In choosing which patient to bathe first she can choose either a male or female patient. She has 2 choices of a male patient and 5 of a female and thus has $2 + 5 = 7$ ways of choosing her first patient.

The restriction in the addition principle that the 2 events cannot occur together can be rephrased to state that the events must be *mutually exclusive.* It should be clear for any of several reasons that this applies to the student nurse and her bathees. To show why this restriction is necessary, let us go back to the card table.

EXAMPLE

Suppose we want to select a single card from a deck of 52 cards. We know that there are 4 ways of selecting a jack from the deck and 13 ways of selecting a heart. There are not 17 ways of selecting a jack or a heart, however, because these 2 events (select a jack, select a heart) are not mutually exclusive. In fact, only 16 cards have the property of being either a jack or a heart. If we apply the addition principle in this situation, we count the jack of hearts twice—once as a jack, once as a heart.

As in the case of the multiplication principle, the addition principle is readily extendable to more than 2 events. For example, if 3 events can occur individually in a, b, and c ways, respectively, then the first or the second or the third can occur in $a + b + c$ ways, provided the events are mutually exclusive.

4-4-3 A Semantic Aid

In the multiplication principle we spoke of the occurrence of event 1 *and* event 2, whereas in the addition principle we were interested in event 1 *or* event 2. Speaking loosely, we can say that enumeration involving the connection of basic events by the word *and* implies multiplication, whereas when the basic events are combined by the word *or* we need to add. Do not lose sight of the restrictions of independence and mutual exclusiveness, which apply, respectively, to the 2 principles.

4-4-4 Examples Combining the Multiplication and Addition Principles

In discussing the example of serum cholesterol and blood glucose determinations we considered 4 methods, I, II, III, and IV, for cholesterol and 2 methods, 1 and 2, for glucose. We then examined the effect of supposing that if method IV is used for cholesterol then method 2 must be used for glucose and determined that 7 ways of carrying out the combined procedure were available. This can be determined by joint application of the 2 enumeration principles. Calling the first 3 cholesterol methods *unrestrictive* and the fourth method *restrictive*, we can say that the combined event is: Choose an unrestrictive cholesterol method and then choose a glucose method, or choose a restrictive cholesterol method and then choose a glucose method. This is of the form E_1 and E_2 or E_3 and E_4, where the E_i's are events.[2] Letting n_i be the number of ways in which the event E_i can occur and making the association between *and* and *multiply* and between *or* and *add* we have $n_1 \cdot n_2 + n_3 \cdot n_4 = $ number of ways of completing the composite event. But this becomes $3 \cdot 2 + 1 \cdot 1 = 6 + 1 = 7$, the result we obtained earlier.

In this example we have quite clearly used a hard method to do something easy, but in other cases it will be convenient, if not necessary, to break down composite events in this way. For example, consider the number of possible radio stations that could be formed and identified in the United States. The call letters of a radio station may consist of either 3 or 4 letters beginning with either W or K. A useful decomposition of the compound event into

[2] Note that E_2 and E_4, although given in the same wording, are different events, since they follow different choices of cholesterol method.

subevents is as follows:

E_1:   Pick 1st letter of 3-letter call sequence. $n_1 = 2$
E_2:   Pick 2nd letter of 3-letter call sequence. $n_2 = 26$
E_3:   Pick 3rd letter of 3-letter call sequence. $n_3 = 26$
E_4:   Pick 1st letter of 4-letter call sequence. $n_1 = 2$
E_5:   Pick 2nd letter of 4-letter call sequence. $n_2 = 26$
E_6:   Pick 3rd letter of 4-letter call sequence. $n_3 = 26$
E_7:   Pick 4th letter of 4-letter call sequence. $n_4 = 26$

The composite event is E_1 and E_2 and E_3 or E_4 and E_5 and E_6 and E_7. The corresponding number of choices is

$$n_1 \cdot n_2 \cdot n_3 + n_4 \cdot n_5 \cdot n_6 \cdot n_7 = 2 \cdot 26 \cdot 26 + 2 \cdot 26 \cdot 26 \cdot 26 = 36{,}504.$$

4-5   The Factorial Notation and Some of Its Properties

In dealing with some enumeration problems a particular notational device will be useful. This is defined by $k! = k(k - 1)(k - 2) \ldots (3)(2)(1)$, where k is a positive whole number. The symbol $k!$ is read "k factorial," and we advise readers when rendering aloud a mathematical passage containing this symbol quickly to overcome the natural tendency to pronounce the k with greater volume or feeling than is used for the rest of the passage. Mathematicians and statisticians are as a group more given to understatement than overstatement, and this fact neatly avoids the apparent ambiguity that might otherwise arise from using the exclamation point in a dual context.

Numerically we have

$$4! = 4 \cdot 3 \cdot 2 \cdot 1 = 24 \qquad 1! = 1, \quad \text{etc.}$$

We notice that $k! = k(k - 1)(k - 2) \ldots (3)(2)(1) = k(k - 1)!$, and if we wish this relationship to hold when $k = 1$ we have $1! = 1(1 - 1)! = 1 \cdot 0!$, but since we had not originally defined $0!$ at all, we can use this to impute such a definition. Dividing both sides by 1, we have

$$\frac{1!}{1} = 0! \quad \text{or} \quad 0! = 1.$$

Now this may rub you the wrong way at first, but remember that $0!$ and 0 are not necessarily the same and, in fact, we can define $0!$ any old way we choose. This definition has the advantage of generalizing the relation

$k! = k(k - 1)!$ to the case $k = 1$ and may partially rationalize the use of the exclamation point. After all, any operation that can make something out of nothing probably deserves an exclamation point. Furthermore, when you see in Sec. 4-6 how fast these factorials increase in size you may feel that an exclamation point is, if anything, too mild a symbol.

4-6   Permutations—Counting Ordered Sequences

An example will perhaps introduce this topic better than a generalization. Suppose that at the time of graduation of a certain medical school class there is no doubt which students are the top 5 members of the class scholastically, but because differences within this group are slight the dean appoints a faculty committee to select the graduates for the individual places. In how many ways can the individual places be determined? Any one of the 5 might be selected for first place, but having selected that individual, there are only 4 choices remaining for second place. Under the multiplication principle, there are $5 \cdot 4 = 20$ ways of filling these 2 places. Then there are 3 choices remaining for third place, after that, 2 choices for fourth place, and finally only a single remaining choice for fifth place. Applying the extension of the multiplication principle to more than 2 events, we finally find $5 \cdot 4 \cdot 3 \cdot 2 \cdot 1 = 5! = 120$ ways of selecting the order of finish of the top 5 students.

More generally, if k objects are to be arranged in order, there are $k!$ possible arrangements, since k choices are available to fill the first position; once it has been filled, there are $k - 1$ choices to fill the second position; etc. Thus $k(k - 1)(k - 2) \ldots (3)(2)(1)$ ordered arrangements are possible.

We mentioned earlier that this quantity $k!$ grows rapidly as k increases. Consider the problem of arranging 15 books on a shelf in all possible orders. By the reasoning we have used, the number of such arrangements is 15!. Now suppose we set about arranging the books in all possible ways and that we work very rapidly, making one rearrangement every second. How long would it take to complete the job? (Here interject a long pause for thought and reflection, to say nothing of the dramatic effect.) Well, if you were very industrious, worked 24 hours a day, 7 days a week, 365 days a year and had begun in 4000 B.C., you'd be roughly 15 per cent finished by now. This may establish for you our earlier point about the rate of increase in size of factorials. [If you think 15! is big, how about (15!)! ?]

Incidentally, an arrangement of objects in a particular order is called a *permutation* of the objects. We can say that the number of permutations of k objects is $k!$.

Now suppose we don't wish to permute all k of the objects but only wish to arrange r of them in all possible ways. For instance, if we consider the medical students at the time they are entering medical school as a class of 100 freshmen, the top 5 finishers might come out as follows: any of the 100 might finish first; having filled first place, any of the remaining 99 might finish second, etc. The multiplication principle then shows that there are $100 \cdot 99 \cdot 98 \cdot 97 \cdot 96$ ways of selecting the first 5 finishers in order from among an entering class of 100 students.

With k objects[3] from which r are to be selected and arranged in all possible orders, we have k choices for the first position; once that position has been filled, any of the remaining $k - 1$ objects can occupy the second position; finally, any of the remaining $k - (r - 1) = k - r + 1$ objects can be selected to fill the rth position. (Notice that in the medical student example when k was 100 and r was 5, the last or rth factor in the product was $96 = 100 - 5 + 1$. The ith factor in any such expression is $k - i + 1$.) The final result, then, is $k(k - 1)(k - 2) \ldots (k - r + 1)$. We shall, in the interest of economy of words, name this expression "the number of permutations of k objects taken r at a time" and denote it by the symbol $_kP_r$. That is, $_kP_r = k(k - 1)(k - 2) \ldots (k - r + 1)$. The factorial notation can give a further economy if we notice that

$$k! = k(k - 1)(k - 2) \ldots (k - r + 1)(k - r)(k - r - 1) \ldots (3)(2)(1)$$

$$= k(k - 1)(k - 2) \ldots (k - r + 1) \cdot (k - r)! = {}_kP_r \cdot (k - r)!.$$

Dividing both sides by $(k - r)!$, we have

$$_kP_r = \frac{k!}{(k - r)!}.$$

Now just to see if we are being consistent, suppose $r = k$. We are again permuting all the objects and by the earlier result we should have $_kP_k = k!$. But examining the formula for $_kP_r$ when $r = k$, we find

$$_kP_k = \frac{k!}{(k - k)!} = \frac{k!}{0!} = \frac{k!}{1} = k!,$$

which is consistent, provided that $0!$ is defined to be 1. Luckily we already thought to do that. Whew![4]

---

[3] We do not here imply that medical students are objects. We might have used the more general term "entities," but that somehow seems a bit formal.

[4] To be read "whew factorial."

4-7   Combinations—Sets of Objects Without Regard to Order

Now suppose that we do not care about order any more. That is, we wish merely to determine the number of ways in which r objects may be selected from the k objects, but order is irrelevant. Such an unordered set of r objects is called a combination. Back to the medical students. Suppose again that the 5 top finishers are known and we wish to select the top 2 finishers, but for the moment we do not care about order among those 2. Let the 5 students be denoted by A, B, C, D, and E. Then if order is ignored, there are 10 possible pairs as follows:

$$AB \quad AC \quad AD \quad AE \quad BC$$

$$BD \quad BE \quad CD \quad CE \quad DE.$$

For a small problem like this, enumeration of the possibilities works well, but what if we had the 100 students as entering freshmen and wished to select all possible sets of 5 top finishers ignoring order? Here enumeration is distinctly less attractive and we seek a general result.

Suppose that we want to find the number of combinations of k objects taken r at a time. We shall use an apparently devious route. Let X be the required number of combinations. Now return momentarily to the number of *permutations* of the k objects taken r at a time, i.e., $_kP_r$. This number can be obtained by applying the multiplication principle to the composite event E_1 and E_2, where

E_1:   Select a *combination* of r of the k objects.

E_2:   Rearrange the r objects selected in E_1 in all possible orders.

But the number of ways of doing E_1 is X and the number of ways of doing E_2 is $r!$, since we are permuting all r of the selected objects. On the other hand, E_1 *and* E_2, the composite event, can be completed in $_kP_r$ ways, since the completion of *both* E_1 and E_2 determines a particular permutation of r of the k objects. But then the multiplication principle says: Number of ways for E_1 and E_2 = (number of ways for E_1) · (number of ways for E_2), i.e.,

$$_kP_r = X \cdot r!,$$

whence, dividing through by $r!$ we obtain

$$X = \frac{{}_kP_r}{r!} = \frac{k!/(k-r)!}{r!} = \frac{k!}{(k-r)!\,r!}$$

and X is determined. It seems only fair to elevate the status of X, however, by giving it a more suitable name, say ${}_kC_r$, the number of combinations of k objects taken r at a time.

In summary

$$ {}_kC_r = \frac{k!}{(k-r)!\,r!} \qquad {}^5 $$

Before moving on we should check the medical student example that introduced this section. If the formula disagrees with our result of 10 determined by enumeration, by all means let us reexamine the formula. Here $k = 5$ and $r = 2$. The formula says

$$ {}_kC_r = {}_5C_2 = \frac{5!}{(5-2)!\,2!} = \frac{5!}{3!\,2!} $$

$$ = \frac{5 \cdot 4 \cdot 3 \cdot 2 \cdot 1}{3 \cdot 2 \cdot 1 \cdot 2 \cdot 1} = \frac{5 \cdot 4}{2 \cdot 1} = 10. $$

Although this, of course, does not prove the correctness of the formula for all possible values of k and r (the above argument based on E_1 and E_2, however, is such a proof), it should bolster our confidence a wee bit. Furthermore, we can now handle the problem of determining the number of ways of selecting the top 5 finishers from the class of 100, ignoring order. This becomes

$$ {}_kC_r = {}_{100}C_5 = \frac{100!}{95!\,5!} $$

$$ = \frac{100 \cdot 99 \cdot 98 \cdot 97 \cdot 96}{5 \cdot 4 \cdot 3 \cdot 2 \cdot 1} = 75{,}287{,}520. $$

[5] Other notations are sometimes used for ${}_kP_r$ and ${}_kC_r$. These are

For ${}_kP_r$: $P_{k,r}$ or $P(k, r)$,

For ${}_kC_r$: $C_{k,r}$ or $C(k, r)$ or $\binom{k}{r}$.

This latter notation for ${}_kC_r$ is quite common, particularly in mathematical literature. We have avoided it to achieve notational symmetry with ${}_kP_r$.

The quantity $_kC_r$ has several interesting and important properties:

1. $_kC_k = 1.$

   Proof:

   $$_kC_k = \frac{k!}{0!\, k!} = \frac{k!}{1 \cdot k!} = 1.$$

   This is reasonable, since there is just one way to take all k of the objects, i.e., just one set consisting of all the objects.

2. $_kC_0 = 1.$

   Proof:

   $$_kC_0 = \frac{k!}{k!\, 0!} = 1.$$

   This is a little curious and seems to say that there is one way of taking exactly none of the objects. On the other hand, if we take none then we leave them all, and there should be just one way of doing that. This is in accord with property 1.

3. $_kC_r = {_kC_{k-r}}$

   Proof:

   $$_kC_r = \frac{k!}{(k-r)!\, r!} = \frac{k!}{r!\,(k-r)!} = \frac{k!}{[k-(k-r)]!\,(k-r)!} = {_kC_{k-r}}.$$

   If you prefer a prose proof to a symbolic one, consider that whenever we select a combination of r objects we leave behind a combination of $k-r$ unselected objects. Thus for every combination of r objects there corresponds a unique combination of $k-r$ objects. Since these can be matched up one to one, it follows that the number of combinations of r from the k and the number of combinations of $k-r$ from the k are the same.

4. $_kC_1 = {_kC_{k-1}} = k.$

   Proof:

   $_kC_1 = {_kC_{k-1}}$ from property 3.
   But $_kC_1 = k!/(k-1)!\, 1! = k$, completing the proof. Again the number of ways of selecting a single object from among k must be k.

4-8 Number of Distinguishable Permutations of Objects Not All Different

It is sometimes necessary to count the number of distinguishable ordered arrangements of objects not all different. Considering the word MAMA, you cannot tell when you see the rearrangement MAMA that we have in fact interchanged the 2 M's. The rearrangement MAAM, on the other hand, is distinguishably different. By enumeration we find that there are 6 distinguishable permutations:

MAMA   MAAM   MMAA

AMAM   AMMA   AAMM

If we try to do the same thing to the word BIOSTATISTICS we may tire of the job of writing a complete list and seek a general formulation.

Suppose that there are k objects, k_1 of which are alike and of the first kind, k_2 of which are alike and of the second kind, etc., finally k_t of which are alike and of the tth kind. For example, in the word MAMA, $k = 4$, $k_1 = 2$ (the number of M's), $k_2 = 2$ (the number of A's), and $t = 2$. Allowing some of the k_i to take the value 1 if necessary, we can insure that $\sum_{i=1}^{t} k_i = k$, since every object in the set of k objects will be of some type, possibly unlike any other object.

In approaching the general solution we temporarily consider the case that all k objects are different. We might, for instance, subscript all the identical items, e.g., $M_1 A_1 M_2 A_2$. Notice that the rearrangement $M_2 A_1 M_1 A_2$ is now distinguishable. In this case the total number of distinguishable permutations is just the total number of permutations of k different objects, i.e., $_kP_k = k!$ But making use of the multiplication principle, we can obtain a particular one of these permutations as the composite events E_0 and E_1 and E_2 and . . . and E_t, where the E_i are

E_0:   Pick a distinguishable permutation in the original sense, ignoring subscripts.

E_1:   Rearrange the k_1 objects of the first kind.

E_2:   Rearrange the k_2 objects of the second kind.

. . .

E_t: Rearrange the k_t objects of the tth kind.

The number of ways of completing the composite event, which we know to be $k!$, is the product of the numbers of ways of completing the individual subevents. But the number of ways of completing E_0 is X, the number we seek; the number of ways of completing E_1 is $k_1!$; the number of ways of completing E_2 is $k_2!$; ...; finally, the number of ways of completing E_t is $k_t!$. Then we have

$$k! = Xk_1!\,k_2!\ldots k_t!,$$

whence

$$X = \frac{k!}{k_1!\,k_2!\ldots k_t!}.$$

Now back to MAMA. We know from enumerating all the cases that $X = 6$, but the formula says $X = 4!/(2!\,2!) = 6$. In the case of BIOSTATISTICS we have $k = 13$ objects, $k_1 = 1$ of which is a B, $k_2 = 3$ of which are I's, $k_3 = 1$ of which is an O, $k_4 = 3$ of which are S's, $k_5 = 3$ of which are T's, $k_6 = 1$ of which is an A, and $k_7 = 1$ of which is a C. Notice that $\sum_{i=1}^{t} k_i = 13 = k$. We have

$$X = \frac{13!}{1!\,3!\,1!\,3!\,3!\,1!\,1!} = 28{,}828{,}800,$$

a number large enough to justify our earlier reluctance to enumerate all the cases. Paraphrasing O. Henry's Mrs. Sampson, we think formulas are just as lovely as they can be. Don't you?

The MAMA example can be made to yield one more point. In this case the result 6 is in fact $_4C_2$. Is this a coincidence? No, the case $t = 2$ always produces the result $_kC_{k_1}$, since

$$\frac{k!}{k_1!\,k_2!} = \frac{k!}{k_1!\,(k - k_1)!} = {_kC_{k_1}},$$

the first equal sign arising from the fact that $k_1 + k_2 = k$ or $k_2 = k - k_1$. Notice that a distinguishable permutation of MAMA is determined as soon as we fix the positions of the M's without regard to order. In other words, from the set of 4 objects (positions) we pick 2 objects, ignoring their order. The remaining positions are, of course, occupied by A's. The generalization to any values of k and k_1 so long as $t = 2$ is immediate.

4-9   Probability of Composite Events

The multiplication and addition principles are useful in determining the probability of certain composite events. We suppose that a procedure can result in n equally likely outcomes, n_A having attribute A and n_B with attribute B. Then $P(A) = n_A/n$ and $P(B) = n_B/n$. We shall call A and B events with the understanding that by the event A we mean that the procedure has resulted in an outcome having attribute A and similarly for B.

Now consider performing the procedure twice. On the second trial suppose that the total number of equally likely outcomes n and the total number n_B having attribute B are not influenced by the outcome of the first trial. Then the multiplication principle says that the 2 trials can result in $n \cdot n = n^2$ equally likely outcomes and that $n_A \cdot n_B$ of these show attribute A on the first trial and attribute B on the second. Then the definition of probability can be applied to the composite event A and B to produce

$$P(A \text{ and } B) = \frac{n_A \cdot n_B}{n^2} = \frac{n_A}{n} \cdot \frac{n_B}{n} = P(A) \cdot P(B).$$

Notice, however, that in order for this result to hold, neither the structure of the procedure nor the outcome of the second trial must depend upon the first trial. When this is true we say that *the events A and B are independent*. Because of the basic character of this relation between probabilities we shall use it as the definition of independent events.

Definition

Two events A and B are said to be *independent in the probability sense* if $P(A \text{ and } B) = P(A) \cdot P(B)$.

EXAMPLE

Suppose a fair coin is tossed twice. Find the probability of getting 2 heads.

The outcome 2 heads can be considered as the composite event: head on the first trial and head on the second.

$$P(\text{head on first}) = \tfrac{1}{2}, \qquad P(\text{head on second}) = \tfrac{1}{2},$$

and

$$P(\text{head on first}) \cdot P(\text{head on second}) = \tfrac{1}{2} \cdot \tfrac{1}{2} = \tfrac{1}{4}.$$

But if the coin is fair, then the 4 outcomes HH, HT, TH, and TT will be equally likely and thus

$$P(2 \text{ heads}) = P(\text{HH}) = \tfrac{1}{4}.$$

Since this is equal to $P(\text{head on first}) \cdot P(\text{head on second})$, these 2 basic events are independent. In fact, the independence here is a natural and direct consequence of the assumption of a fair coin.

The definition of independence can be extended to more than 2 events as follows:

Definition

The events $A_1, A_2, \ldots, A_k$ are said to be *mutually independent* if

$$P(A_1 \text{ and } A_2 \text{ and } \ldots \text{ and } A_k) = P(A_1) \cdot P(A_2) \cdots P(A_k).$$

The addition principle can help in calculating probabilities for events of the form "*A* or *B*." If *A* can occur in n_A ways and if *B* can occur in n_B

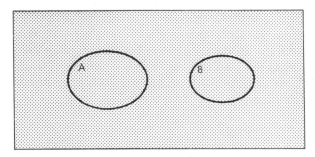

Figure 4-1   Two mutually exclusive events.

ways and if *A* and *B* are mutually exclusive, then the number of equally likely outcomes favorable to *A* or *B* is $n_A + n_B$ and

$$P(A \text{ or } B) = \frac{n_A + n_B}{n} = \frac{n_A}{n} + \frac{n_B}{n} = P(A) + P(B).$$

If the *n* outcomes are represented by points within a rectangular region and the outcomes showing attributes *A* and *B* are represented by points within oval shaped subregions, the situation diagrammatically is shown in Fig. 4-1. Quite clearly a similar result holds for more than 2 events. If $A_1, A_2, \ldots, A_k$ are mutually exclusive events, then

$$P(A_1 \text{ or } A_2 \ldots \text{ or } A_k) = P(A_1) + P(A_2) + \cdots + P(A_k).$$

A more general formula eliminating the assumption of mutual exclusiveness can be derived. Figure 4-2 shows this situation; the dots are eliminated for clarity.

Here the ovals for *A* and *B* overlap in the shaded lens-shaped region.

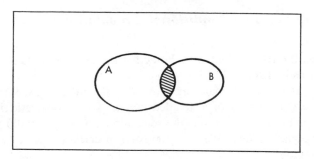

Figure 4-2 Two events which are not mutually exclusive.

In this region both *A* and *B* occur. In determining the number of outcomes favorable to *A* or *B*, we note that the count $n_A + n_B$ is too large, since it counts outcomes in the shaded region twice. But the number of points in this region may be denoted by $n_{A \text{ and } B}$, since it is the number of outcomes favorable to the joint event *A* and *B*. As a result the number of points favorable to *A* or *B* is

$$n_A + n_B - n_{A \text{ and } B}$$

and

$$P(A \text{ or } B) = \frac{n_A + n_B - n_{A \text{ and } B}}{n} = \frac{n_A}{n} + \frac{n_B}{n} - \frac{n_{A \text{ and } B}}{n}$$

or

$$P(A \text{ or } B) = P(A) + P(B) - P(A \text{ and } B).$$

This reduces to the earlier case if *A* and *B* are mutually exclusive, since then $n_{A \text{ and } B} = 0$ and $P(A \text{ and } B) = 0$.

EXAMPLE

Find the probability of drawing a jack or a heart from a well shuffled deck of 52 cards.

Here *A*: drawing a jack.

 B: drawing a heart.

A and *B*: drawing a jack and a heart, i.e., drawing the jack of hearts.

$$P(A) = \tfrac{4}{52} = \tfrac{1}{13}; \qquad P(B) = \tfrac{13}{52} = \tfrac{1}{4}; \qquad P(A \text{ and } B) = \tfrac{1}{52}.$$

From the formula, $P(A \text{ or } B) = P(A) + P(B) - P(A \text{ and } B) = \tfrac{1}{13} + \tfrac{1}{4} - \tfrac{1}{52} = \tfrac{16}{52}$. This conforms with the result of the second example of Sec. 4-4-2.

4-10   *Marginal and Conditional Probability*

Suppose the 100 residents of a home for the aged are cross-classified by age and sex as in Table 4-1.

That is, there are 20 males aged 70–79, 10 females aged 90 and over, etc. Suppose that as part of a detailed clinical study, one resident is to be selected at random—i.e., so that the selection of any of the 100 individual residents is equally likely. We are interested in certain types of events and their probabilities.

Table 4-1

CROSS-CLASSIFICATION BY AGE AND SEX OF 100 RESIDENTS
OF A HOME FOR THE AGED

Sex	70–79	80–89	90 *and over*	Total
Male	20	10	10	40
Female	30	20	10	60
Total	50	30	20	100

For example, what is the probability of selecting a male? Since 40 of the 100 subjects are male, we have

$$P(\text{selecting a male}) = \tfrac{40}{100} = 0.4.$$

Similarly, $P(\text{selecting an 80–89 year old}) = \tfrac{30}{100} = 0.3$. Such events are called marginal events and their probabilities are called marginal probabilities. The terminology is suggested by the fact that the frequencies in the numerators of such probabilities are marginal totals in the cross-classification. For example, in $P(\text{selecting a male}) = \tfrac{40}{100}$, the 40 is the marginal row total of the male row.

Definition

If the n equally likely outcomes of a procedure are cross-classified according to 2 or more classification variables and if the specification of an event fails to mention one or more of the classification variables, then such an event is called a *marginal event* and its probability is called a *marginal probability.*

The 2 classification variables in the present example are age and sex, and the event "male"—the words "selecting a" being omitted for brevity—fails to mention the age classification; the event "male" is a marginal event.

Suppose that for a special substudy we wish to select our subject only from among the female residents. Now consider within this restricted group the probability of selecting an 80–89 year-old. In this case there are only 60 equally likely outcomes—the 60 women in the home—and the required probability is $\frac{20}{60} = \frac{1}{3}$. In this case the event refers to a reduced set of equally likely outcomes, and a special notational device is needed to make this clear. We write $P(80\text{–}89 \mid \text{female}) = \frac{1}{3}$ and we say, "The conditional probability of selecting an 80–89 year-old, given that we select a female, is $\frac{1}{3}$."

Notice that $P(80\text{–}89 \mid \text{female})$ and $P(\text{female} \mid 80\text{–}89)$ refer to different events. In the latter case we are limiting selection to 80–89 year-olds and asking the probability of selecting a female. In fact, $P(\text{female} \mid 80\text{–}89) = \frac{2}{3}$. Similarly, $P(\text{male} \mid 90 \text{ and over}) = \frac{1}{2}$ and $P(90 \text{ and over} \mid \text{male}) = \frac{1}{4}$.

Definition

Events of the form "A given B" which state the occurrence of A given that B must occur are called *conditional events*. Their probabilities are called *conditional probabilities* and are denoted by $P(A \mid B)$, provided $P(B) \neq 0$.

The restriction that $P(B)$ must be different from zero is important. Suppose we ask for $P(\text{male} \mid 60\text{–}69)$, that is, the probability of selecting a male given that a 60–69 year-old is selected. We might say, "I don't understand the question. There aren't any 60–69 year-olds." On the other hand, there are no males in that age group, and thus the probability must be zero. Yet, all the 60–69 year-olds are males (or anything else for that matter, since there are none of them), and so the probability must be 1.[6] Clearly, we are in a dilemma. In fact it does not make sense to talk about the probability of such an event, and as a result we do not define $P(A \mid B)$ unless $P(B) \neq 0$.

Considering again the event 80–89 given female, notice that

$$P(80\text{–}89 \mid \text{female}) = \frac{1}{3} = \frac{20/100}{60/100} = \frac{P(80\text{–}89 \text{ and female})}{P(\text{female})}.$$

The event in the numerator specifies jointly both age (80–89) and sex (female) and thus is called a *joint event*. It is true in general that for any 2 events A and B

$$P(A \mid B) = \frac{P(A \text{ and } B)}{P(B)}.$$

[6] Logically speaking, any statement about a group consisting of no objects is true—vacuously true, as the logicians say. Thus, we can say that all the 60–69 year-olds are male or that they are all female, both statements being vacuously true.

This expression, whose denominator is $P(B)$, may make clearer the reasons we do not define $P(A \mid B)$ when $P(B) = 0$. Division by zero is an undefined operation.

To see why this expression for $P(A \mid B)$ holds, suppose the procedure has n equally likely outcomes with n_A favorable to A, n_B favorable to B, and $n_{A \text{ and } B}$ favorable to both A and B. $P(A \mid B)$ is the probability that A occurs given that B occurs—out of the n_B outcomes favorable to B we want the relative number favorable to A as well, i.e., $P(A \mid B) = (n_{A \text{ and } B})/n_B$.

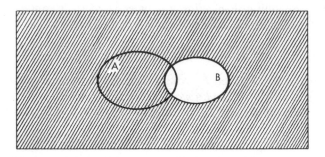

Figure 4-3   **Elimination of certain outcomes in the calculation of** $P(A \mid B)$.

But dividing numerator and denominator by n, we have

$$P(A \mid B) = \frac{n_{A \text{ and } B}/n}{n_B/n} = \frac{P(A \text{ and } B)}{P(B)},$$

the required result. The conditional probability eliminates all outcomes not favorable to B (shaded area) in Fig. 4-3 and asks for the ratio of the number of outcomes in the lens-shaped (A and B) region to the number in the B region.

If A and B are independent, we have

$$P(A \mid B) = \frac{P(A \text{ and } B)}{P(B)} = \frac{P(A) \cdot P(B)}{P(B)} = P(A).$$

This casts more light on the definition of independence. If A and B are independent, then the probability of A is unaffected by the occurrence or nonoccurrence of B. That is, the conditional probability of A given the occurrence of B is the same as the probability of A without mention of B.

4-11 Bayes's Rule

As we have seen, the two quantities $P(A \mid B)$ and $P(B \mid A)$ refer to different events and in general are not equal. For example, if B is some kind of effect and A is a cause, we often have information concerning $P(B \mid A)$ when in fact what we want is $P(A \mid B)$. The basic problem of medical diagnosis is of this type, for the physician may know the probability of a particular complex of signs and symptoms given a particular disease, but he usually wants for the individual patient the probability of the disease given the observed symptom complex.

We shall consider first a simple case involving determination of the presence or absence of a particular disease D based upon an observed symptom complex S. Let D denote presence of disease and $\bar{D}$ denote its absence. Suppose we know $P(D)$ in the general population. Property 3 of Sec. 4-1 then says that $P(\bar{D}) = 1 - P(D)$. Furthermore, suppose we know $P(S \mid D)$ and $P(S \mid \bar{D})$, the probabilities, respectively, that subjects with and without the disease will exhibit the symptom complex. We wish to find $P(D \mid S)$. Notice first that all individuals with S either have D or fail to have it; i.e., S can be written "S and D or S and $\bar{D}$." Since the 2 events linked by "or" are mutually exclusive, we have

$$P(S) = P(S \text{ and } D) + P(S \text{ and } \bar{D}).$$

Furthermore, since

$$P(S \mid D) = \frac{P(S \text{ and } D)}{P(D)} \quad \text{and} \quad P(S \mid \bar{D}) = \frac{P(S \text{ and } \bar{D})}{P(\bar{D})},$$

it follows, if we multiply by $P(D)$ and $P(\bar{D})$, respectively, that $P(S \text{ and } D) = P(D) \cdot P(S \mid D)$ and $P(S \text{ and } \bar{D}) = P(\bar{D}) \cdot P(S \mid \bar{D})$. But

$$P(D \mid S) = \frac{P(D \text{ and } S)}{P(S)} = \frac{P(S \text{ and } D)}{P(S \text{ and } D) + P(S \text{ and } \bar{D})}$$

or finally

$$P(D \mid S) = \frac{P(D) \cdot P(S \mid D)}{P(D) \cdot P(S \mid D) + P(\bar{D}) \cdot P(S \mid \bar{D})},$$

the expression commonly known as Bayes's rule after the Rev. Thomas Bayes (1702–1761) (3). Notice that the expression on the right involves only quantities that we have assumed to be known.

EXAMPLE

A new screening test is proposed for the detection of a particular form of cancer in women over 40 years of age. The prevalence of the disease in the general population of such women has been estimated by a nationwide survey to be 1 per cent; i.e., 1 per cent of women over 40 have the disease. The test has been investigated in women with confirmed disease and is found to give a positive result in 95 per cent of such cases; i.e., in the terminology applied to such screening devices this test has a sensitivity of 95 per cent. When given to women known to be free of the malignancy it gives a positive result in only 3 per cent; i.e., the specificity of the test or the probability of a negative result in disease-free subjects is 97 per cent. Suppose we let D denote presence of the particular form of cancer and let S denote a positive response to the screening test. The epidemiologists and public health administrators considering widespread use of this screening procedure in large populations want to know the proportion of women with positive tests who when followed up by confirmatory biopsy and histologic examination, will actually be found to have disease. Translating prose into probability statements, we have

$$P(D) = 0.01,$$

whence

$$P(\bar{D}) = 0.99; \quad P(S \mid D) = 0.95, \quad \text{and} \quad P(S \mid \bar{D}) = 0.03.$$

We seek $P(D \mid S)$. But Bayes's rule says

$$P(D \mid S) = \frac{(0.01)(0.95)}{(0.01)(0.95) + (0.99)(0.03)} = 0.24.$$

In other words, after application of the screening test in the general population with follow-up of all positives, only 24 per cent of these positives will be found actually to have disease. There is a moral here, and we cannot resist stating it. Even a screening test with high sensitivity and specificity, when applied to a population in which the disease in question is of low frequency, may produce a low yield of true positives among those who are screened out as positive. The possible resulting inconvenience, cost, pain, or actual risk of applying the confirmatory procedure to many subjects who do not need it should be weighed carefully.

There is a more general form of Bayes's rule which we shall now state without proof, although the proof follows similar lines to that given for the more restricted form. Suppose we again have a symptom complex S but wish to choose between t distinct disease states $D_1, D_2, \ldots, D_t$, possibly including complete absence of disease as one state, such that $\sum\limits_{i=1}^{t} P(D_i) = 1$ and such that $P(D_i)$ and $P(S \mid D_i)$ are known for each i. Then

$$P(D_j \mid S) = \frac{P(D_j) \cdot P(S \mid D_j)}{\sum\limits_{i=1}^{t} P(D_i) \cdot P(S \mid D_i)}$$

for any $j = 1, 2, \ldots, t$. Notice that the restricted case established earlier in this section corresponds to $t = 2$. The more general form can be used in some situations to make a differential diagnosis between the t disease states by selecting that D_j for which $P(D_j \mid S)$ is a maximum. Do not get the idea that Bayes's rule in its more general form is a kind of statistical panacea for all kinds of diagnostic situations. Notice that all the $P(D_i)$ must be known— that is, the disease prevalence for each of the t disease states must be known or estimated for the population under investigation. This is often a very tough bill to fill. Furthermore, the $P(S \mid D_i)$ must also be available; this means that a series of reliable clinical investigations concerning the manifestations of each disease state giving results in quantitative form applicable to the patient group under investigation must have been completed. Even so, there are instances in which this formulation has proved useful.

Bayes's rule applies to many other situations than those illustrated here. This will be brought out more clearly by the exercises at the end of the chapter.

EXERCISES

4-1 Suppose a dietitian has available the following foods listed by their main vitamin content:

Vitamin A	Vitamin B Complex	Vitamin C
lettuce	peanuts	oranges
carrots	peas	lemons
squash	lean meat	
egg yolk	egg white	
butter	liver	
	milk	
	cereal	

a. How many meals are possible containing one food from each vitamin group?

b. How many meals are possible containing 3 foods from the vitamin A group (and none from vitamin B or C groups)?

c. How many meals are possible containing 2 foods from the vitamin A group, 3 from the vitamin B group, and none from vitamin C?

    d. How many meals are possible containing 2 foods from group A, 3 from group B, and one from group C?

    e. How many meals are possible containing 4 A foods, 4 B foods, and 1 C food?

4-2 As chairman of the nominations committee of the local medical society you are to select:

    a. One candidate for president, one for vice-president, and one for secretary. If 100 members of the society are eligible, how many slates are possible?

    b. Three members to go as representatives to the national convention in Miami. If all 100 members are eligible, how many sets of three representatives are possible?

4-3 How many distinguishable permutations are there of the letters in the words:

    a. MEDICINE?

    b. STETHOSCOPE?

    c. PHARMACEUTICAL?

4-4 In a particular hospital 50 patients are eligible for study of long-term hospital care. As the administrator you have 50 cards, one for each patient, on your desk with the information side face down. The pertinent distributions are

Sex		Age		Length of Illness	
Male	20	20–40	10	<6 months	15
Female	30	40–60	10	6–12 months	15
		60–80	30	12–36 months	10
				>36 months	10

    a. You randomly select one card. What is the probability that:

      1. It is a female's card?

      2. It belongs to a patient whose length of illness has been 12–36 months?

      3. It belongs to a patient whose age interval is either 20–40 or 60–80?

    b. You randomly select 5 cards with replacement. What is the probability that:

      1. All 5 are males' cards?

      2. All 5 cards are females'?

    c. You randomly select 3 cards without replacement. What is the probability that:

      1. All 3 are in the age interval 40–60?

      2. The first 2 are in the age interval 20–40 and the third is in the interval 60–80?

    d. Assuming that the three variables, sex, age, and length of illness, are independent, you randomly select 4 cards without replacement. What is the probability for these 4 that: 2 are males and 2 females, 3 are in the age interval 20–40 and one is in the interval 60–80, and one is in each of the 4 length of illness categories?

4-5 Let A and B denote two genetic characteristics and suppose that the probability is $\frac{1}{2}$ that an individual chosen at random will exhibit A, $\frac{3}{4}$ that he will exhibit B. Assume that these characteristics occur independently.

    a. What is the probability that an individual chosen at random will exhibit both?

    b. Neither?

    c. Exactly one?

4-6 Five identical rabbits are in a cage. Some have been inoculated against a particular virus. Find the probability that you select the inoculated rabbit(s) if:

    a. You select one and only one was inoculated.

    b. You select three and two were inoculated.

    c. You select two and two were inoculated.

    d. You select two and want them in the order they were inoculated; two were inoculated, one on Wednesday and one on Thursday.

4-7 A geneticist studying a certain hybrid flower is able to grow plants producing flowers of one of three colors, red, white, or blue, with probabilities $\frac{1}{2}$, $\frac{1}{3}$, and $\frac{1}{6}$. If he has several plants with unknown flower colors, what is the probability that if:

    a. He selects two, one is red and the other white?

    b. He selects two, at least one is red?

    c. He selects one, it is red or white?

    d. He selects three, one is red, one white, and one blue?

4-8 Suppose a certain opthalmic trait is associated with eye color. Three hundred randomly selected individuals are studied with results as follows:

Eye Color

Trait	Blue	Brown	Other
Yes	70	30	20
No	20	110	50

Using these data, find:

    a. P (trait).

    b. P (blue eyes and trait).

    c. P (blue or brown eyes and trait).

    d. P (brown eyes | trait).

4-9 Four hundred adult males with angina pectoris are classified by age and weight as follows:

Weight (in pounds)

Age (Years)	130–149	150–169	170–189	≥ 190
30–39	10	20	20	40
40–49	10	15	50	70
50–59	5	15	50	40
60–69	5	10	15	25

Using the above table, find for an individual the probability that he:

a. Is in the age interval 40–49.
b. Is in the age interval 40–49 and weighs 170–189 lb.
c. Is in the age interval 40–49 or 60–69.
d. Is in the age interval 40–49 or 60–69 and weighs 150–169 lb.
e. Is in the age interval 40–49 given that he weighs 150–169 lb.
f. Weighs less than 170 lb.
g. Weighs less than 170 lb and is less than 50 years old.
h. Weighs less than 170 lb given that he is less than 50 years old.

4-10 You and 3 other intern friends set your identical (except for initials inside the cup) stethoscopes on a table while drinking coffee. You each pick up a stethoscope at random when you leave. What is the probability that:

a. Each picks up his own stethoscope if you all leave together?
b. You and one friend will have your stethoscopes between you (i.e., you may have your own or your friend's and vice versa) if both of you leave before the others?

4-11 A medical investigator is studying three drugs identical in appearance numbered 1, 2, or 3. When the drugs are injected in guinea pigs the probabilities that an antitoxin will develop are $\frac{1}{4}$ for number 1, $\frac{1}{8}$ for number 2, and $\frac{1}{3}$ for number 3. There are 2 bottles of number 1, 3 bottles of number 2, and one bottle of number 3. The investigator, in hurrying to give a demonstration of these drugs, takes one bottle without noting its number. In transit the label is torn off. When a guinea pig is injected with this drug, no antitoxin forms. What is the probability that it was drug number 2?

4-12 In an epidemiological experiment involving rats, cages holding one healthy rat each are used. A second rat infected with one of the organisms A, B, or C, is introduced into each cage. The probability is $\frac{1}{3}$ that the rat introduced is infected with A, $\frac{1}{3}$ that it is infected with B, and $\frac{1}{3}$ that it is infected with C. If an A-infected rat is introduced, the probability is $\frac{1}{2}$ that the healthy rat will become ill; the corresponding probabilities for B-infected rats and C-infected rats are $\frac{1}{3}$ and $\frac{1}{4}$, respectively. In one cage, the technician loses the information concerning the type of rat that was introduced. However, the healthy rat in the cage did become ill. What is the probability that the rat introduced was infected with A?

4-13 Show how the multiplication principle can be used alone, i.e., without the addition principle, to arrive at the total number (36,504) of possible United States radio stations as discussed in Sec. 4-4-4.

5 Populations, Samples, and Inference

We now elaborate some of the ideas of Chap. 1 concerning populations, samples, and methods for making inferences from a sample to a population. Such inferences are called statistical inferences, and the specifics of the procedures involved will be the subject of most of the remainder of this book. A series of rather formal definitions and less formal figures will open our discussion.

5-1 Definitions and Basic Ideas Related to Populations and Samples

Definition

A *population of units* is a group of entities having some quantifiable characteristic in common.

The units may be people, machines, animals, bacteria, families, or any other entities. They may even be individual times at which measurements on a single person are taken. They may either be infinite or finite in number. In this book most such populations will be assumed to be either infinite or at least extremely large. You will see later that this is not such a severe restriction as it seems at first glance. The quantifiable characteristic may be a continuous or a discrete variable and might even be a coded or nominal variable such as sex or color. Figure 5-1 is a schematic representation of a population of units. Again, the sets of 3 dots are used to represent continuation and to indicate that the population may be very large.

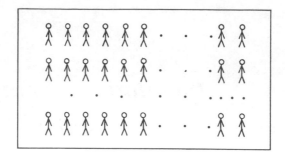

Figure 5-1   A population of units.

Definition

A *population of observations* is a group consisting of the numerical values of a quantifiable characteristic determined on each member of a population of units.

The same population of units will often have more than one population of observations associated with it. For instance, in studying blood pressure, systolic and diastolic blood pressure are ordinarily both determined. Thus, the same population of units—in this case, people—yields 2 populations of observations. Figure 5-2 represents a population of observations, using x as a generic notation for the values of the characteristic under study. As we mentioned in Chap. 1, the gap between the population of units and the population of observations may be large, and while the investigator really wishes to study the units, he does so through the observations with the attendant risk of error of measurement and possible misinterpretation.

Definition

A *sample of units* is some finite number of the units from a population of individuals.

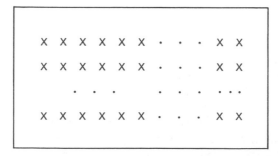

Figure 5-2   A population of observations.

Figure 5-3   A sample of units.

Definition

A *sample of observations* is some finite number of the observations from a population of observations.

In other words, a sample is some part of a population. Figures 5-3 and 5-4 illustrate the situation.

In practice, although the populations of units and observations both exist, they are not available to the investigator for detailed study. For example, if the population of units consists of all adult males in the United States and if the observation of interest is height, then no investigator will be able to study all individuals in the population or to measure their heights. He must be content with measuring the heights of a sample of individuals. From this he hopes to learn something about the population of men. If he is interested in the arithmetic mean height of all the men in the population, then he might examine the arithmetic mean height of all the men in the sample. Of course, he will never be able to assert that these arithmetic means are identical, but if the sample is selected appropriately, he might be able to make precise statements concerning the mean of the population of heights. This procedure is inductive in the sense that the investigator reasons from the specific (the sample) to the general (the population). It is the business of statistics to develop procedures for making such inductions or inferences from sample to population and to study the influence of such things as changes in sample selection procedure, sample size, and choice of sample characteristic providing maximum information about a given population characteristic.

It is rare in the biological sciences that the investigator has access to information on the entire population. He is more often interested in generalizing his conclusions to individuals not specifically measured or studied. The clinician, for example, who is studying the relative merits of new and

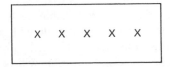

Figure 5-4   A sample of observations.

old forms of treatment against a particular disease will ordinarily wish to learn from his therapeutic trial what to expect later when treating similar patients. His colleagues in other centers will hope to apply his findings successfully to their own patients. Finally, the basic scientist seeking to refine and improve the treatment procedure still further will hope that the individual trial results can be depended upon to hold generally and thus to provide the basis for his continuing biochemical, physiological, or pharmacological studies. In fact, the patients in the clinical trial are a sample of units in the sense that they are providing information about a much larger group of patients to whom inferences need to be made. This is not to say that the patients in the trial are unimportant either in their own right as individuals or as providers of information of benefit to other similar patients.

Though this example is stated in the context of clinical medicine, readers from other fields will recognize the role of sample and population in their own areas of interest. The attempt to recognize order and system in the universe, when based upon quantitative observation, almost inevitably depends upon this process of generalization from sample to population. Such generalizations can, of course, be made without applying formal statistical methodology, but most scientists today would agree that systematic study of procedures for making such generalizations is valuable. That systematic study is statistics.

The statistician is occasionally confronted with the argument that the data from the study are an end in themselves and that no generalizations are either possible or desirable. For example, the clinician may claim that his sole interest is in the individual patient currently under treatment. To that argument the statistician can only ask if the relevant decisions concerning patient management are not based upon training, prior personal experience with similar patients, or familiarity with the appropriate literature. If any one of these bases applies, then a generalization of an inductive nature is in fact being applied to the current patient. Furthermore, records of the patient's status are often kept for the purpose of making an inference either to the same patient at a later time or to other similar patients, again inductive processes of a statistical nature.

Lest we seem to give the impression that it is inevitably less desirable to study a sample rather than the entire population, let us emphatically deny the point right now. It is our view that for large populations study of every population member is rarely desirable even if it is possible. Most studies must operate within fixed financial and other resources, and in such cases a greater scope and depth of study are possible for a sample. The attempt to spread resources over the entire population will result in a thin, if not emaciated, study. On the other hand, if fixed information is required on each individual, then lower cost and less time will be required to complete a sample study. Generally, fewer observers or technicians will be required if we do not study the entire population. This results in less observer training, lower cost, and more uniformity of recorded observations.

Studies rarely have as an objective the determination of a population characteristic down to the last iota of precision. Even if we are interested in the mean height of all American men, it is hard to imagine a purpose for which that mean must be determined to the nearest micron. In most cases, after the sample size increases beyond a certain point, depending both upon the way the sample is selected and the characteristic being studied, additional increases produce a useless or unnecessary increase in precision—a kind of diminishing returns effect. A properly selected sample can yield precision sufficient for almost any reasonable purpose.

We will now give a more formal definition of statistical inference.

Definition

A conclusion concerning a population of observations made on the basis of a sample of observations is called a *statistical inference.*

Frequently a sample of observations will be used to provide information concerning some summary descriptive characteristic of the population of observations, such as the population mean, median, or variance. Such information will often be based upon a similar summary descriptive characteristic of the sample. The frequency of such usage might be expected to produce special terminology, as indicated in the following definitions.

Definition

A summary descriptive characteristic of a population of observations is called a *parameter.*

Definition

A summary descriptive characteristic of a sample of observations is called a *descriptive statistic* or, more briefly, a *statistic.*

In the sense of these definitions parameters and statistics are given in numerical form. One kind of statistical inference, then, though not the only kind, seeks information about a parameter on the basis of a statistic. Whereas the numerical value of a statistic can be calculated, since it is based solely upon sample information, the numerical value of the corresponding parameter is ordinarily unavailable—it depends upon information from the entire population. A mnemonic device useful in fixing terminology takes note of the correspondence between initial letters in sample and statistic, and population and parameter.

5-2   *Random and Nonrandom Samples*

So far we have not attempted to compare samples or to introduce standards for evaluating samples. We now proceed to do this. We shall ordinarily assume that the population being sampled is either infinite or very large relative to the number of observations in the sample. In Sec. 5-5 we shall temporarily eliminate this assumption and examine the effect of a smaller-sized population.

Definition

A *random sample* is a sample drawn from a population of units in such a way that every member of the population has the same probability of selection and different units are selected independently.

The technical advantages of random samples are extensive. The random sample plays no favorites, but assigns the same selection probability to every member of the population. In doing so it assures that the population is fully known and defined by the investigator. This avoids biases and oversights, which can lead to faulty inference. A further advantage is that the majority of statistical theory available for making statistical inferences assumes this type of sampling. It is not the purpose of this book to expound that theory in detail, partly because of the level of mathematical proficiency required to do so, but primarily because our emphasis is upon the statistical techniques themselves rather than upon their mathematical basis. Anyone who wishes to go into this matter in greater depth may study such references as Freund (19), Mood and Graybill (38), or Hoel (26).

Random sampling permits the quantitative assessment of sampling error, and this is another of its advantages. We know that the sample is not an exact reflection of its population. For example, a sample mean will not exactly equal its corresponding population mean. The difference between a sample measure or statistic and its corresponding population measure or parameter is called sampling error. It is not error in the sense of blunder or mistake, but a calculated error we make because we do not collect data on the entire population. It is the statistician's goal to measure sampling error in quantitative terms, and random sampling permits him to do so.

Random sampling has some disadvantages as well. In practice it can be a very difficult and expensive technique to implement, particularly if the individuals in the population are widely scattered. For example, it would be wasteful to attempt to select a random sample of all adults in the United States. Better methods have been worked out for sampling naturally oc-curring human populations; these will be discussed briefly in Sec. 5-5.

In some situations it will be impossible to obtain a list of all members of the population; thus random sampling is ruled out. In such cases we may be presented with a "sample" and be faced with the problem of deciding to what population statistical inferences can reasonably be applied. For example, the patients in a clinical trial to compare different treatments will not ordinarily be selected by any random method. The investigator must then be wary about his statistical inferences. He should search for obvious and hidden biases that tend to make his patient group unusual in any sense—for example, in its sex, race, or other demographic characteristics; in terms of stage or severity of disease, or in terms of prior medical history. In order to make a statistical inference he will be forced to ask, "From what population might this group constitute a random sample?" and limit his purely statistical inferences accordingly. His experience with the subject matter may make it possible for him to widen the scope of his scientific inference, but he does so without the assistance of formal statistical methodology, and his inferences must be to that extent nonquantitative.

It would be presumptuous of any statistician to claim that the only kind of scientific inference is statistical inference and that conclusions can be based only upon samples selected by some method based upon probability, such as random sampling. Such a dictum would serve to eliminate the majority of scientific research in all fields. Certainly human geneticists have learned a great deal from studies of fruit flies, but the inference from a sample of fruit flies to a population of human beings is not entirely statistical. It is often wise to check models based upon fruit fly observations by observing samples of humans, since men differ from flies in several important respects. So you see, we are not really too worried about the fact that nonstatistical inferences are made and that statistical inferences are often made on the basis of non-random or nonprobability samples. This is inevitable to the progress of science. It is important, we feel, to emphasize both the strengths and limitations of statistical methods in order that they might be applied with something approximating good judgment.

It is not necessary that all individuals in the population have the same probability of entering the sample. The crucial point is not whether the selection probabilities are different, but rather whether they are known. For some purposes conscious variation in selection probabilities can be useful. In a study of blood pressure levels in Nassau, Bahamas, for example,[1] an important purpose was the comparison of levels between the races. In order to achieve this comparison the white residents were sampled at a higher rate than the blacks, but for both groups the selection probabilities were known.

Trouble arises either when selection probabilities for some members of the population are unknown or when some individuals in the population have zero probability of being sampled. Community immunization levels

[1] Johnson and Remington (29).

against diphtheria in children could hardly be determined by studying records in pediatricians' offices, since a segment of the population will not receive routine preventive medical care from a pediatrician. This latter group may be the very segment of the population in which a diphtheria outbreak could be initiated and maintained and could as a result be the most important group to study.

One type of sampling that deserves special attention is the so-called quota sample, used by some public opinion polling groups. In this method interviewers are given interviewing quotas to fill. They must, say, obtain a specified number of interviews from married white women aged 40–49, a specified number from single black men aged 30–39, etc. Beyond this the interviewers are given no instructions but can interview any people within their interviewing area who fall into these categories. The quotas are so arranged that the sample will mirror the population in its age, sex, and racial composition and possibly with respect to other variables as well, depending upon the definition of the initial quota groups. That is, if the population is 10 per cent black, then the sample can be selected so that 10 per cent of the respondents are black, etc. This is an attempt to select a "representative sample" from the population. Indeed, it is possible to select a sample that will represent the population with respect to any variable whose distribution in the population is known. That is, if we know the percentage of females in the population, we can guarantee by quota sampling that the sample will contain the same percentage of females. We can, if we know the percentage of single white women aged 40–49 who have completed a high school education and who reside in one of the midwestern states, even guarantee that the sample matches that percentage. But notice that it is never possible to represent the population in this way with respect to the variables that we are really interested in studying—say the percentage of adults using a fluoride-containing dentifrice. For these variables the distribution in the population is unknown. In fact, if it were known we would not be doing the study. The key point is that it is possible to guarantee that the quota sample will represent the population with respect to all sorts of variables, except those in which we are really interested.

With respect to these variables the only hope is for a kind of representation based upon probability, such as is provided by random sampling or by the other probability sampling techniques of Sec. 5-5. This representation cannot guarantee that sample information on these variables is identical to the corresponding information in the population, but it does permit us to say with a specified level of confidence—a concept to be defined precisely in Chap. 7—that a sample statistic is not farther than a predetermined distance from the corresponding parameter in the population. This deviation can be specified numerically in advance of selecting the sample and the sample designed to provide for it. The technique of stratification (see Sec. 5-5) does permit representation of the population by the sample with respect to variables of known distribution in the population.

The essential difficulty with the quota sample is that, except for the restrictions imposed by the quotas, the interviewer is free to select anyone as a respondent. As human beings these interviewers are not likely to approach potential respondents whom they might expect to be uncooperative or even actively hostile. Investigations in this area show that quota-sampling interviewers have a tendency to obtain interviews from people in many respects similar to themselves. For these methods the sampled population is not necessarily the one about which inferences are wanted, and without some known element of probability in the selection process no assessment of sampling error is possible. The quota sample has unknown selection probabilities for every member in the population.

5-3   Random Numbers and Their Uses

5-3-1   PROPERTIES OF A RANDOM NUMBER TABLE

Table A-1 consists of a list of digits constructed by a process designed to satisfy the following 2 conditions:

1. In any predetermined geographic position in the table any one of the 10 digits 0 through 9 has probability $\frac{1}{10}$ of occurring.
2. The occurrence of any specific digit in a predetermined geographic position in the table is independent in the probability sense of the occurrence of specific digits in other positions in the table.

Notice that these 2 conditions—equal probability and independence—correspond to the 2 conditions in the definition of a random sample given in Sec. 5-2. Although a process thought to be essentially random is used to produce such a table, the actual sequence of digits produced is subjected to a series of statistical tests to determine that the 2 conditions are satisfied. The impact of these 2 properties is important. They imply that in a sequence of 5 consecutive predetermined positions any one of the 100,000 possible sequences—applying the multiplication principle of Sec. 4-4-1—has probability 1/100,000 of occurring. That is, the sequence 3, 3, 3, 3, 3 is just as likely as the sequence 1, 8, 3, 9, 6. Careful, though; this is very different from saying that regular sequences are as likely as irregular ones. For instance, there are only 10 possible sequences consisting of 5 identical digits: 0, 0, 0, 0, 0; 1, 1, 1, 1, 1; . . .; 9, 9, 9, 9, 9, and they are only 1/10,000th of the possible sequences.

5-3-2   USE OF THE RANDOM NUMBER TABLE
TO SELECT A RANDOM SAMPLE

As an example, suppose that we have a population of 9,367 units from which we wish to select a sample of 50. It is necessary first to assign a unique identification number to each member of the population. These numbers would ordinarily be taken as 0001, 0002, . . . , 9,367. A starting point in the table is selected by some device designed to avoid a purposive selection of a particular unit from the population and to guarantee that if the table is used to select repeated random samples from the same or different populations, an identical starting point is not used repeatedly. Flipping the book open to one of the 4 pages of Table A-1, staring off into space, plopping the dominant index finger down on that page, and agreeing to start with some unseen digit in a determined position covered by the finger is one method of starting. This method is not as bad as it sounds at first. A sense of distaste for this may drive some people to flip a coin twice to select a page, using, say, the sequence TH for the first page of the table, HT for page 2, HH for page 3, and TT for page 4. A similar method can be used to pick a row and a column on the selected page.

The starting point provides an initial digit for a sequence of 4 digits. Four digits are used because the 9,367 units in the population require a maximum of 4 digits per identification number. When such a sequence is selected, the unit with that identification number is elected to the sample. A sequence like 9842 to which no unit corresponds is rejected and the next sequence is picked. Successive sequences are selected by proceeding in a regular predetermined geometric manner, e.g., by moving across the rows of the table or down its columns. If a sequence corresponding to a unit previously selected occurs again, it is skipped and the following sequence is used. Strictly speaking, a random sample requires that the same individual can be selected more than once, but in practice this is wasteful of information. Furthermore, when the population is large relative to the sample, as in this case, the effect of nonreplacement is negligible. This will be discussed again in Sec. 5-5.

Let us illustrate this process of selecting a random sample by use of a table of random numbers. We have a population of 9,367 units, and we wish to select 50 units at random. Suppose that our starting point in Table A-1 is on the first page of that table 15 lines from the bottom beginning with the first digit on the left edge of the page. Also suppose that we decide to use a design wherein we proceed across a row, then return to the start of the next row, as one would read a book. Recall that we must group our digits into sets of 4 and discard both those sets beyond 9,367 and those sets which already have occurred. The selected numbers indicate that those members of the population whose identification numbers correspond will be our

random sample of units. In this case these random identification numbers
are

6968	5785	1304	9332	3542
2191	5549	8614	4587	0981
0994	2161	0741	6487	9175
1899	3368	0566	2744	0274
4626	6663	0510	3103	3105
0929	1301	6572	2069	1427
9106	0404	7664	8455	7950
4955	1250	8812	7053	2517
4055	1393	3047	4418	4698
4571	0769	0011	0326	1843

5-3-3   OTHER USES OF THE RANDOM NUMBER TABLE

The probabilistic properties of a random number table are useful for
purposes other than the selection of random samples. For example, if a group
of patients is to be divided into 2 similar subgroups of equal size, one group
receiving a treatment procedure and the other a control procedure, the
division can be effected through the use of Table A-1. In this case the
individuals to receive the treatment might be selected from the full group in
exactly the manner described in Sec. 5-3-2 for selecting a random sample.
If the groups are reasonably large, this kind of randomization can be expected
to produce treatment and control groups that are similar in all respects except
for the fact of treatment and its consequences. The randomization can also
be carried out within individual patient subgroups. That is, male patients
might be randomized separately from female patients to guarantee an
identical sex ratio in treatment and control groups. Stage of disease, age,
race, occurrence of prior treatment, or other variables could be used to
divide the patients into similar subgroups prior to randomization.
Patients frequently enter a clinical trial serially in time. In this case the
random number table can again be useful, but in a slightly different way. We
may wish to guarantee that every sequence of 10 successive patients will be
divided into 5 treatment and 5 control patients. In the random number
table even and odd digits occur with equal probability. Thus we might
identify even digits with treatment, odd with control, and select a sequence
of single digits from the table. If the sequence 3, 4, 4, 1, 4, 8, 2, 1, 5, 7 occurs,
this means that the first 7 successive patients are assigned as follows, T being
used for treatment, C for control: C, T, T, C, T, T, T. At this stage the
following 3 patients must be assigned to the control group because of the
restriction that exactly 5 patients in each successive group of 10 must be in
each group, the 5 patients in the T group having already been determined.
This restricted randomization procedure has the property that a patient

arriving for allocation in any predetermined position in the patient sequence has equal probability of being assigned to either group. Stratification of patients into homogeneous subgroups can also be accomplished with restricted randomization.

Randomizations into more than 2 groups follow from a natural generalization of these procedures. For example, with 3 groups, one-third of the patients can be selected at random for group 1 and one-half the remaining subjects (one-third the original group) then can be selected for group 2, the remainder going to group 3. For patients arriving serially in time a restricted randomization might guarantee that of every 9 successive patients 3 are assigned to each group. Here the digit 0 might be discarded when it arises from the random number table and the remaining digits might be grouped as 1–3, 4–6, 7–9 to correspond to the 3 allocation groups.

5-4 *Random Variables, Distributions, and Sampling Distributions*

5-4-1 DEFINITIONS AND GENERAL BACKGROUND

A fundamental concept in statistics is that of random variable.

Definition

A *random variable* is a quantity that takes various values or sets of values with various probabilities.

In this definition 2 characteristics of a random variable are important— its values and the probabilities associated with those values or sets of values.

EXAMPLE

When a fair penny and nickel are tossed together we might be interested in the total number of heads. This quantity takes the values 0, 1, and 2 with probabilities $\frac{1}{4}$, $\frac{1}{2}$, and $\frac{1}{4}$, respectively, as we found in an example in Sec. 4-1. If we let x denote the number of heads, then according to our definition x is a random variable.

EXAMPLE

An adult male is to be selected at random from the population of American men, and his standing height is to be measured. There is some probability that his height will be between 69 and 73 inches. Similarly, there will be probabilities that his height will be over 72 inches, between 70 and 71 inches, etc. Thus, his height is a random variable.

The random variable in the first example, number of heads, is a discrete variable according to the terminology of Sec. 3-6 and thus is called a discrete random variable, whereas the second random variable, standing height in inches, is a continuous random variable.

Definition

A table, graph, or mathematical expression giving the probabilities with which a random variable takes different values or sets of values is called the *distribution of the random variable*.

A distribution indicates for a population of observations the relative frequencies with which different values or sets of values of the random variable occur. When we speak of the mean, variance, standard deviation, or any other descriptive measure of a distribution, we are referring to that measure on the entire population of observations.

Much scientific investigation is concerned with the determination of characteristics of the distributions of random variables. For example, we may be interested in the mean of the distribution of serum cholesterol level of middle-aged American men, in the variance of the distribution of incubation periods of animals infected with the hog cholera organism, in the mean of the distribution of length of hospitalization of patients undergoing cholecystectomy, or in the standard deviation of the distribution of Lactobacillus acidophilus counts in saliva specimens of children. In comparative studies such as the clinical trial we may be interested in the difference between the means of the distributions of some physiologic response variable in treated and control patients. In other applications we may need to know whether some distribution follows a predetermined pattern or shape.

Statistical inference is concerned with the conclusions that can be drawn about a population of observations based upon a sample of observations. If we limit our attention to random samples, then probability enters the sample selection process; here we wish to learn something about a distribution upon the basis of a random sample from that distribution. In practice we have only one specific random sample upon which to base inferences, but in studying statistical methods we are interested in the behavior of random samples in general. The process by which the sample is selected might have resulted in the selection of any one of many different random samples. The statistician is concerned with regularities and similarities shared by all these potentially occurring random samples as well as with the amount of variation or difference between random samples. For example, he wants to know the probability of a really aberrant or unusual random sample.

One of the most important ideas to get fixed in your mind right now is that even though the data we work with in numerical form constitute only a single random sample, that sample is but one of very many different samples

which might have been produced by the selection process. This concept is
the turning point in statistics. More potential students of statistics run
aground here than at any other place. If you fix this idea and grasp it through
text, examples, diagrams, exercises, hook or crook, the rest of this subject
will be much easier. If you do not, will not, or cannot understand it, then
there really is not much point in proceeding farther.

In the hope of assisting you over this rough spot we provide another
series of diagrams. Figure 5-5 shows a random sampling population of
units. Notice that in this population the units are random samples of the

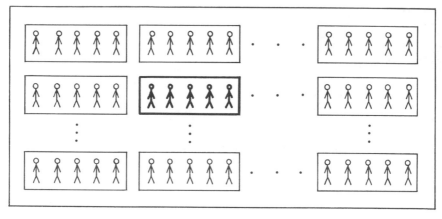

Figure 5-5   A random sampling population of units.

underlying population of units shown in Fig. 5-1. Think of this figure as
showing all possible random samples of a given size—size 5 in this case—
that could be selected from the underlying population. A given unit from
the underlying population may, of course, be selected in many different
random samples and thus will be in many of the units of Fig. 5-5. There are
a large number of such samples, as indicated by the dots in Fig. 5-5. The
single sample indicated in boldface is the only one that will actually provide
numerical information for our study. That does not mean that all the others
do not exist. They do—we just do not get observations from them.

Corresponding to the random sampling population of units there is a
random sampling population of observations. But here an observation is
somewhat different. It is an observation on a random sample rather than on
a single member of the underlying population. In other words, it is a statistic
such as the sample mean $\bar{x}$, the sample variance s^2, or the sample standard
deviation s. For example, instead of being a single systolic blood pressure
it is the mean of a random sample of systolic blood pressures. Figure 5-6
shows a random sampling population of observations, using $\bar{x}$ as the observation. Since we only have data from a single random sample, we only have in

numerical form one value of $\bar{x}$ shown in boldface in Fig. 5-6. That does not mean that all the other values do not exist. They do—we just do not have access to their numerical values, since they correspond to unselected samples.

Just as to the underlying population of units (Fig. 5-1) there correspond many underlying populations of observations like Fig. 5-2, depending upon the variable that is measured, so corresponding to the random sampling population of units (Fig. 5-5), there are many random sampling populations of observations like Fig. 5-6, depending upon the statistic that is measured.

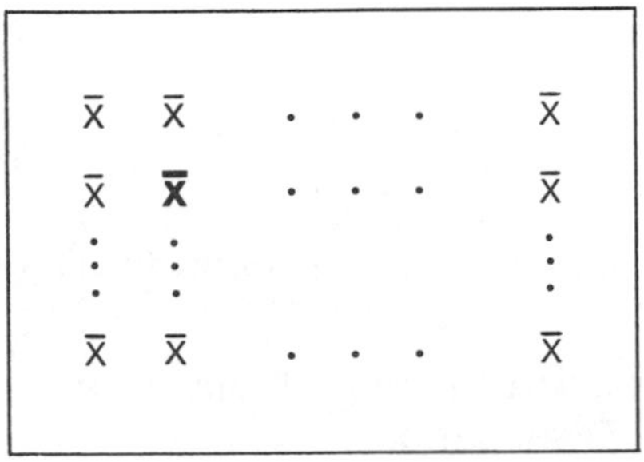

Figure 5-6 A random sampling population of observations.

The statistic, $\bar{x}$ in our example, now satisfies the definition of a random variable. It varies from random sample to random sample, and it takes on various values or sets of values with various probabilities. That is, just as there is some probability that the height of an American male selected at random from the population lies between 69 and 73 inches, so there is some probability that the mean height of a random sample of, say, 50 American men lies between 69 and 73 inches. The 2 probabilities are not the same in numerical value. In fact, a little reflection will perhaps convince you that the second probability will be higher. Since $\bar{x}$ is a random variable, it has a distribution, called its sampling distribution.

Definition

The distribution of a statistic over all random samples of size n from an underlying population is called the *sampling distribution of the statistic for random samples of size n.*

When there is no fear of ambiguity concerning random sampling or sample size we will, for example, just speak of the sampling distribution of $\bar{x}$, eliminating the phrase "for random samples of size n."

It has become common practice in statistics to observe certain notational conventions. In general, we will denote parameters by lower case Greek letters and statistics by lower case Latin letters. Table 5-1 lists some of this notation.

Table 5-1

SOME NOTATIONAL CONVENTIONS

Quantity	Notation	Terminology
Population mean	μ	mu
Population variance	σ^2	sigma-squared
Population standard variation	σ	sigma
Sample mean	$\bar{x}$	x-bar*
Sample variance	s^2	s-squared
Sample standard deviation	s	s

* Readers from the Western or Southwestern United States may prefer to say "bar-x."

5-4-2 SOME IMPORTANT GENERAL RESULTS ON SAMPLING DISTRIBUTIONS

A surprising number of general statements can be made concerning sampling distributions—general in the sense that the statements hold for any random variable on the underlying population whatever its distribution, provided only that we have random samples of size n.

Statement 1. For random samples of size n from any underlying population the mean of the sampling distribution of $\bar{x}$, the sample mean, is μ, the mean of the underlying population.

Statement 2. For random samples of size n from any underlying population, the variance of the sampling distribution of $\bar{x}$, the sample mean, is σ^2/n, the variance of the underlying population divided by the sample size.

Statement 3. For random samples of size n from any underlying population, the mean of the sampling distribution of s^2, the sample variance, is σ^2, the variance of the underlying population.

The variance of the sampling distribution of the sample variance depends upon the shape of the underlying population distribution, and thus no statement as general as these 3 statements can be made about it.

At first the 3 statements may seem somewhat artificial, but in fact they conform remarkably well to common sense. The first and third statements say, loosely speaking, that the statistics $\bar{x}$ and s^2 "on the average" equal their corresponding parameter values, μ and σ^2. This kind of "on the average" is rather special, however, and refers to the average value of the statistic over all random samples of size n. You will agree that that is a terrific amount of averaging. It would be mildly surprising if the sample mean did not in some sense tend to vary around the population mean, and the first statement says that it does. Notice that the sample mean might never exactly equal the population mean for any single sample and that the statement only says that it does so "on the average."

The second statement indicates that the sample mean is less variable than a single observation. The variability of a single observation x can be measured by the size of σ^2—larger σ^2, more variable x; smaller σ^2, less variable x.[2]

An example will show that $\bar{x}$ should intuitively be less variable than x. Again considering heights of American men, ask yourself: with what probability do extreme values of x and $\bar{x}$, e.g., values greater than 76 inches, occur? To get a value of x over 76 inches we must merely select one person of such a height. On the other hand, a value of $\bar{x}$ in this range requires that n men have been randomly selected from the population and found to have an *average* height over 76 inches—an unusual event if n is of even moderate size. Thus extreme values of $\bar{x}$ are rarer than correspondingly extreme values of x, and this is just another way of saying that $\bar{x}$ is less variable. As we have noted in this example, the reduction in variability should depend upon n: the larger the value of n, the less probable extreme values of $\bar{x}$ become. Notice again that statement 2 agrees with this and quantitates this dependence upon the sample size by stating that the variance of $\bar{x}$ is σ^2/n.

Definition

The standard deviation of the sampling distribution of a statistic for random samples of size n is called the *standard error* of the statistic.

In Chap. 3 the standard deviation was defined to be the positive square root of the variance. Since the variance of the sampling distribution of $\bar{x}$

---

[2] It is necessary to make a slight modification in the formula for the variance when it is calculated from an entire finite population of size N. The formulas of Sec. 3-4-5 now contain N instead of $N - 1$, as a divisor; i.e.,

$$\sigma^2 = \sum_{i=1}^{N} \frac{(x_i - \bar{x})^2}{N} = \frac{N \sum_{i=1}^{N} x_i^2 - \left(\sum_{i=1}^{N} x_i \right)^2}{N^2}$$

You will recall that the divisor $n - 1$ in the definition of s^2 was (and still is) a little unnatural. The advantages this divisor holds for the statistic s^2—to be delineated in Chap. 7—do not hold for the parameter σ^2.

is σ^2/n, the standard deviation must be $\sqrt{\sigma^2/n} = \sigma/\sqrt{n}$. We say that the standard error of $\bar{x}$ is $\sigma/\sqrt{n}$.

The adequately curious reader will wonder where results like those in statements 1 through 3 originate. The statements are really theorems whose proofs require an understanding of mathematics at least through integral calculus, a level of preparation we are not assuming. We believe that the results are intuitively reasonable, however, and anyone who wants to see formal proofs may consult such references as Freund (19), Mood and Graybill (38), or Hoel (26).

EXAMPLE

Suppose the pH of serum of arterial blood in American men has a distribution with $\mu = 7.42$ and $\sigma = 0.016$ ($\sigma^2 = 0.00026$). Then the sampling distribution of $\bar{x}$ for random samples of size 25 from this population will have mean 7.42 and variance $0.00026/25 = 0.000010$. The standard error of $\bar{x}$ will be $\sqrt{0.000010} = 0.0032$. The mean of the sampling distribution of s^2 will be 0.00026.

A notational convention helpful in discussing sampling distributions uses a parameter symbol subscripted by the appropriate statistic to indicate the parameters of sampling distributions, i.e., $\mu_{\bar{x}}$ for the mean of the sampling distribution of $\bar{x}$, $\sigma^2_{\bar{x}}$ for the variance of the sampling distribution of $\bar{x}$, and μ_{s^2} for the mean of the sampling distribution of s^2. With this convention and with the assumption of random samples of size n from an underlying population with mean μ and variance σ^2, a considerable compression of statements 1 through 3 of this section is possible; i.e.,

Statement 1: $\mu_{\bar{x}} = \mu$.

Statement 2: $\sigma^2_{\bar{x}} = \dfrac{\sigma^2}{n}$.

Statement 3: $\mu_{s^2} = \sigma^2$.

5-5 *Sampling Finite Populations*

The results of the preceding section assume random sampling, a method basically applicable to populations that are either infinite or very large relative to the sample size. The definition of random sampling requires that the selection of different individuals from the population be independent. This implies, as we saw in Sec. 4-10, that the conditional probability of selecting individual A given the selection of individual B must be the same as the probability of selecting individual A; i.e., $P(A \mid B) = P(A)$. In sampling from a finite population, unless we replace each individual after selection, thereby allowing the same individual to be sampled more than once, this property will not hold exactly, even though we insure that every member of the population has equal probability of selection. In fact we shall show in this section that for nonreplacement equal probability sampling from a finite population, the probability of selecting any given individual is n/N, where n is the sample size and N is the population size. Thus $P(A) = n/N$. On the other hand, we shall also show that in this case $P(A \mid B) = (n - 1)/(N - 1)$. For large N these probabilities will be close to each other in value. If N is large relative to n, it is intuitively clear that the question of replacement or nonreplacement is unimportant, since the probability of selecting the same individual more than once will be small even with replacement. We mentioned earlier the inefficiency in most practical circumstances of using replacement sampling.

In a sense, replacement sampling has the effect of generating an infinite population. For example, with replacement sampling there is nothing to stop us from drawing a sample of size 100 from a population of size 50—nothing except the excessive stupidity of doing so. In fact, if we were sufficiently motivated, financed, and cerebrally ischemic, we could draw a sample of any fixed size, however large, from such a population, provided we replaced each individual after selection. This then amounts to producing an infinite population.

The issue of replacement versus nonreplacement leads to the definition of a type of sample very similar to random sampling but eliminating the restriction of independence. This latter restriction, as we have seen, does not strictly hold for nonreplacement sampling.

Definition

A *simple random sample* of size n from a population of size N is a sample selected in such a way that every group of n different units has the same probability of being selected as the sample.

From the results of Chap. 4 we see that there are ${}_N C_n$ such samples, ignoring order of selection. To calculate the probability that any unit will be selected into the sample, we notice that the number of samples containing that unit is ${}_{N-1} C_{n-1}$, since from the remaining $N - 1$ units in the population we must select the remaining $n - 1$ units in the sample, again without regard to order. But the probability of selecting a given unit say $P(A)$, is then ${}_{N-1} C_{n-1}/{}_N C_n$, if we use the basic definition of probability given in Sec. 4-1. This follows since of the ${}_N C_n$ equally likely outcomes, ${}_{N-1} C_{n-1}$ are favorable to the selection of the given unit. But

$$P(A) = \frac{_{N-1}C_{n-1}}{_NC_n} = \frac{(N-1)!/[(N-1)-(n-1)]!\,(n-1)!}{N!/(N-n)!\,n!}$$

$$= \frac{(N-1)!\,(N-n)!\,n!}{N!\,(N-n)!\,(n-1)!} = \frac{n}{N}\,.$$

The probability of selection is the same for any unit, and thus a simple random sample satisfies the first part of the definition of a random sample. Since by definition a simple random sample consists of n different units, such a sample is essentially a nonreplacement sample. Of course, the same unit will belong to many of the different simple random samples that could occur.

To see that $P(A \mid B)$, the probability of selecting individual A given the selection of B, is equal to $n - 1/N - 1$, recall that

$$P(A \mid B) = P(A \text{ and } B)/P(B).$$

But as we have seen, $P(B) = n/N$. The number of samples containing both A and B is $_{N-2}C_{n-2}$, since the remaining $n - 2$ places in the sample must be filled by individuals selected from the remaining $N - 2$ population members. Thus,

$$P(A \text{ and } B) = \frac{_{N-2}C_{n-2}}{_NC_n} = \frac{(N-2)!/(N-n)!\,(n-2)!}{N!/(N-n)!\,n!} = \frac{n(n-1)}{N(N-1)}$$

and

$$P(A \mid B) = \frac{n(n-1)/N(N-1)}{n/N} = \frac{n-1}{N-1}\,.$$

But since $P(A) = n/N$, the selections for a simple random, i.e., nonreplacement, sample are not independent.

It is interesting that for simple random samples statement 1 of Sec. 5-4-2 still holds. That is, for simple random sampling $\mu_{\bar{x}} = \mu$. However, statements 2 and 3 are changed and now become:

Statement 2a. For simple random samples of size n from an underlying population of size N, the variance of the sampling distribution of $\bar{x}$, the sample mean, is $[(N-n)/(N-1)] \cdot \sigma^2/n$, where σ^2 is the variance of the underlying population. In other words,

$$\sigma_{\bar{x}}^2 = \frac{N-n}{N-1} \cdot \frac{\sigma^2}{n}\,.$$

Statement 3a. For simple random samples of size n from an underlying population of size N, the mean of the sampling distribution of s^2, the sample variance, is $\dfrac{N}{N-1} \cdot \sigma^2$. In other words,

$$\mu_{s^2} = \frac{N}{N-1}\,\sigma^2.$$

The change from the original form of statement 2 only consists of multiplication by the factor $(N-n)/(N-1)$. If the sample size is small relative to the population size, then this factor will be close to 1 and σ^2 will be close to σ^2/n. This agrees with our observation that replacement and nonreplacement sampling approximately agree in this case.

Statement 3a is identical to statement 3 except for the factor $N/N-1$, which will be close to 1 if N is large.

We will illustrate the operation of these 3 statements under simple random sampling by an example in which we list all the 15 possible simple random samples of size 2 from a small artificial population of size 6 and observe that the properties hold in the stated form. This is, of course, not a proof that the statements will always hold—such a proof is beyond the mathematical level we are using.

EXAMPLE

Suppose the population of units consists of 6 families—that is, each family is a unit and the relevant observation is the number of children in the family. Suppose further that the families are denoted by letters A through F and that their respective numbers of children are as shown in Table 5-2.

Table 5-2

FAMILIES AND NUMBERS OF CHILDREN IN THE
UNDERLYING POPULATION

Family	A	B	C	D	E	F
No. of Children	2	1	2	0	5	2

For this population,

$$\mu = \frac{\sum\limits_{i=1}^{N} x_i}{N} = \frac{12}{6} = 2$$

and

$$\sigma^2 = \frac{N\sum\limits_{i=1}^{N} x_i^2 - \left(\sum\limits_{i=1}^{N} x_i\right)^2}{N^2} = \frac{(6)(38) - (12)^2}{36} = \frac{7}{3}.$$

There are $_6C_2 = 15$ simple random samples of size 2 from this population; Table 5-3 presents each sample with its sample mean and sample variance.

Table 5-3

ALL SIMPLE RANDOM SAMPLES OF SIZE 2 FROM THE
POPULATION OF FAMILIES

Sample of Units	Sample of Observations	Sample Mean $\bar{x}$	Sample Variance s^2
A, B	2, 1	1.5	0.5
A, C	2, 2	2.0	0.0
A, D	2, 0	1.0	2.0
A, E	2, 5	3.5	4.5
A, F	2, 2	2.0	0.0
B, C	1, 2	1.5	0.5
B, D	1, 0	0.5	0.5
B, E	1, 5	3.0	8.0
B, F	1, 2	1.5	0.5
C, D	2, 0	1.0	2.0
C, E	2, 5	3.5	4.5
C, F	2, 2	2.0	0.0
D, E	0, 5	2.5	12.5
D, F	0, 2	1.0	2.0
E, F	5, 2	3.5	4.5
Total		30.0	42.0
Mean		2.0	2.8

From Table 5-3 it can be seen that

$$\mu_{\bar{x}} = 2 = \mu \quad \text{and} \quad \mu_s^2 = 2.8 = \frac{6}{5} \cdot \frac{7}{3} = \frac{N}{N-1} \sigma^2.$$

Also,

$$\sigma_{\bar{x}}^2 = \frac{(15)(74) - (30)^2}{(15)^2} = \frac{74 - 60}{15} = \frac{14}{15} = \left(\frac{4}{5}\right) \cdot \left(\frac{7/3}{2}\right) = \left(\frac{N-n}{N-1}\right) \cdot \frac{\sigma^2}{n},$$

since $\sum_{i=1}^{K} \bar{x}_i^2 = 74$ and $\sum_{i=1}^{K} \bar{x}_i = 30$; $K = {}_NC_n$, the number of simple random samples.

Random sampling and simple random sampling are examples of a more general type of sampling called probability sampling.

Definition

A *probability sample* is a sample selected from a population in such a way that every member of the population has a known non-zero probability of being selected into the sample.

Notice that this definition does not require that every member of the population have the same probability of entering the sample—only that the probability be known. An example of a kind of probability sample that does not, in general, have equal selection probabilities is the stratified random sample.

Definition

Suppose a population is divided into subpopulations or strata. If a simple random sample is selected from each of these strata, the aggregated sample is called a *stratified random sample*.

There is nothing in this definition to prevent the investigator from using a different sampling fraction in each stratum. In the Nassau blood pressure survey[3] referred to in Sec. 5-2 the population was initially stratified into white and black subpopulations, and the smaller white stratum was sampled at a higher rate to provide an adequate number of whites for the racial comparison needed in the study. An advantage of stratified random sampling is that information about the composition of the population with respect to a number of stratifying variables can be taken into account. If, for example, the age, sex, and race composition of the population is known, then the sample can be selected to conform exactly to this composition by using these variables as the basis for stratification. Thus the stratified random sample shares the ability of the quota sample to represent the population with respect to variables with known distributions, without the disadvantages of non-probability sampling. Ideally strata should be as internally homogeneous as possible with respect to the variables under study. In fact, if the observations are relatively homogeneous within strata as compared to their variability between strata, a stratified random sample can show real gains in precision. That is, the sampling error compared to that of a simple random sample of the same total size may be considerably smaller.

Another type of probability sampling particularly useful in sampling large human populations is the *multistage probability sample*. In this sample the population is first divided into large units called primary sampling units, or P.S.U.'s. In selecting a probability sample of the population of the United States, these are often counties. A simple random sample, or more

[3] Johnson and Remington (29).

often a stratified random sample of P.S.U.'s, is selected. The selected P.S.U.'s are then enumerated or canvassed to produce a complete listing of second-stage units, possibly households or groups of households. A probability sample, say a simple random sample of second-stage units, is then selected. Additional stages of enumeration and sampling may or may not be used.

The systematic random sample is useful in sampling units that are arranged in sequence, e.g., hospital patient charts, library catalog cards, or death certificates in a file. A one in k systematic random sample is selected by picking every kth member of the population in order after selecting the first unit at random from the first k units in the sequence. If the units are arranged in an order unrelated to the variables under investigation, this can be an effective type of sampling. However, unexpected regularities or cyclic patterns may exist and upset such a study. In case of doubt, it is probably best to avoid this type of sample.

The statistical theory and practice of probability sampling is today a well-developed subspecialty of statistics that has proved useful in many areas of application.

EXERCISES

5-1 Develop an example requiring statistical inference from your own field of interest.

5-2 Consider the first 25 rows on the second page of random numbers. Since each row contains 50 digits, 1250 digits will be considered in the following experiment to test for "randomness," i.e., equal and independent probabilities for each digit.

    a.   Count the number of times each digit occurs. Check to see if the distribution of these digits is reasonable (in Chap. 9 we shall develop a statistical test to reinforce our intuition). If each digit is equally likely, it should occur about 125 times.

    b.   Consider the digit 5. Count the number of times a 5 is followed by another 5. If digits are independently generated, approximately one-tenth of the time a 5 occurs it will be followed by another 5.

    c.   In similar fashion consider 00, 11, . . . , 99 pairs. Do these pairs occur with about the frequency expected if we have equal and independent probabilities? What frequencies would you expect?

5-3 Using the data of Table 3-4, select a simple random sample (without replacement) of 10 kidney weights. Describe in detail how you used the table of random numbers.

5-4 Suppose that the 100 individuals whose serum cholinesterase values are presented in Exercise 3-10 are to be randomly assigned to one of two regimes.

Fifty will be on a study drug and 50 on a placebo. Assign the individuals randomly to the two groups and describe completely the system you use.

5-5 Suppose a study is structured to allow 100 subjects. Subjects enter the study across time and at the time of entry are to be assigned to one of five treatment groups. To insure balance the study is to be designed so that ultimately 20 subjects will be in each group. A restricted randomization is to be used so that after every 20 subjects who enter, exactly 4 will be in each treatment group.

    a. Devise and describe a randomization scheme so that these requirements are satisfied.

    b. Suppose that the study can accommodate 107 subjects; how will this affect your scheme? What will you do with the last few subjects?

5-6 Consider the population of observations: 3, 4, 5.

    a. Compute μ and σ^2 for this population.

    b. List all possible random samples (i.e., with replacement) of size $n = 2$.

    c. For each sample find $\bar{x}$ and s^2.

    d. Find $\mu_{\bar{x}}$ and $\sigma_{\bar{x}}^2$ and note that their relationships to μ and σ^2 satisfy properties 1 and 2 of Sec. 5-4-2.

    e. Find μ_{s^2} and note that its relationship to σ^2 satisfies property 3 of Sec. 5-4-2.

5-7 Repeat Exercise 5-6 for simple random sampling, noticing that properties 2a and 3a of Sec. 5-5 are satisfied.

5-8 How do repeat measurements on the same subject fit into our definitions and terminology for samples and populations? Describe the population and sample of units, population and sample of observations, and the random sampling populations of units and observations.

6

Some
Important
Distributions

A catalog listing of all the distributions that have been discovered and investigated by statisticians would be lengthy. Some of these distributions have been used for special purposes or in limited areas of application, while others are of primary interest because they have unusual mathematical properties. Still others are of rather general interest and importance and have wide areas of relevant application. These latter distributions include the binomial, Poisson, and normal distributions—important descriptors of many sets of data, in the sense that a number of random variables measured on individual units from the underlying population of units tend approximately to follow one or another of them. These distributions also have a number of general properties that make them of particular importance in statistical inference.

6-1 The Binomial Distribution

We will attempt to set the stage for the binomial distribution by presenting an example. Suppose a treatment for a particular disease has 0.6 probability of alleviating all signs and symptoms in an individual patient. For brevity we will say that such a patient is cured. Find the probability that in a series of 4 treated patients exactly 2 will be cured, assuming that the outcomes cure or noncure in the individual patients are mutually independent.

In attempting to solve this problem we can dissect it into several fundamental parts. If the patients are denoted by A, B, C, D then the outcome "exactly 2 cures" can occur in several different ways, e.g., A and B cured, C and D not cured; A and C cured, B and D not cured; etc. In fact, using $E_1, E_2, \ldots, E_6$ to denote the specific ways of obtaining exactly 2 cures, the situation is as follows:

Patient

	A	B	C	D
E_1:	cured	cured	not cured	not cured
E_2:	cured	not cured	cured	not cured
E_3:	cured	not cured	not cured	cured
E_4:	not cured	cured	cured	not cured
E_5:	not cured	cured	not cured	cured
E_6:	not cured	not cured	cured	cured

The fact that there are 6 results with the attribute "exactly 2 cures" is no accident. In fact, $6 = {}_4C_2$, and arises from the fact that from the 4 patients we must pick a set of 2 patients to be cured. Order is unimportant, since, e.g., "A and B cured" is identical to "B and A cured."

The required probability is

$$P(\text{exactly 2 cures}) - P(E_1 \text{ or } E_2 \text{ or } \ldots \text{ or } E_6),$$

since one or another of the E_i must occur if we are to have exactly 2 cures. The E_i are mutually exclusive, and thus

$$P(E_1 \text{ or } E_2 \text{ or } \ldots E_6) = P(E_1) + P(E_2) + \cdots + P(E_6).$$

But E_1 is the event: A cured and B cured and C not cured and D not cured, and since the outcomes for the individual patients are mutually independent, we have

$$P(E_1) = P(\text{A cured and B cured and C not cured and D not cured})$$

$$- P(\text{A cured}) \cdot P(\text{B cured}) \cdot P(\text{C not cured}) \cdot P(\text{D not cured})$$

$$= (0.6)(0.6)(0.4)(0.4) = (0.6)^2(0.4)^2.$$

The numerical result follows from the fact that any individual patient is cured with probability 0.6, and thus any individual patient is not cured with probability $1 - 0.6 = 0.4$.

Similarly, $P(E_2) = (0.6)(0.4)(0.6)(0.4) = (0.6)^2(0.4)^2$. In fact, $P(E_i) = (0.6)^2(0.4)^2$ for any $i = 1, 2, \ldots, 6$. Therefore,

$$P(\text{exactly 2 cures}) = \sum_{i=1}^{6} P(E_i) = \sum_{i=1}^{6} (0.6)^2(0.4)^2 = 6 \cdot (0.6)^2(0.4)^2$$

$$= 0.3456.$$

Similarly, $P(\text{exactly 1 cure}) = 4(0.6)(0.4)^3 = 0.1536$, since the result "exactly one cure" can occur in 4 equiprobable mutually exclusive ways corresponding to the cure of any one of the 4 patients. A complete listing of the probabilities of any outcomes of treating the 4 patients is

$P(\text{exactly 0 cures}) = 1 \cdot (0.4)^4 = 0.0256$

$P(\text{exactly 1 cure}) = 4 \cdot (0.6)(0.4)^3 = 0.1536$

$P(\text{exactly 2 cures}) = 6 \cdot (0.6)^2(0.4)^2 = 0.3456$

$P(\text{exactly 3 cures}) = 4 \cdot (0.6)^3(0.4) = 0.3456$

$P(\text{exactly 4 cures}) = 1 \cdot (0.6)^4 = 0.1296.$

It is instructive to examine several properties of these probabilities. First, $P(\text{exactly } r \text{ cures}) = {}_4C_r(0.6)^r(0.4)^{4-r}$ for $r = 0, 1, 2, 3, 4$; recall that any nonzero quantity raised to the zero power equals 1; e.g., $(0.6)^0 = 1$. Second, since the event (exactly 0 cures) or (exactly 1 cure) or ... or (exactly 4 cures) includes all possible results of treatment, and since the subevents connected by "or" are mutually exclusive, the sum of the probabilities of the subevents should be 1. That is,

$P(\text{exactly 0 cures or exactly 1 cure or ... or exactly 4 cures})$

$= P(\text{exactly 0 cures}) + P(\text{exactly 1 cure}) + \cdots + P(\text{exactly 4 cures}) = 1.$

A little addition verifies that this property holds numerically in the present example.

Let us see how far we can generalize the results of this example. Suppose we replace "treatment of a single patient" by "trial," "cured" by "success," "not cured" by "failure," "4 patients treated" by "*n* trials," "exactly 2 cures" by "exactly *x* successes," and "probability 0.6 that any individual patient is cured" by "probability *p* that any individual trial results in a success." In this new frame of reference our problem is altered as follows.

Suppose that a trial can result in either a success or a failure with probabilities p and $1 - p$, respectively. Suppose further that n mutually independent trials are performed. Find the probability of exactly x successes.

The solution follows the lines of the example. The x trials resulting in successes can be selected from the total of n trials in ${}_nC_x$ mutually exclusive ways. Each of these ways has probability $p^x(1 - p)^{n-x}$, and thus:

$P(\text{exactly } x \text{ successes}) = {}_nC_x p^x(1 - p)^{n-x}, \qquad x = 0, 1, \ldots, n.$ Such probabilities are called *binomial probabilities*. The number of successes x takes the various values $0, 1, 2, \ldots, n$ with various probabilities and thus is a random variable; its distribution is called the *binomial distribution*.

The probabilities and therefore the distribution depends upon the quantities n and p. In fact, there are many different binomial distributions, corresponding to the various numerical values of n and p. We use the term "binomial distribution" in 2 senses—generically, to indicate any of the distributions with probabilities $_nC_x p^x(1 - p)^{n-x}$, whatever the values of n and p, and specifically, to indicate a distribution with particular numerical values of these 2 quantities.

In practice n and p play quite different roles. The number of trials n is ordinarily known. The success probability p is usually unknown, and an important class of problems in statistical inference concerns attempts to gain information about its value on the basis of x, the observed number of successes in a series of n independent trials. You can see that in the example of treating patients, estimation of the probability that an individual patient will be cured is of potential interest and importance. The figure 0.6 attached to that probability for our numerical example will frequently be unavailable in practice.

The word "success" is intended to be very general. In fact, the outcome described as a success might be death, worsening of symptoms, being a female, tossing a head with a coin, or any other attribute. The only requirement is that a single trial must result in either of 2 outcomes, one of which is labeled "success," the other "failure."

With respect to the basic ideas of sampling set forth in Chap. 5, the n independent trials are considered to be a random sample from an infinite underlying population of units in which the basic observation takes values 1 and 0 corresponding to "success" and "failure," respectively. Then the random variable x is a statistic taking a value equal to the sum of the sample observations. For example, with our 4 patients the result E_1 gives a value of 1 to patients A and B and 0 to patients C and D and $x = 1 + 1 + 0 + 0 = 2$. The distribution of x—the binomial distribution—is then a sampling distribution.

The distribution of the basic random variable taking the values 1 and 0 is sometimes called the *point binomial distribution*. It is of a very simple form with probabilities p and $1 - p$ corresponding to the values 1 and 0, respectively. The mean μ of the point binomial distribution can be determined by using the ideas of Sec. 4-2. If we perform a series of trials and compute the proportion of successes or equivalently the average number of successes per trial, then it is both intuitively reasonable and true that as the number of trials increases this mean number of successes should approach the population mean. But the mean number of successes is the relative frequency of successes, i.e., total number of successes divided by total number of trials, which will approach the probability of success, p. Thus $\mu = p$.

It is more difficult to establish an intuitive basis for the value of σ^2 for the point binomial, but in fact $\sigma^2 = p(1 - p) = pq$, if we let $q = 1 - p$. Certainly if p is small, say 0.1, we would expect mainly to observe 0's and seldom to observe 1's; i.e., the variance should be fairly small. In this case

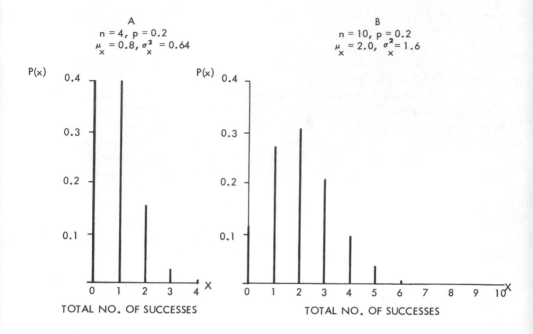

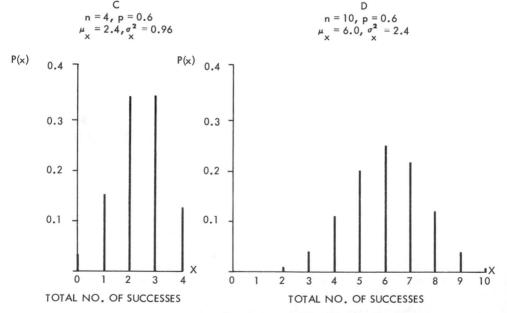

Figure 6-1 Examples of binomial distributions.

$\sigma^2 = (0.1)(0.9) = 0.09$. Similarly, if p is large, say 0.9, we would expect mainly to observe 1's and again should have a small variance. Again $\sigma^2 = (0.9)(0.1) = 0.09$. However, if p takes an intermediate value such as 0.5, we would expect about equal numbers of 1's and 0's, i.e., a relatively large variance. Here $\sigma^2 = (0.5)(0.5) = 0.25$, in conformity with our expectations. In fact σ^2 takes its maximum value when $p = 0.5$.

We will use the mean and variance of the point binomial distribution to determine μ_x and σ_x^2, the mean and variance of the binomial distribution, for x, the total number of successes in n trials. We must do this through a somewhat indirect route. Statements 1 and 2 of Sec. 5-4-2 enable us to find $\mu_{\bar{x}}$ and $\sigma_{\bar{x}}^2$, the mean and variance of the sampling distribution of $\bar{x}$. Using those results, we obtain

$$\mu_{\bar{x}} = \mu = p; \qquad \sigma_{\bar{x}}^2 = \frac{\sigma^2}{n} = \frac{pq}{n}.$$

But $\bar{x} = x/n$ since, as we have observed, x is the sample total, the sum of all the basic observations (0's and 1's) in the sample. Hence $x = n\bar{x}$. The effects upon the mean and variance of uniform changes in observations, discussed in Sec. 3-5, hold for populations and samples alike. Thus multiplying $\bar{x}$ by the constant n multiplies its mean by n and its variance by n^2. Formally,

$$\mu_x = n \cdot \mu_{\bar{x}} = np \quad \text{and} \quad \sigma_x^2 = n^2 \sigma_{\bar{x}}^2 = npq.$$

These 2 results are very important, and we restate them for emphasis.

Conclusion: The mean and variance of the binomial distribution of x, the total number of successes in n independent trials, with probability p of success on an individual trial are np and $np(1 - p) = npq$, respectively.

A little reflection may convince you that the mean should depend directly upon both n and p. Certainly for a fixed number of trials, a larger value of p, the single-trial success probability, should be associated with a larger mean number of successes. Also for fixed p an increase in n, the total number of trials, should result in an increased mean. The quantity np exhibits both these properties.

Figure 6-1 shows 4 specimen binomial distributions for $n = 4$ and 10 and $p = 0.2$ and 0.6. The vertical scale of each graph gives $P(x)$, the probability of x successes. Several observations concerning the relative shapes of these 4 distributions are instructive. The distributions for $p = 0.6$ are better balanced, i.e., more symmetric, than are the distributions for $p = 0.2$. In general, the nearer p is to 0.5, the more symmetric is the corresponding distribution. In fact, when $p = 0.5$ the binomial distribution is perfectly symmetric for any value of n. We also notice that the distributions for

$p = 0.6$ are more spread than those for $p = 0.2$. This conforms with the fact that the corresponding variances are larger. There is also more spread in the distributions for $n = 10$ than for $n = 4$, again in conformity with larger variances.

A useful notational device for the binomial probabilities uses the symbol $B(x; n, p)$ for the probability that the total number of successes is x in n independent trials with success probability p on a single trial. For example, with 4 patients and single patient cure probability 0.6, the probability of exactly 2 cures is $B(2; 4, 0.6) = 0.3456$. Formally,

$$B(x; n, p) = {_nC_x}p^x(1 - p)^{n-x}.$$

The construction of tables giving binomial probabilities can be simplified by taking advantage of the property that

$$B(x; n, p) = B(n - x; n, 1 - p).$$

It is as easy to establish this relation semantically as it is analytically, so we shall do it both ways. The statement says that the probability of x successes in n independent trials when the probability of success on a single trial is p is the same as the probability of $n - x$ failures in n independent trials when the probability of failure on a single trial is $1 - p$. The designation of one kind of outcome as success and the other as failure is, as we have noted, purely arbitrary. The labels can be interchanged, provided we remember that if the probability of one outcome is p then the probability of the other is $1 - p$. This verbal or semantic argument establishes the relation.

Analytically we need to recall that ${_nC_x} = {_nC_{n-x}}$, a fact established in Sec. 4-7. Then

$$B(x; n, p) = {_nC_x}p^x(1 - p)^{n-x} = {_nC_{n-x}}(1 - p)^{n-x}[1 - (1 - p)]^{n-(n-x)}$$

$$= B(n - x; n, 1 - p).$$

This relation makes it unnecessary to tabulate binomial probabilities for values of p larger than 0.5. If, for example, we want $B(6; 10, 0.8)$, we notice that it is numerically equal to $B(4; 10, 0.2)$. Table A-2 gives binomial probabilities for several values of n and p.

EXAMPLE

Suppose that the probability of a single human birth producing a male infant is 0.5, ignoring twin and other multiple births. (In fact, this probability is slightly greater than 0.5.) Assuming that different deliveries are independent with respect to sex, find the probability that a family of

5 children will include: (a) exactly 3 boys, (b) at least 1 boy, (c) at most 1 boy.

Solutions:

(a) P(exactly 3 boys) = $B(3; 5, 0.5)$ = 0.3125.

(b) P(at least 1 boy) = $1 - P$(no boys) = $1 - B(0; 5, 0.5)$

$$= 1 - 0.0312 = 0.9688.$$

(c) P(at most 1 boy) = P(exactly 0 boys or exactly 1 boy)

$$= B(0; 5, 0.5) + B(1; 5, 0.5)$$

$$= 0.0312 + 0.1562 = 0.1874.$$

6-2   The Poisson Distribution

Another important discrete distribution is the Poisson distribution. It is the distribution of a random variable taking values 0, 1, 2, ... , and is useful in a wide variety of applications dealing with counts. Of course, not all variables arising from counting applications follow this distribution.

If x is a particular value, then the probability from the Poisson distribution that this value will occur is

$$\frac{e^{-\mu}\mu^x}{x!} \qquad \text{for } x = 0, 1, 2, \ldots.$$

The quantity e in the above formula is a constant with value approximately equal to 2.71828. Its actual value is

$$\frac{1}{0!} + \frac{1}{1!} + \frac{1}{2!} + \frac{1}{3!} + \cdots = 1 + 1 + \frac{1}{2} + \frac{1}{6} + \cdots,$$

where the dots indicate that reciprocals of successive factorials are to be added indefinitely, depending upon the degree of accuracy needed in the numerical value of e. As we observed in Sec. 4-6, successive factorials increase in size very rapidly, and as a result their reciprocals decrease rapidly. Thus the amount of change in the approximate value of e by adding successive terms is, after the first few terms, very slight. The surprising thing is that this is true even though there are at any stage infinitely many terms still to be added. We say that the expression $\frac{1}{0!} + \frac{1}{1!} + \frac{1}{2!} + \frac{1}{3!} + \cdots$ converges to the value e.

In the expression for the Poisson probability the quantity μ is a parameter that can take any value greater than zero. We have used the symbol μ because it is, in fact, the mean of the distribution. Notice that in contrast to the binomial distribution, which required knowledge of the values of n and p before probabilities could be calculated, the Poisson distribution requires knowledge of μ alone. We sometimes say that the binomial is a 2-parameter family of distributions, while the Poisson is a one-parameter family.

The Poisson distribution can be a reasonably adequate representative of such diverse quantities as radioactive counts per unit time, the number of calls arriving at a busy telephone switchboard during certain time periods, the number of raisins per cookie in a batch of cookies from a bakery, the number of plankton per aliquot of sea water, the number of deaths by horse kick per year of Prussian army officers, or the count of bacterial colonies per Petri plate in a microbiological study.

Two general models can be shown to lead directly to the Poisson probability. The first considers a large quantity of some medium such as sea water, air, serum, or cookie dough in which are found a large total number of discrete small entities, such as plankton, bacteria, granulocytes, or raisins. The material is thoroughly mixed to overcome clumping or social instincts. The mixing must guarantee a uniform density of entities throughout the medium. Then when a small quantity of the medium is examined, the probability that x of the entities will be found in the examined portion is the Poisson probability. The demonstration of this fact requires a mathematical argument involving limits, which we will omit.

The other model producing Poisson probabilities concerns events occurring in time. These events might be emissions of radioactive particles, calls arriving at a switchboard, or patients entering a waiting line at a hospital outpatient registration desk. Suppose that the events occur independently and that the probability that an event will occur in a short interval of length Δt is proportional to the length of the interval—in other words, the probability is $c\,\Delta t$, where c is some constant. Suppose further that Δt is short enough that the probability that more than one event occurs in an interval of such a length is negligible. Then the probability that x events occur in an interval of some fixed length t is the Poisson probability. Again we omit the proof.

We are concerned that you might think nearly everything is Poisson-distributed. Notice that both models require a kind of independence—the first by speaking of thorough mixing or uniform density, the second by stating that the events occur independently in time. This is an important and essential assumption. The Poisson distribution cannot be expected to fit counts of entities that tend to aggregate or clump, and it similarly cannot be expected to fit arrivals at a waiting line when the length of the line tends to stimulate new arrivals perhaps in search of a bargain or alternatively to repel them because of the possibility of a long wait. Similarly, the number of cases of streptococcal infection in households should not follow the Poisson distribution because of the effect of contagion.

The Poisson probabilities can be derived as limits of binomial probabilities when p gets very small (i.e., close to zero) and n gets very large in such a way that np remains constant. A demonstration that the first model dealing with entities in a medium produces Poisson probabilities proceeds by examining the binomial probability under these circumstances. The probability p that a given entity enters the examined aliquot is very small under our assumptions of a large total volume of medium compared to the sampled volume. We assume that n, the total number of entities, is large, and the assumption of thorough mixing without clumping or aggregation guarantees that the long-term average number of entities per aliquot np will remain constant.

These considerations suggest that np, the binomial mean, should equal μ, the parameter of the Poisson distribution, which, as we have stated, is its mean. They also allow us to determine the variance of the distribution. The variance of the corresponding binomial distribution is $np(1 - p)$ $= \mu \cdot (1 - p)$. But as p approaches 0, $1 - p$ approaches 1 and the variance approaches μ. An interesting fact about the Poisson distribution emerges: its mean and variance are both equal to the quantity μ appearing in the expression for the Poisson probability. Remember that the Poisson distribution is important in its own right and should not be regarded *solely* as a limiting form of the binomial. The purpose of discussing this relationship is to show the connection between these 2 important distributions.

As an example of a set of data to which the Poisson distribution provides a reasonably good fit, we show in Table 6-1 observed and Poisson expected frequencies for a set of 100, 0.1 minute radiologic counts from a single

Table 6-1

OBSERVED AND POISSON EXPECTED FREQUENCIES FOR 100, 0.1 MINUTE RADIOLOGIC COUNTS

Count	Observed Frequency	Poisson Expected Frequency
0	11	11.1
1	20	24.4
2	28	26.8
3	24	19.7
4	12	10.8
5	5	4.8
6 and over	0	2.5

source. The counts were made by Dr. John Nehemias. The expected frequencies are obtained by calculating the Poisson probabilities for a distribution having the same mean, 2.2, as the observed data. These probabilities are tabulated in Table A-3 for a variety of values of μ. If any member of the Poisson family of distributions is to fit the data, we would expect the best fit from the distribution with $\mu = \bar{x}$, the sample mean of the data.

The probabilities needed for the data in Table 6-1 are $P(x = 0) = 0.110803$, $P(x = 1) = 0.243767$, $P(x = 2) = 0.268144$, $P(x = 3) = 0.196639$, $P(x = 4) = 0.108151$, $P(x = 5) = 0.047587$, and $P(x \geq 6) = 0.024909$. These values are found from an extended version of Table A-3 which includes $\mu = 2.2$.

Since a probability can be regarded as a kind of long-term relative frequency (see Sec. 4-2), we multiply each probability by 100, the sample size, to get the expected frequency. These expected frequencies will not necessarily be whole numbers, since they represent the long-run average frequencies for repeated samples of size 100 from a Poisson distribution with $\mu = 2.2$. The general agreement of observed and expected frequencies can be seen from the table. We do not expect perfect agreement because of sampling variation and because the underlying distribution is probably not *exactly* Poisson. Remember that the Poisson distribution is simply a mathematical model useful as an idealization of natural phenomena. We never expect an *exact* fit of any such model. In Chap. 9 we will examine a class of statistical procedures for investigating the adequacy of fit of a set of expected or theoretical frequencies to a set of observed frequencies.

6-3   Probability Density

In the case of discrete random variables it is frequently possible to write down a mathematical expression that can be evaluated to determine the probability of a particular occurrence. For instance, for a binomially distributed random variable the probability of exactly x successes is given by the expression $_nC_x p^x q^{n-x}$ and for a Poisson random variable the probability of a count of x is $e^{-\mu}\mu^x/x!$. For continuous random variables it is not reasonable to expect a similar situation to hold. In Sec. 3-6 we said that a continuous variable (and therefore a continuous random variable) has the property that between any 2 potentially observable values there lies another potentially observable value. In other words, a continuous random variable may take any value along a whole interval of measurement, possibly an interval of infinite length. The fact is that a continuous random variable can take a fantastic number of different values. It is clear enough that it can take infinitely many different values, but it is true that the order of this infinity is much larger than, say, the number of whole numbers. A Poisson random variable potentially takes an infinite number of values, too; any nonnegative whole number occurs with some probability, however small. But the order of infinity of values of a continuous random variable is the order of the continuum, a much larger value. You might be inclined to say, "That's nonsense; infinity is infinity is infinity." 'Tain't so, my friend. Again, the demonstration of this interesting fact is at a mathematical level beyond this book, but we shall try to make the problem clear by an example.

Suppose we ask you to consider the probability that a man selected at random from the U.S. population will have a height of *exactly* 69 inches. Now be clear about what we are saying. We do not mean $69\frac{1}{2}$ inches or $69\frac{1}{1000}$ inches or even $69 + \frac{1}{15!}$ inches. We mean exactly 69.00000000... inches and those 3 dots merely indicate that we got tired of writing zeros. We mean the infinite decimal representation of *exactly* 69 inches. Now you might say that the probability will be fairly small. But we are persistent— how small? You might say, 1/1,000,000. But that is not nearly small enough. Suppose we agree that values within a small region of 69—say, between 69 and $69\frac{1}{4}$ inches—are about equally probable, so that you would be willing to assign on the average the same small value such as 1/1,000,000 to each. Just how many values are in that little interval of $\frac{1}{4}$ inch in length? Well, any value of the form $69 + 1/4k$ for $k = 1, 2, 3, \ldots$ is in there, and there are a lot more than a million of those. For instance, for $k = 10,000,000$ we have $69 + 1/40,000,000$, and although that is fairly close to 69, it is still in the interval. Just considering the first 10,000,000 values of k (not nearly all of them), we find that there are well over 10,000,000 different values, each supposed to have probability 1/1,000,000. But since we have mutually exclusive values,

$$P\left[69 + \frac{1}{(4)(1)} \quad \text{or} \quad 69 + \frac{1}{(4)(2)} \quad \text{or} \ldots \text{or} \quad 69 + \frac{1}{(4)(10,000,000)}\right]$$

$$= P\left[69 + \frac{1}{(4)(1)}\right] + P\left[69 + \frac{1}{(4)(2)}\right] + \cdots + P\left[69 + \frac{1}{(4)(10,000,000)}\right]$$

$$= \underbrace{\frac{1}{1,000,000} + \frac{1}{1,000,000} + \cdots + \frac{1}{1,000,000}}_{\text{10,000,000 terms}} = 10.$$

That is a little awkward, is it not? Here we are with only some, not all, of the values within a little interval of length $\frac{1}{4}$ inch occurring with a total "probability" of 10. That is not even sensible. We know that a probability larger than 1 just does not happen. It is inconsistent with our definition of probability.

What is the problem? The value 1/1,000,000 that we assigned to the probability of a single height is *grossly* too large. You will convince yourself after a little reflection that no matter how small a positive value you chose for that probability, we could give a similar argument to show that *it* was grossly too large. That drives us to the wall, and we are forced to conclude that the only feasible value to assign to the probability of a height of exactly 69 inches is zero. Now, cheer up; things are not as bad as they seem. When we say that this probability is zero, we mean to say only that the value, exactly 69, gets thoroughly lost among its near neighbors and that we should be prepared, given unlimited capital and an adventuresome spirit, to bet against its occurrence and give whatever odds are necessary. Of course, there is nothing special about the value 69. Our conclusion applies to any prespecified value. Furthermore, there is nothing special about height. Our conclusion applies to any continuous random variable.

But what is the real difficulty here? Simply that it is not important or even very sensible to be interested in the probability of a single value of a continuous random variable. In practice we cannot measure things that precisely, anyway. We will inevitably be more interested in probabilities for specific intervals of values—probabilities that a height will fall between 69 and 69$\frac{1}{4}$ inches, that a height will be over 72 inches, or that a height will be less than 67 inches. In these cases nonzero probabilities are appropriate and can be computed with the aid of a mathematical expression giving probability density.

An expression for probability density of a continuous random variable indicates for a given value x of the random variable how frequently values within the immediate vicinity of x occur. Specifically, if we make a graph showing probability density on the vertical scale and x values on the horizontal scale, then the area under the resulting curve over an interval of x values gives the probability that the value of the random variable will lie in that interval.

6-4   The Normal Distribution

6-4-1   GENERAL PROPERTIES OF THE NORMAL CURVE

The distribution occupying the central position in statistical theory and practice is the normal distribution. The distribution is remarkable and of great importance, not because naturally occurring continuous random variables follow it exactly, and not even because it is a useful model in all but abnormal circumstances. In fact, no observable random variable follows the normal distribution exactly, and many variables occurring in normal experience do not tend approximately to follow the distribution. The pervasiveness and importance of the distribution lie in its convenient mathematical properties leading directly to much of the theory of statistics available as a basis for practice, in its availability as an approximation to other distributions, in its direct relationship to sample means from virtually any distribution, and in its application to many random variables that either are approximately normally distributed or can be easily transformed to approximate normal variables.

The term "normal" as used in describing the normal distribution should not be construed as meaning "usual," "typical," "physiological," or "most common." In particular, a distribution that does not follow this form should be termed "nonnormal" rather than "abnormal." This problem of terminology has led many writers to call the distribution the Gaussian distribution, but this substitutes a historical inaccuracy for a possible misinterpretation of the word *normal*. In fact, the distribution should probably be called the De Moivrian distribution, but we will stick with the word *normal* in deference to the force of common usage.[1]

The mathematical expression for the normal density is

$$\frac{1}{\sqrt{2\pi\sigma^2}} \cdot e^{-\frac{1}{2}\left(\frac{x-\mu}{\sigma}\right)^2}$$

This gives the probability density at any value x. The quantity e is the constant discussed in Sec. 6-2, with a value of approximately 2.71828, while π is the familiar constant expressing the ratio of the circumference of a circle to its diameter and has approximate value 3.14159. The quantities μ and σ^2 are, respectively, the mean and variance of the distribution. These, of course, are parameters, and since they are the only quantities that must be specified in order to determine numerically the value of the density, the normal distribution is a 2-parameter family of distributions. The normal density or distribution is represented graphically in Fig. 6-2.

The curve is bell-shaped and has a single peak or mode at $x = \mu$. It is symmetric about $x = \mu$ in the sense that the value of the density is identical

---

[1] Professor Karl Pearson (42), in discussing the history of the normal distribution, commented upon this point as follows: "When we remember how common a process it is even today to deduce the normal curve from the limit to a binomial, I think it is only fitting to drop the association of that curve with the name of either Laplace or *a fortiori* Gauss and to credit De Moivre with its discovery." De Moivre's discovery of the mathematical expression for the normal density was published in 1733. Although the work of Gauss extended the available information about the distribution, all this work first appeared after 1800.

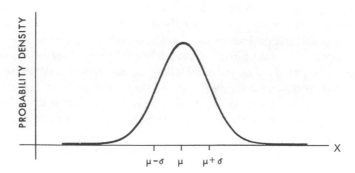

Figure 6-2 A normal distribution.

at $\mu + c$ and $\mu - c$ for any value of c. There is a change in the direction of curvature at 2 places. Whereas the curve is concave downward at and near μ, it becomes concave upward when x gets far enough from μ in either direction. The points at which the direction of concavity changes are called inflection points; these are located at a distance of σ units above and below the mean μ. The total area under the curve is 1 in accord with the fact that areas under densities give probabilities, and that some value of x between plus and minus infinity must occur. In other words, the probability that x falls somewhere must be 1.

We can observe the effect of changing the value of μ or σ. For fixed σ, changing μ has the effect of shifting the whole curve intact to the right or left a distance corresponding to the amount of the change. For fixed μ,

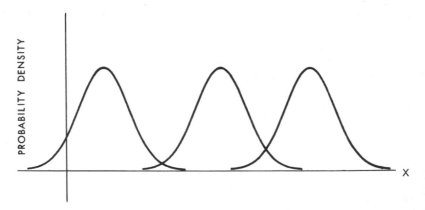

Figure 6-3 The effect of changing μ in the normal distribution while keeping σ constant.

changing σ has the effect of locating the inflection points closer to or farther from μ, and since the total area is fixed at 1, this results in values clustered more closely or less closely about the mean. This agrees with the fact that σ, the standard deviation, measures the amount of dispersion or spread in the curve—smaller σ, less spread; larger σ, more spread. Figures 6-3 and 6-4 show the effects of changes in μ and σ, respectively.

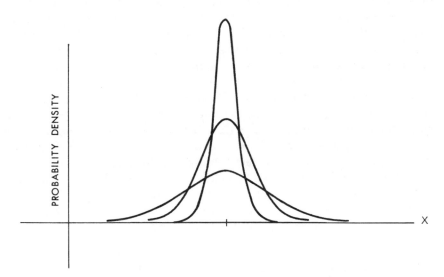

Figure 6-4   **The effect of changing σ in the normal distribution while keeping μ constant.**

Since x can take any value between plus and minus infinity, it is reasonable that μ would also have this property, and that is the case. The variance σ^2 must, of course, be positive, and σ, the standard deviation, is the positive square root of σ^2 and is, therefore, also positive.

Interesting features of the normal distribution are its long tapering tails. In fact, these tails extend indefinitely far in either direction, corresponding to the infinite range of x values in the distribution. If the total area is to be 1, it follows that the elevation of the curve above the x axis must decrease rapidly as the value of x gets farther and farther from μ. As a result, the area under the curve for intervals of x values farther and farther from μ must also decrease. In other words, intervals of x values become very improbable as their distance from μ increases. To illustrate this fact quantitatively, for the normal distribution the set of all values of x farther than 3 standard deviation units from the mean occurs with probability 0.0027; the set of all values farther than 4 standard deviations occurs with probability 0.00006, and the set of values farther than 5 standard deviations occurs with probability 0.0000006.

6-4-2 THE STANDARD NORMAL DISTRIBUTION

The normal distribution is, as we have noted, really a large family of distributions corresponding to the many different values of μ and σ. In attempting to tabulate the normal probabilities for various parameter values some simplification is essential. Fortunately, such a simplification is readily available through a device known as standardization. Geometrically this amounts to an agreement to convert the basic scale of x values in order that we measure on a standard scale with zero value corresponding to μ and with a measurement unit of 1 standard deviation—in other words, we agree to convert our measurements to numbers of standard deviation units above or below the mean. If we define $z = (x - \mu)/\sigma$, then z has this property. To check this, note that when $x = \mu$, $z = 0$; when $x = \mu + \sigma$, $z = 1$; when $x = \mu - 2\sigma$, $z = -2$, etc. In fact, multiplying by σ and adding μ in the expression relating x and z, we find that $x = \mu + z\sigma$. This states that z is the number of standard deviations from μ at which the value x is located; z is positive if x is greater than μ and negative if x is less than μ. Figure 6-5 shows the relation between the measurement scales.

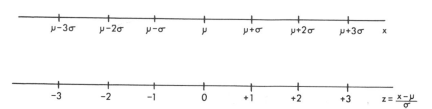

Figure 6-5   Conversion of scale of measurement to standardized z scale.

Since x is a random variable, so is z. It seems natural that the mean of z's distribution will be 0 and its standard deviation will be 1. We can determine the mean and variance of z by making use of Sec. 3-5 and again noting that the results of that section apply to populations and samples alike. Table 6-2 shows the situation.

This kind of standardization will produce a derived random variable z with mean 0 and standard deviation 1 for any random variable x with mean μ and standard deviation σ, whatever the distribution of x—normal or nonnormal, continuous or discrete. Naturally, if x is nonnormal, z is also.

If x is normally distributed, z will also be normal. In fact, if x is normally distributed with mean μ, and standard deviation σ and if c and d are constants,

Table 6-2

DETERMINATION OF MEAN AND STANDARD DEVIATION OF z

Random Variable	Mean	Standard Deviation
x	μ	σ
$y = x - \mu$	$\mu - \mu = 0$	σ
$z = \dfrac{x - \mu}{\sigma} = \dfrac{y}{\sigma}$	$\dfrac{0}{\sigma} = 0$	$\dfrac{\sigma}{\sigma} = 1$

then the random variable $w = c + dx$ is normally distributed with mean $c + d\mu$ and standard deviation $|d|\sigma$. The variable z is a special case of this general situation, for which $c = -\mu/\sigma$ and $d = 1/\sigma$. Since z is a normal variable with $\mu = 0$, $\sigma = 1$, its probability density is obtained by substituting these values in the general expression of Sec. 6-4-1. This produces

$$\frac{1}{\sqrt{2\pi}} \cdot e^{-z^2/2}.$$

6-4-3   INEQUALITIES

The treatment and manipulation of inequalities is important to many of the statistical ideas and techniques we wish to develop. Inequalities were mentioned briefly in Sec. 4-1 but will now be discussed at greater length. The basic symbols used to express inequalities are $<$ and $>$. The symbols used between 2 algebraic or numerical quantities are read, respectively "is less than" and "is greater than." The notation is sensible and easy to remember if you notice that the sign points toward the smaller of the 2 quantities; i.e., the pointed side is directed toward the smaller and the open side toward the larger of the quantities. The following statements are all true: $2 < 4$, $8 > 6$, $-5 < 7$, $-2 > -5$.

Notice that any inequality is equivalent to a reverse inequality in the following sense: $2 < 4$ is equivalent to $4 > 2$; $-2 > -5$ is equivalent to $-5 < -2$. Since any inequality involving the symbol $>$ can, therefore, be converted to an equivalent inequality involving the symbol $<$, we shall tend to limit formal discussions and presentations to the latter symbol, rather than giving separate presentations for inequalities in both directions. Inequalities may be chained or strung together as follows:

$$2 < 4 < 6, \quad 1 > -1 > -10,$$

but notice that chain inequalities are so arranged that the quantities involved either increase or decrease in sequence. That is, all inequality signs in a chain must point in the same direction. We do not write $2 < 4 > 1$, but we can transmit the same information by the proper form: $1 < 2 < 4$. Inequalities have a number of algebraic properties.

Property 1

The same quantity may be added to or subtracted from both members of an inequality without affecting the sense, i.e., the direction of the inequality. In other words, if $a < b$ then $a + c < b + c$ and $a - c < b - c$. As an example, $2 < 4$ and $2 + 3 < 4 + 3$, i.e., $5 < 7$ and $2 - 3 < 4 - 3$, i.e., $-1 < 1$.

Property 2

Both members of an inequality may be multiplied or divided by the same positive quantity without affecting the sense of the inequality. In other words, if $a < b$ and $c > 0$, then $ac < bc$ and $a/c < b/c$. As an example, $5 < 8$ and $2 > 0$. Thus $(5)(2) < (8)(2)$, i.e., $10 < 16$ and $5/2 < 8/2$, i.e., $2\frac{1}{2} < 4$.

Property 3

If both members of an inequality are multiplied or divided by the same negative quantity, the sense of the resulting inequality is reversed. In other words, if $a < b$ and $c < 0$, then $ac > bc$ and $a/c > b/c$. As an example, $6 < 9$ and $-3 < 0$. Thus $(6)(-3) > (9)(-3)$, i.e., $-18 > -27$ and $6/-3 > 9/-3$, i.e., $-2 > -3$.

Property 4

If 2 positive quantities are related by an inequality, their reciprocals are related by the reverse inequality. In other words, if $a < b$, $a > 0$, and $b > 0$, then $1/a > 1/b$. As an example, $3 < 6$, $3 > 0$, and $6 > 0$. Thus, $1/3 > 1/6$.

Two additional symbols are used to connect algebraic quantities that are either unequal in a certain direction or are possibly equal. These symbols are $\leq$ and $\geq$ and are read "is less than or equal to" and "is greater than or equal to," respectively. Equivalently they may be read "is at most equal to" and "is at least equal to," respectively. The same 4 properties apply to these inequalities as well, and we restate them in abbreviated form.

Property 1a

If $a \leq b$, then $a + c \leq b + c$ and $a - c \leq b - c$.

Property 2a

If $a \leq b$ and $c > 0$, then $ac \leq bc$ and $a/c \leq b/c$.

Property 3a

If $a \le b$ and $c < 0$, then $ac \ge bc$ and $a/c \ge b/c$.

Property 4a

If $a \le b$, $a > 0$ and $b > 0$, then $1/a \ge 1/b$.

Again notice that although these properties are stated in terms of the relation $\le$, an equivalent property involving the relation $\ge$ is always available. That is, whenever $a \le b$, the dual relation $b \ge a$ is implicit.

6-4-4 THE USE OF TABLES FOR THE STANDARD NORMAL DISTRIBUTION

Table A-4 gives the height or ordinate h of the standard normal distribution at the value z, as well as a number of useful areas or probabilities under the curve, denoted by the letters A through E.[2] Examination of the table shows that regular or "nice" values of z, e.g., $z = 0.1, 0.2, 1.0$, etc., are accompanied by irregular values of heights or areas. For that reason, the table is also set up to include regular values of the areas, resulting, of course, in irregular values of z. This accounts for the variation in numbers of decimal places shown for z and the areas. The table gives sufficient accuracy so that interpolation is unnecessary in most applications. Uses of the table will be illustrated by a series of examples.

EXAMPLE 1

Find, for the standard normal distribution, (a) $P(0 < z < 1.20)$, (b) $P(z > 1.20)$, (c) $P(-1.20 < z < 1.20)$, (d) $P(z < -1.20$ or $z > 1.20)$, (e) $P(z < 1.20)$, (f) the height or ordinate at $z = 1.20$, and finally, (g) $P(1.50 < z < 2.00)$.

As we have noted, probabilities for a continuous random variable are given by areas under the density or distribution curve. The probability for part (a) is the area under the curve between 0 and 1.20, an area designated in Table A-4 as A corresponding to $z = 1.20$. From the column of A values we read the result, 0.3849. The probability for part (b) is the table area designated as B corresponding to $z = 1.20$, that is, 0.1151. For part (c) the required probability is table area C corresponding to $z = 1.20$, that is, 0.7699. Part (d) asks for area D corresponding to $z = 1.20$, that is, 0.2301. Part (e) requires area E corresponding to $z = 1.20$, that is, 0.8849. Notice that the column of E

---

[2] h is the value of the probability density at z.

values with the corresponding z values give various percentiles and percentile ranks (see Sec. 3-8) of the standard normal distribution. For example, to find the 80th percentile we set $E = 0.80$ and read $z = 0.8416$; to find the percentile rank of 1.00 we set $z = 1.00$ and read $E = 0.8413$ or 84.13 per cent. Use of the table to determine percentiles and percentile ranks corresponding to percentages less than 50 will be discussed in a subsequent example. The solution to part (f) of this example is the tabulated value of h for $z = 1.20$ or 0.1942. Part (g) asks for the area under the curve between $z = 1.50$ and $z = 2.00$. Using the column of E values, if we subtract the value of E for $z = 1.50$ from that for $z = 2.00$, then we will get the required solution. That is, $P(1.50 < z < 2.00) = 0.9772 - 0.9332 = 0.0440$.

EXAMPLE 2

For the standard normal distribution: (a) Between which 2 z values does the central 95 per cent of the area under the curve lie? (b) What value of z cuts off 95 per cent of the area to its left? (c) What value of z has the property that 1 per cent of the area lies outside the interval from $-z$ to $+z$?

With respect to part (a) we want the quantity C to take the value 0.95. We find this value opposite a z value of 1.9600, which implies that the central 95 per cent of the area lies between a z value of -1.9600 and $+1.9600$. In part (b) the given area is a value of E. Corresponding to $E = 0.95$ we have $z = 1.6449$. Part (c) gives the value $D = 0.01$. We find here that $z = 2.5758$.

EXAMPLE 3

For the standard normal distribution, find (a) $P(z < -0.85)$, (b) $P(z > -1.16)$, (c) the value z_0 such that $P(z < z_0) = 0.3$, (d) the value z_0 such that $P(-z_0 < z < 0) = 0.42$, (e) the 30th percentile, (f) the percentile rank of -1.28.

In attacking these problems we notice that Table A-4 provides no negative values of z. This is not because negative values of z do not occur. In fact, as we have seen, any value of z along the entire negative and positive axis can occur. The reason for eliminating negative values from the table is that the symmetry of the standard normal distribution makes it easy to convert probability statements involving negative values of z to corresponding statements involving positive values. Letting z_0 denote a particular positive value for z, we have, using the letters A through E to denote the various areas in Table A-4:

1. $P(-z_0 < z < 0) = P(0 < z < z_0) = A.$

2. $P(z < -z_0) = P(z > z_0) = B.$

3. $P(z > -z_0) = P(z < z_0) = E.$

If you have trouble seeing that these relationships hold, try drawing a standard normal curve and shading the various areas involved.

Returning to the specific problem at hand (Example 3), we have for part (a) $P(z < -0.85) = P(z > 0.85) = B$, corresponding to a z of 0.85: that is, 0.1977. In part (b), $P(z > -1.16) = P(z < 1.16) = E$, corresponding to a z of 1.16: that is, 0.8770. In part (c), the required value z_0 must be negative, since the area to its left is less than 0.5. Again using the symmetry of the normal curve, we find that z_0 is the negative of a number z_1 for which $P(z > z_1) = 0.3$. But 0.3 is thus a B value to which corresponds $z_1 = 0.5244$, whence $z_0 = -0.5244$. In part (d), $P(-z_0 < z < 0) = P(0 < z < z_0) = 0.42 = A$. Therefore, $z_0 = 1.4051$. The 30th percentile of the distribution is the value z_0 for which $P(z < z_0) = 0.30$. But we found in part (c) of this problem that $z_0 = -0.5244$, and this is the required percentile. The percentile rank of -1.28 is the numerical value of $P(z < -1.28)$. But $P(z < -1.28) = P(z > 1.28) = 0.1003$ or 10.03 per cent, from the column of B values.

EXAMPLE 4

Suppose that in a certain pediatric population, casual, sitting systolic blood pressure is normally distributed with $\mu = 115$ mm Hg, $\sigma^2 = 225$ (mm Hg)^2. Find the probability that a child randomly selected from this population will have: (a) a systolic pressure less than 140 mm Hg, (b) a pressure greater than 100 mm Hg, (c) a pressure between 110 and 120. Also find (d) the systolic pressure below which the pressures of 99 per cent of the population lie, (e) the 2 systolic pressures between which the central 50 per cent of the pressures in the population lie. Find (f) the quintiles of the distribution, (g) the percentile rank of 100 mm Hg.

In order to use Table A-4 to solve this set of problems we must first standardize the distribution. That is, if x denotes systolic blood pressure then we must convert numerical values of x to values of z by the relation $z = (x - \mu)/\sigma$. If we require as a solution a value of x, we shall obtain directly from the table a value of z from which x is determined by the relation $x = \mu + z\sigma$.

Solutions:

(a) We must convert the x value 140 to a z value.

$$\frac{140 - \mu}{\sigma} = \frac{140 - 115}{15} = 1.67.$$

We find $P(x < 140) = P(z < 1.67) = 0.9525$.

(b) Here $x = 100$ and $z = \dfrac{100 - 115}{15} = -1$.

$$P(x > 100) = P(z > -1) = P(z < 1) = 0.8413.$$

(c) $$x_1 = 110, \qquad z_1 = \frac{110 - 115}{15} = -0.33;$$

$$x_2 = 120, \qquad z_2 = \frac{120 - 115}{15} = 0.33.$$

$$P(110 < x < 120) = P(-0.33 < z < 0.33) = 0.2586.$$

(d) We have $P(x < x_0) = P(z < z_0) = 0.99$ whence $z_0 = 2.3263$ and $x_0 = 115 + (2.3263)(15) = 149.8945$ or approximately 150 mm Hg.

(e) We know that $P(-z_0 < z < z_0) = 0.50$ and therefore $z_0 = 0.6745$. Thus, between the values $z_1 = -0.6745$ and $z_2 = 0.6745$ lie the central 50 per cent of pressures. The x's corresponding to these z's are

$$x_1 = 115 + (-0.6745)(15) = 104.8825$$

and

$$x_2 = 115 + (0.6745)(15) = 125.1175.$$

The central 50 per cent of pressures are found between the approximate values 105 and 125 mm Hg.

(f) The quintiles of the distribution divide the total area or probability into 5 equal parts. Again we must approach the problem by finding first the quintiles of the standard normal distribution. These are quantities z_1, z_2, z_3, z_4 such that $P(z < z_1) = 0.20$, $P(z < z_2) = 0.40$, $P(z < z_3) = 0.60$, and $P(z < z_4) = 0.80$. From Table A-4 we find $z_1 = -0.8416$, $z_2 = -0.2533$, $z_3 = 0.2533$, and $z_4 = 0.8416$. These z values are converted to x's with the following results. $x_1 = 102.3760$, $x_2 = 111.2005$, $x_3 = 118.7995$, and $x_4 = 127.6240$, or, in rounded form, $x_1 = 102$, $x_2 = 111$, $x_3 = 119$, $x_4 = 128$, the required quintiles of the distribution of systolic blood pressure.

(g) The given value, 100 mm Hg, is a blood pressure and therefore an x value. Its corresponding z value is -1 and $P(z < -1)$ $= P(z > 1) = 0.1587$. The percentile rank of 100 is 15.87 per cent.

6-4-5 FITTING A NORMAL DISTRIBUTION TO GROUPED DATA

When data have been grouped in a frequency table, it is often of interest to note the extent to which they conform to the general shape of the normal distribution. An operation known as *fitting a normal distribution* can be carried out to investigate this question. In this section we shall describe the fitting process itself. Questions of testing whether an observed random sample comes from an underlying population that is normally distributed will be treated in quantitative terms later in the book.

The fitting process calculates a set of expected frequencies for the classes of the frequency table. Each expected frequency is obtained by multiplying the total frequency by the calculated probability from a normal distribution having the same mean and variance as the grouped data. If any normal distribution fits the data, we expect the distribution with the same mean and variance as the data to fit best. The fitting process can be carried through for any grouped set of observations, no matter how far the set apparently deviates from the normal form. The adequacy or goodness of fit is another matter to be treated quantitatively later.

We will illustrate the fitting process by examining a set of heights of 195 adult male students enrolled in biostatistics courses at the University of Michigan. The heights were taken by self-report rather than actual measurement and were given to the nearest inch. Table 6-3 is a frequency table of the heights, grouped in 2-inch intervals.

Table 6-3

FREQUENCY TABLE OF HEIGHTS IN INCHES
OF 195 MALE BIOSTATISTICS STUDENTS
AT THE UNIVERSITY OF MICHIGAN

Class Boundaries	Class Limits	Class Midpoints	Observed Frequency
61.5–63.5	62–63	62.5	2
63.5–65.5	64–65	64.5	16
65.5–67.5	66–67	66.5	30
67.5–69.5	68–69	68.5	48
69.5–71.5	70–71	70.5	48
71.5–73.5	72–73	72.5	39
73.5–75.5	74–75	74.5	11
75.5–77.5	76–77	76.5	1

The mean height calculated from the grouped data is 69.47 inches, and the standard deviation is 2.8164 inches. In order to calculate the expected frequencies for the normal distribution with this mean and standard deviation, we must determine the area or probability under the normal curve for each class interval. To avoid gaps in the measurement scale, we must get the area between the class boundaries rather than between the limits. A problem arises with the first and last intervals. Since the normal distribution has infinitely long tails in both directions, the total area or probability will not add to 1 unless we take this into account. As an adjustment we shall take for the probability corresponding to the first class interval the entire area under the normal curve to the left of the upper class boundary, 63.5 in this example. For the last interval we shall use the entire area to the right of the lower class boundary 75.5. Unless some such adjustment is made to include tail areas, the total expected frequency will not equal the total observed frequency, and this might distort our notion of the adequacy of fit.

To use Table A-4 to provide the relevant areas or probabilities corresponding to each class interval we must first convert the class boundaries to values of z, the standard normal variable. In this case $z = (x - \bar{x})/s$, where x is a boundary. Remember that we are fitting a normal distribution with μ and σ equal to the observed $\bar{x}$ and s of the sample. We shall convert the upper boundary of each class except for the last class, for which the entire remaining upper tail area under the normal curve will be used. Cumulative areas are computed for each upper boundary; adjacent areas are then subtracted from each other successively to give areas in the intervals. The cumulative area through the upper class boundary of the next to the last class is subtracted from 1 to give the area for the last class. The areas are

Table 6-4

FITTING A NORMAL DISTRIBUTION TO HEIGHTS OF 195 MALE
BIOSTATISTICS STUDENTS AT THE UNIVERSITY OF MICHIGAN

Class Boundaries	Observed Frequency	z at Upper Class Boundary	Cumulative Area through Upper Class Boundary	Area in Class	Expected Frequency
61.5–63.5	2	−2.12	0.0170	0.0170	3.32
63.5–65.5	16	−1.41	0.0793	0.0623	12.15
65.5–67.5	30	−0.70	0.2420	0.1627	31.73
67.5–69.5	48	+0.01	0.5040	0.2620	51.09
69.5–71.5	48	+0.72	0.7642	0.2602	50.74
71.5–73.5	39	+1.43	0.9236	0.1594	31.08
73.5–75.5	11	+2.14	0.9838	0.0602	11.74
75.5–77.5	1			0.0162	3.16

then multiplied by the total frequency to find the expected number of cases in each class. Table 6-4 shows the results of these computations for the 195 heights. The general agreement between the observed and expected frequencies is an indication that the underlying population of heights may be fairly well modeled by the normal distribution.

Figure 6-6 shows histograms for observed and expected frequencies plotted on the same graph.

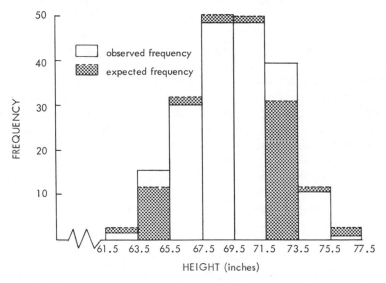

Figure 6-6 Histograms of observed and expected frequencies for fit of a normal distribution to heights of 195 biostatistics students.

6-4-6 THE NORMAL APPROXIMATION TO THE BINOMIAL DISTRIBUTION

In 1733, when De Moivre discovered the normal curve, arithmetic computation was a tedious chore. Modern computers and desk calculators were, of course, unavailable, and a large part of mathematical research was devoted to the search for expressions to use as approximations to other expressions that were more tractable computationally. It was such a search for a simplified approximation to the binomial probabilities that led to De Moivre's discovery.

It seems curious that the continuous normal distribution could approximate the discrete binomial distribution, and we must examine the sense in which such an approximation is possible. The binomial distribution with parameters n and p gives probabilities that a random variable x takes any of its possible values, $0, 1, 2, \ldots, n$. As we have seen, a continuous random variable takes any single value with probability 0. The dilemma can be resolved by assuming that the value of the binomial variable results from a measurement process for an underlying continuous variable and that the recorded value is the result of rounding to the nearest whole number. Of course, this is not ordinarily true, but it provides the bridge we need between the discrete and continuous distributions. Under this assumption an observed value, say $x = 2$, really arises from some underlying value between 1.5 and 2.5. This procedure of adjusting by a half-unit on either side of the observed value is called a *correction for continuity*. An additional difficulty similar to one encountered in the preceding section arises here. This is the problem caused by the long tails of the normal distribution. The smallest and largest values of a binomial variable are 0 and n, respectively. In approximating the corresponding probabilities by areas under the normal curve we shall take the entire area to the left of 0.5 to approximate the binomial probability that $x = 0$ and the entire area to the right of $n - 0.5$ for the probability that $x = n$. You will recognize that this corresponds to our practice for the first and last class intervals in fitting a normal distribution to grouped data.

Which normal distribution would we expect to provide the best approximation to the binomial probabilities? Again, that distribution whose mean and variance are the same as the mean and variance of the distribution being approximated; that is, the normal distribution with $\mu = np$ and $\sigma^2 = np(1 - p) = npq$.

The adequacy of this approximation depends jointly on the values of n and p. In general terms, the larger the value of n, the better the approximation, but if p is close to 0.5, a smaller value of n can be used than if p is near 0 or 1. A rule of thumb that will provide a guide is that the normal approximation is adequate for most purposes so long as both np and nq are greater than 5. It will be instructive to evaluate some of the approximating probabilities for $n = 10$, $p = 0.5$. Note that this is a "breakpoint" case that just violates our rule of thumb, since here $np = nq = 5$.

We shall illustrate the computational method by finding the normal approximate probability that $x = 2$. The relevant normal area lies, as we have seen, between the values $x_1 = 1.5$ and $x_2 = 2.5$. Converting these to z values, noting that $\mu = np = 5$, $\sigma^2 = npq = 2.5000$, and $\sigma = 1.5811$, we have

$$z_1 = \frac{1.5 - 5}{1.5811} = -2.21 \quad \text{and} \quad z_2 = \frac{2.5 - 5}{1.5811} = -1.58.$$

From Table A-4 we find that the corresponding area under the normal curve is $0.9864 - 0.9429 = 0.0435$. Table A-2 shows the exact binomial probability to be 0.0439. In Table 6-5 we show the exact and approximate probabilities for the full distribution. Although there is fairly close agreement between the exact and approximate probabilities in this case, remember that the normal approximation is ordinarily used for much larger values of np and nq, and in these cases the agreement will be better.

Table 6-5

EXACT AND APPROXIMATE VALUES BASED ON THE NORMAL
DISTRIBUTION FOR BINOMIAL PROBABILITIES WITH
$n = 10, p = 0.5$

Value of x	Exact Probability of x Value	Approximate Probability of x Value
0	0.0010	0.0022
1	0.0098	0.0114
2	0.0439	0.0435
3	0.1172	0.1140
4	0.2051	0.2034
5	0.2461	0.2510
6	0.2051	0.2034
7	0.1172	0.1140
8	0.0439	0.0435
9	0.0098	0.0114
10	0.0010	0.0022

EXAMPLE

The rate of operative complications in' a certain complex vascular reconstructive procedure is 20 per cent. This includes all complications, ranging in severity from wound separation or infection to death, and is the proportion of patients showing some complication. In a series of 50 such procedures, what is the probability that there will be at most 5 patients with operative complication?

We shall assume independence of the occurrence or nonoccurrence of complication in different patients, and we shall also assume that each patient has a complication probability of 0.20. Then the binomial distribution provides the correct solution as follows:

$$P(\text{at most 5 complications}) = P(x \leq 5) = \sum_{x=0}^{5} B(x; 50, 0.2)$$

$$= \sum_{x=0}^{5} \binom{50}{x}(0.2)^x(0.8)^{50-x}.$$

Although large tables of the binomial distribution are available and could be used in this case, the normal approximation should be satisfactory here, since $np = 10$ and $nq = 40$. In applying the approximation we want all the area under the normal curve to the left of the point representing the upper boundary of the interval corresponding to $x = 5$. This boundary is 5.5. Here $\mu = np = 10$, $\sigma^2 = npq = 8$ and $\sigma = 2.8284$. Accordingly, the z value corresponding to 5.5 is $z = (5.5 - 10)/2.8284 = -1.59$. The area to the left of this value is, from Table A-4, 0.0559.

6-4-7 THE CENTRAL LIMIT THEOREM

One of the most interesting and important mathematical results in this field deals with the normal distribution and gives one reason for its central importance. This result says that for random samples from any underlying distribution with mean μ and variance σ^2, however nonnormal it may be, the sampling distribution of $\bar{x}$, the sample mean for random samples of size n, is approximately normal, the approximation improving as n increases. The approximating normal distribution will, of course, have mean μ and variance σ^2/n, as follows from statements 1 and 2 of Sec. 5-4-2.

This result is remarkable because we start with very few assumptions—in fact, we assume only random sampling and the existence of a mean and variance—and arrive at a very specific conclusion.[3] This is a kind of "all roads lead to Rome" result in that all distributions from a very wide class, including continuous, discrete, skewed, symmetric, unimodal, and multimodal distributions, produce sample means that tend to be normally distributed.

Some investigators, in attempting to apply this result to real data, have stretched it a bit out of shape. Their reasoning is that any natural phenomenon in which many small effects, some acting positively and some negatively, combine to produce a single numerical value will result in data that are normally distributed. They reason that the observed value is a kind of arithmetic mean resulting from all these effects. An example is adult height, a variable subject to a wide variety of hereditary, environmental, endocrinologic, nutritional, and developmental factors. When this variable is examined, for a single sex, of course, it ordinarily tends to follow the normal distributional form rather closely.

This cannot be used as a general argument that all such variables will be normal. For example, adult weight is similarly the result of many small effects, but in the U.S. population it has a positively skewed distribution. This is perhaps because the majority of those small effects in this case are things

---

[3] Some distributions, of interest primarily in mathematical statistics, do not have finite means or variances. We say for brevity that the mean or variance of such a distribution does not exist. None of the distributions considered in this text will exhibit such pathologies.

like cream puffs, french fries, milk shakes, and hours logged in the TV chair, which tend to operate on weight in only one direction. Certainly weight is freer to vary on one end than the other (no pun intended). It is relatively easy to produce very large positive effects on weight, but not so easy in the other direction. Spreading adult weight along a continuum, we see that the right-hand direction is relatively free and open, whereas the left is quite sharply limited. A living adult must carry a certain minimum weight consistent with survival and including the weights of heart, blood, lungs, skeleton, and other essential organs and tissues.

The central limit theorem can be used very explicitly in certain circumstances to improve the degree to which a set of data will follow the normal form. Sets of individual observations, grouped together at random, can be formed. The observations within each such set can be averaged and the final analyses based upon the set of averages. The central limit theorem applies here and provides assurance that the averages will be more nearly normally distributed than were the individual observations. This technique is of rather limited usefulness, however, since it requires a sufficiently large collection of data to permit the formation of random subsets. Furthermore, our primary interest may not be in averages at all, and we may have by this procedure constructed an artificial situation.

Some individuals tend to become panic-stricken whenever they are confronted with a set of nonnormal data. By way of balm and comfort, although it is true that many techniques in statistical inference assume normality, some of the most important of these have been found to be quite robust against departures from normality. This is not to say that normal-based techniques can be applied willy-nilly to any old set of observations. Like a package insert for a drug, all we can say is when in doubt, consult your statistician.

It is natural to inquire relative to the central limit theorem about the distribution of the sample mean for random samples from a normal distribution. If means from nonnormal distributions tend toward normality as the sample size increases, how about distributions that are already there to begin with? As you might expect, the sample mean for random samples of size n from a normal distribution with mean μ and variance σ^2 has a sampling distribution that is *exactly* normal with mean μ and variance σ^2/n.

EXAMPLE

In Example 4 of Sec. 6-4-4 we considered a pediatric population in which systolic blood pressure was normally distributed with $\mu = 115$ and $\sigma = 15$. If a random sample of size 25 is selected from this population, find $P(110 < \bar{x} < 120)$.

The sample mean is normally distributed with mean 115 and standard deviation (standard error) $\sigma/\sqrt{n} = \frac{15}{5} = 3$. The z values corresponding to 110 and 120 are -1.67 and $+1.67$. The required probability is 0.9051. In part (c) of Example 4 we determined that the probability that a single blood pressure would lie between these limits is 0.2586. The difference is, of course, due to the fact that the variability of the sample mean is less than that of a single observation, and therefore the probability that the sample mean will fall within 5 units of the population mean should be larger than the corresponding probability for a single observation.

EXERCISES

6-1 The quoted figure for the 5-year mortality rate for a particular form of leukemia is 80 per cent. In the hospital where you are a resident interested in malignant neoplasm research, of the last 5 cases with this form of leukemia 4 are cured and 1 died. Do you feel you should check to see if some new procedure was used on the patients or that they were special in some other way, or pass off the cures to chance?

6-2 Suppose 60 per cent of the voting population in a city, about to have a referendum on adding sodium fluoride to the drinking water, favor fluoridation.

    a. A sample of 10 persons are interviewed. What is the probability that 5, 6, or 7 favor fluoridation?

    b. A sample of 100 persons are interviewed. What is the probability that between 55 and 65 inclusive favor fluoridation?

6-3 Under microscopic investigation, on the average 5 particular microorganisms are found on a one square centimeter untreated specimen. One such specimen was chemically treated. If it is assumed that the treatment was ineffective and if the Poisson distribution is used, what is the probability of finding: (a) less than 3 organisms? (b) exactly 5? (c) more than 6? (d) 2 or 3?

6-4 A student in quantitative analysis has been told that the probability of obtaining a successful end point in a particular titration is 0.7. This student carries out 5 such titrations and obtains only one successful end point. Should he think of a career that does not involve chemistry?

6-5 a. Suppose a dentist engaged in research states at a dental convention that in a survey on "What makes an article acceptable to a journal?" it was concluded that "Two out of every three published articles summarized the major findings in the first lines of the paper." Assuming that journals do not dictate the format of articles, if you were to write an article, would you structure it with the findings in the first lines? Discuss.

b. Assume now that the conclusion of the survey was "Two out of every three articles that had the findings in the first lines were published." Suppose you read drafts of 6 articles, all with the findings in the first lines. What is the probability that all will be published?

6-6 Let x be a random variable assuming the values 1 and 0 for the presence and absence of amblyopia. Consider the results of ocular tests for 500 children:

Amblyopia

Present   *Absent*

50   450

a. Using the grouped computing formula, where $x_i = 0$ or 1 and f_i represents the associated frequencies, find

$$\mu = \frac{\sum\limits_{i=1}^{2} f_i \cdot x_i}{\sum\limits_{i=1}^{2} f_i} \quad \text{and} \quad \sigma^2 = \frac{N \sum\limits_{i=1}^{2} f_i \cdot x_i^2 - \left(\sum\limits_{i=1}^{2} f_i \cdot x_i\right)^2}{N^2}$$

b. Letting $P(x = 1) = p$, find p and compute μ and σ^2, using the binomial formulae of Sec. 6-1. Verify that these agree with the values found in part a.

6-7 Suppose that a particular strain of staphlococcus produces a certain symptom in 2 per cent of persons infected. At a church picnic 200 persons ate contaminated food and were infected with the organism. What is the probability of the following numbers having this symptom: (a) 10 or fewer? (b) none? (c) more than 4? (d) exactly 4?

6-8 *Multinomial distribution.* Suppose that there are k mutually exclusive and exhaustive ways for a trial to result so that the probability that the trial ends in the first way is p_1, in the second way is p_2, ..., in the kth way is p_k. Recalling the formula for computing the number of distinguishable permutations of objects not all different (Sec. 4-8),

a. Develop the following formula for the probability that in n such trials, t_1 are of the first type, t_2 of the second, ..., t_k of the kth type:

$$P(t_1, t_2, \ldots, t_k) = \frac{n!}{t_1! \, t_2! \ldots t_k!} p_1^{t_1} \cdots p_k^{t_k}.$$

b. What restrictions are there on the t_i's and p_i's?

6-9 A certain type of benign tumor responds to appropriate chemotherapy in the following proportions: 60 per cent are improved, 30 per cent are unchanged, and 10 per cent are worse. What is the probability that of the next 10 patients given this drug, 8 improve, one is unchanged, and one is worse?

6-10 *Hypergeometric distribution.* Suppose a population consists of n objects, m of which are of type A and the remainder of type B. In an experiment k objects are selected at random without replacement.

   a. Develop the following formula for the probability that exactly r of the k objects selected are of type A:

$$P(r \text{ are of type A}) = \frac{{}_mC_r \cdot {}_{n-m}C_{k-r}}{{}_nC_k}$$

   b. What restrictions are there on m, r, and k?

6-11 Suppose that 20 rabbits were to be used in a study of blood-coagulating agents. As a first phase of the experiment 10 were given an anticoagulant, but unfortunately all 20, unmarked, were then put in a single pen. Twelve rabbits were needed for the second phase of the study and were selected at random without replacement. What is the probability that of the 12 selected, 6 had the drug and 6 did not?

6-12 *Negative binomial distribution.* Suppose that an experiment consists of a number of trials, each of which can result in success or failure. This experiment continues until k successes are noted. Let p and q be the probabilities of success and failure, respectively, for a given trial.

   a. Develop the following formula for the probability that exactly $k + x$ trials are required to observe k successes:

$$P(k + x \text{ trials needed for } k \text{ successes}) = {}_{k+x-1}C_{k-1}p^k q^x.$$

   Note: $P(k + x$ trials needed for k successes)

$$= P(k - 1 \text{ successes in first } k + x - 1 \text{ trials})$$

$$\cdot P[\text{success on } (k + x)\text{th trial}].$$

   b. Are there any restrictions on x, k, or p? (This formula may be generalized to apply whenever $k > 0$, not necessarily a whole number as we have assumed. In the more general form the resulting distribution is called the negative binomial distribution. In the special form we have considered a distribution for the waiting time to the kth success).

6-13 *Geometric distribution.* As a special case of the negative binomial distribution, suppose that an experiment continues until the first success is noted. Again each trial can result in success or failure with probabilities p and q. Show that the probability that x trials are required for the first success is

$$P(\text{first success on } x\text{th trial}) = q^{x-1}p.$$

6-14 Suppose that in a dental study on eighth graders it is assumed that the probability is $\frac{2}{3}$ that a student's teeth will have at least one cavity. What is the probability that:

a. Exactly 6 students will be examined to find the first decayed tooth?
b. Fifteen students will have to be examined to find 5 students with carious activity?

6-15 Suppose that $\frac{1}{5}$ of a large population of laboratory guinea pigs have been inoculated with a particular toxin causing liver enlargement. For your study you need the livers from five such guinea pigs.

a. What is the probability that you must sacrifice exactly 9 guinea pigs to find 5 with enlarged livers?
b. What is the probability that the sixth pig sacrificed is the first with an enlarged liver?
c. Find x such that

P (exactly x trials are required to reach the first success) $= 0.02684$.

6-16 Suppose that weight x of 6-year-old boys is normally distributed with a mean $\mu = 44$ lb and a standard deviation $\sigma = 5$ lb.

a. Find $P(40 < x < 48)$.
b. Find $P(x < 42)$.
c. Find $P(x > 45)$.
d. Between what 2 values does the middle 90 per cent of weights lie?
e. Your son (first and only) weighs 38 lb and is 6 years old. Should you fear that he is abnormally light and doomed never to become a professional football player?

6-17 For the weights discussed in Exercise 6-16, a random sample of ten 6-year-old boys are selected and weighed.

a. Find $P(42 < \bar{x} < 46)$.
b. Find $P(\bar{x} < 40)$.
c. Find $P(\bar{x} > 48)$.
d. Between what 2 values does the middle 95 per cent lie?
e. If $\bar{x} = 38$ lb, would this indicate an unusual sample of boys?

6-18 Suppose that diastolic blood pressure y in hypertensive women centers about a mean of 100 mm with a standard deviation of 14 mm and is normally distributed. Find:

a. $P(96 < y < 104)$.
b. $P(y < 88)$.
c. $P(y > 115)$.
d. t such that $P(y < t) = 0.95$.
e. t such that $P(y > t) = 0.95$.
f. 2 symmetric values a and b such that the $P(a < y < b) = 0.95$.

6-19 The probability an individual with a rare syndrome will be cured is $p = \frac{1}{100}$.

    a. A random sample of 10 persons with this syndrome is selected; find $P(1 \text{ person is cured})$, using the binomial distribution.

    b. A random sample of 300 persons with this syndrome is selected; find $P(\text{less than 4 are cured})$. Note that this problem requires values beyond the tabled values of the binomial distribution. To solve such problems you may use the Poisson distribution to approximate the binomial distribution. Use $\mu = np$.

    c. A random sample of 600 persons with this syndrome is selected; find $P(\text{between 4 and 12 persons, inclusive, are cured})$ using the normal approximation to the binomial.

6-20 Consider the following example illustrating that the transformation $z = (x - \mu)/\sigma$ does transform a variable x with mean μ and variance σ^2 into a variable z with mean 0 and variance 1. Let the population for the variable x consist of the values 3, 5, 7.

    a. Find the mean and variance of x.

    b. Apply $z = (x - \mu)/\sigma$ to the values of x, generating a population z.

    c. Find the mean and variance of z.

6-21 Suppose that a population of monkeys is given injections of a virus to cause them to develop antibodies preparatory to developing a vaccine for humans. The virus is in 2 lots, only one of which is effective. Eighty per cent of the monkeys receive the potent virus. A lab technician selects 25 monkeys at random. What is the probability that between 15 and 20 inclusive have been effectively injected?

6-22 Using the fact that $(a + b)^n = \sum_{r=0}^{n} {}_nC_r a^r b^{n-r}$, for all n, a, and b, prove $\sum_{r=0}^{n} {}_nC_r = 2^n$.

6-23 *Pascal's triangle.* Consider the following array of numbers:

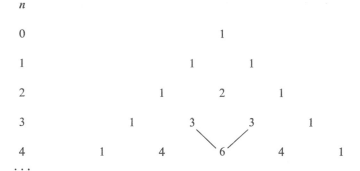

n							
0				1			
1			1		1		
2		1		2		1	
3	1		3		3		1
4	1	4		6		4	1

. . .

Each row begins and ends with a 1, and each intervening value is found from the preceding row by adding the two adjacent values in the row above, e.g., 6 in the row $n = 4$ is found by adding 3 and 3 bordering it from the row $n = 3$ (as illustrated).

a. Label the rows $n = 0, 1, 2$, etc. and let $r = 0, 1, 2, \ldots, n$ denote the various numbers in the nth row (start from left). What seems to be the relationship between the binomial coefficients, $_nC_r$, and the numbers in a given row?

b. As noted, the values in one row are found from those in the row above. Verify that the following equation, an algebraic statement of this relationship, is true, by substituting the definitions of the combination symbols.

$$_nC_r = {}_{n-1}C_{r-1} + {}_{n-1}C_r.$$

7 Estimation

The 2 general subcategories of statistical inference are estimation and hypothesis testing. We have seen that distributions often depend upon parameters. For example, if the mean and variance of a normal distribution are known, then the probabilities of various events can be determined. In general, however, since these parameters are attributes of the underlying population, they will be unknown, and we are faced with the problem of learning as much as possible about them from the information contained in a single random sample; that is, we must estimate their numerical value. Because of sampling variation we cannot say that the exact parameter value is some specific number, but we can make in some sense an optimum single choice based on sample information and then perhaps indicate a range of values within which we are confident the unknown parameter lies. Inferences of this sort fall into the category of *estimation*.

In hypothesis testing we make some statement about the underlying population—a statement about its form or about the numerical value of one or more of its parameters. We then select and inspect a random sample from the population to determine whether the sample is consistent or inconsistent with the stated hypothesis. Problems of hypothesis testing will be considered in detail in Chap. 8.

7-1 Point Estimation

When we use the information in a random sample to determine a single numerical value that will be a good indicator of the value of an underlying parameter, we have a problem of *point estimation*. For example, if we wish to estimate the average height of all American men, the sample mean might be a good estimator. It is not the only way of using the sample information to come to a numerical value, however. We might use the sample median, sample midrange, or some other sample measure of central tendency. We need criteria to assist in choosing from among the several different statistics those that might be "reasonable" estimators of the parameter.

In making such a choice we would prefer an estimator that is always exactly equal to the correct parameter. Of course we would. We might prefer to be a millionaire, discover the fountain of youth, and be President of the United States as well, a joint event scarcely less likely than finding such a paragon of an estimator. We are forced to sample, and therefore we are forced to live with sampling error. Since we cannot wish it away, let us learn to live with it.

It is reasonable to ask that in some sense our estimator should be close to the parameter value with a fairly high probability, or to insist that the estimator should "on the average" equal the correct parameter value. Both these concepts refer to properties of the sampling distribution of the statistic used to estimate the parameter. In general, suppose that the parameter is denoted by θ (the Greek letter theta) and the estimator by $\hat{\theta}$ (read variously as "theta hat" or "theta caret"). Then since $\hat{\theta}$ is calculated from the observations in a random sample, it is a random variable and has a sampling distribution.

We formalize the criterion that an estimator should equal the parameter value "on the average" by saying that the mean of the sampling distribution of $\hat{\theta}$ should be θ. Whenever this statement holds no matter what the numerical value of θ, we say that $\hat{\theta}$ is an *unbiased estimator* of θ. We have already found 2 unbiased estimators. In Sec. 5-4-2, statements 1 and 3 say, respectively, that the sample mean is an unbiased estimator of the population mean and that the sample variance is an unbiased estimator of the population variance.

In Sec. 3-4-5 when the variance was discussed as a descriptive statistic, we promised to clear something up later—the clearing up time is now. Recall that we seemed faintly disquieted at that point at defining the variance with $n - 1$ instead of n for its denominator—that is, at using as a definition for the variance the quantity

$$\frac{\sum_{i=1}^{n}(x_i - \bar{x})^2}{n - 1} \quad \text{instead of} \quad \frac{\sum_{i=1}^{n}(x_i - \bar{x})^2}{n}.$$

If we denote the latter quantity by $\hat{\sigma}^2$ we find that it is on the average equal to $\sigma^2 - (\sigma^2/n)$, in the sense that this expression is the mean of its sampling distribution. The division by n has produced a biased estimator. Thus $\hat{\sigma}^2$ will on the average be less than σ^2. We say that $\hat{\sigma}^2$ is biased to the left or negatively biased, since the quantity $\sigma^2 - (\sigma^2/n)$, when plotted on an axis, lies to the left or in a negative direction from the true parameter value of σ^2. Notice that if n is reasonably large, the amount of the bias is small.

The other criterion we proposed for a good estimator was that it should with high probability be close to the correct parameter value. In attempting to quantitate this idea we might consider the several configurations for the sampling distribution of $\hat{\theta}$ illustrated in Fig. 7-1. Here the correct value of θ is located as a point on the horizontal axis and the distribution of $\hat{\theta}$ is plotted as a probability density (see Sec. 6-3). The horizontal axes on the 4 graphs

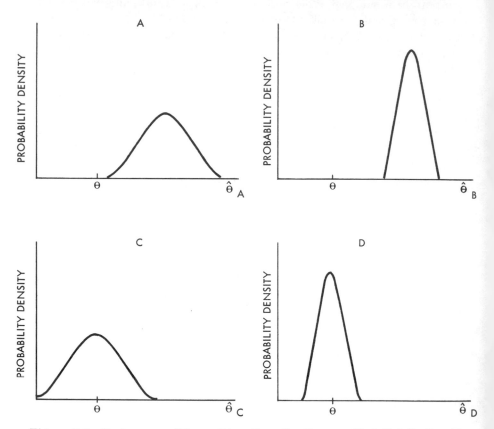

Figure 7-1   Various possible configurations for the sampling distribution of $\hat{\theta}$ as an estimator of θ.

in Fig. 7-1 give values of $\hat{\theta}$ for each of 4 estimators $\hat{\theta}_A$, $\hat{\theta}_B$, $\hat{\theta}_C$, and $\hat{\theta}_D$, with the correct population parameter value θ indicated at the same position on each graph.

Notice that $\hat{\theta}_C$ and $\hat{\theta}_D$ have sampling distributions with means equal to θ and thus are unbiased estimators. Both $\hat{\theta}_A$ and $\hat{\theta}_B$, however, have sampling distributions whose means are considerably to the right of the correct value of θ. Thus $\hat{\theta}_A$ and $\hat{\theta}_B$ are biased to the right or positively biased.

Another important feature of an estimator is evident from Fig.
Notice the difference between θ_A and θ_C, on the one hand, and θ_B and θ
on the other. The sampling distributions of θ_A and θ_C have larger variances
than those of θ_B and θ_D. This is another characteristic of an estimator that
requires consideration. Given 2 unbiased estimators such as θ_C and θ_D, we
would prefer to use the one with the smaller variance, θ_D in this case. In
fact, of the 4 estimators shown in Fig. 7-1, θ_D comes nearest our stated ideals
that a good estimator should be close to the parameter value with high
probability and should equal the parameter value "on the average." Although
we have illustrated the distributions of estimators in the continuous case
only, limiting our attention to probability densities, similar graphs could be
drawn for discrete estimators.

We will formalize these ideas with a definition.

Definition

An estimator θ is said to be a *minimum variance unbiased estimator* of a
parameter θ if it is unbiased and if no other unbiased estimator based on
random samples of size n has a sampling distribution with a smaller variance
than that of θ.

In general, the demonstration that a particular estimator is minimum
variance unbiased is a mathematical problem beyond the level of this book.
However, $\bar{x}$, the sample mean based on random samples of size n from a
normal distribution with mean μ and variance σ^2, is a minimum variance
unbiased estimator of μ. The sample variance s^2 is also a minimum variance
unbiased estimator of σ^2. These results depend heavily upon the normality
of the distribution. It is not true that $\bar{x}$ is a minimum variance unbiased
estimator of the population mean for every distribution.

The symmetry of the normal distribution implies that the mean and the
median are both equal to μ. It is, therefore, natural to examine the characteristics of the sample median as an estimator of the population mean μ
(also the median) of the normal distribution. The sample median is unbiased,
and since we know that the sample mean is minimum variance unbiased, it
follows that the sample median cannot have a sampling distribution with a
smaller variance. In fact, we know from Sec. 5-4-2 that the variance of the
sampling distribution of $\bar{x}$ is σ^2/n. For random samples of size $n = 10$ this
becomes $(0.1)\sigma^2$. By way of contrast, the variance of the sampling distribution of the sample median can be shown to be $(0.138)\,\sigma^2$ in this case, an
increase in variance of 38 per cent. Naturally, the actual value of the variance
of the sampling distribution of the sample median depends upon the sample
size, as does the variance of the sample mean.

or ratio of variances for 2 different unbiased estimators is
ir relative adequacy. In the case of the normal distribution
ie sample mean over the sample median becomes greater
ict, the ratio of the variances approaches a limiting value
;ets larger and larger.

statisticians have suggested and investigated a large
‑‑‑‑‑‑‑ properties of point estimators. From a practical point
of view, however, consideration of the mean and variance of the sampling
distribution of the estimator are fundamental. In practice we usually want
more than a point estimate of a parameter. We know that it is unlikely that
the values of a point estimator and the parameter it is estimating will exactly
coincide—in fact, for a continuous random variable the probability that the
estimator will equal any particular value, including the fixed unknown value
of the parameter, is zero. We almost always wish to include with a point
estimator some notion of the variability to which it would be subject under
repeated sampling. This can be accomplished by giving an interval of values
within which we are quite confident the true parameter lies. This is the topic
of the rest of this chapter.

7-2   *Interval Estimation*

7-2-1   GENERAL IDEAS

Now let us get one thing straight right off the bat. We are sampling
from some fixed population. We are examining values of a random variable
obtained by selecting a random sample from that fixed population. The
random variable has a certain distribution in the fixed population. That
distribution depends upon a parameter that we wish to estimate. Since the
population and the distribution of the random variable we are investigating
are fixed, it follows that the parameter is fixed. It has a numerical value like
3.7 or 18.96 or 0.0022. We shall never know that numerical value, however.
Otherwise, we would not be trying to estimate it by sampling. Now do not
let that parameter get away from you and start slipping around. It is fixed
and it is going to stay that way.

7-2-2   CONFIDENCE INTERVAL FOR THE MEAN OF A NORMAL
DISTRIBUTION WITH KNOWN VARIANCE

It is convenient to start the discussion of interval estimation with an
example that is more important pedagogically than in any other respect.

This case assumes that we are sampling from a normal distribution with a known variance—that is, that the numerical value of σ^2 is available to us. We shall eliminate this latter assumption in the next section.

We might, for example, be studying the potency of a drug. The production process for the drug might have been slightly changed recently and we might wonder whether the mean potency of lots produced under the new process is different from the former mean. We might be willing to assume that the lot-to-lot variance of potency remains fixed. Furthermore, we might have extensive testing records under the former process and feel that we know the value of the variance quite closely.

It is reasonable to use $\bar{x}$ as the basis for an interval estimator of μ. It is a good point estimator, as we saw in Sec. 7-1, and we know the form of its distribution. If $\bar{x}$ is based upon a random sample of size n from a normally distributed population with unknown mean μ and known variance σ^2, then its sampling distribution is normal with mean μ and variance σ^2/n, as discussed in Sec. 6-4-7. The standard normal variable related to $\bar{x}$ is obtained by subtracting μ and dividing by $\sigma/\sqrt{n}$, the standard deviation or standard error of $\bar{x}$. This gives

$$z = \frac{\bar{x} - \mu}{\sigma/\sqrt{n}}.$$

In constructing an interval estimator of μ we want limits between which μ will lie with high probability, say 0.95. These limits will be random variables, and in order to determine them we use z.

We know from Table A-4 that

$$P(-1.96 < z < 1.96) = 0.95$$

and we examine the event $-1.96 < z < 1.96$. This is really the event $-1.96 < (\bar{x} - \mu)/(\sigma/\sqrt{n}) < 1.96$, if we use the definition of z. We seek to recast the form of this event and construct a completely equivalent event of the form $a < \mu < b$. This event will then also have probability 0.95. To do this we will make use of the properties of inequalities discussed in Sec. 6-4-3. Notice that $\bar{x}$ is the only component of the event which is a random variable.

Whenever $-1.96 < (\bar{x} - \mu)/(\sigma/\sqrt{n}) < 1.96$, then multiplying through by $\sigma/\sqrt{n}$, we obtain

$$-1.96\,\frac{\sigma}{\sqrt{n}} < \bar{x} - \mu < 1.96\,\frac{\sigma}{\sqrt{n}}.$$

That is, in exactly those random samples for which $(\bar{x} - \mu)/(\sigma/\sqrt{n})$ is between plus and minus 1.96, $\bar{x} - \mu$ is between plus and minus 1.96 $\sigma/\sqrt{n}$. Recall that multiplying both sides of an inequality by the same positive quantity, $\sigma/\sqrt{n}$ in this case, does not change the sense or direction of the inequality. Now, subtracting $\bar{x}$ from the 3 parts of our chain inequality, we have a completely equivalent event:

$$-\bar{x} - 1.96\,\frac{\sigma}{\sqrt{n}} < -\mu < -\bar{x} + 1.96\,\frac{\sigma}{\sqrt{n}}.$$

Recall that subtracting the same quantity from both sides of an inequality does not change its sense. Now, multiplying through by -1, we obtain the equivalent event

$$\bar{x} + 1.96\,\frac{\sigma}{\sqrt{n}} > \mu > \bar{x} - 1.96\,\frac{\sigma}{\sqrt{n}}.$$

Multiplication of both sides of an inequality by a negative quantity reverses the sense of the inequality. But now reversing the order of the terms of this chain inequality and writing the smallest quantity on the left instead of the right, we have the equivalent event

$$\bar{x} - 1.96\,\frac{\sigma}{\sqrt{n}} < \mu < \bar{x} + 1.96\,\frac{\sigma}{\sqrt{n}},$$

an event of the form we are seeking. Note well that this is the same event as

$$-1.96 < z < 1.96.$$

That is, for the same random samples in which $z = (\bar{x} - \mu)/(\sigma/\sqrt{n})$ is between plus and minus 1.96, μ will be between $\bar{x} - 1.96\,\sigma/\sqrt{n}$ and $\bar{x} + 1.96\,\sigma/\sqrt{n}$. Notice, however, that in the original form of the event the middle expression in the chain inequality was the random variable, since it involved $\bar{x}$, whereas in the revised form the outside expressions are the random variables. Since we have only changed the form of the original event and have not changed its probability, we can write

$$P\!\left(\bar{x} - 1.96\,\frac{\sigma}{\sqrt{n}} < \mu < \bar{x} + 1.96\,\frac{\sigma}{\sqrt{n}}\right) = 0.95.$$

This statement takes into account the fact that $\bar{x}$ is a random variable with a sampling distribution. All we need to do is select a random sample, calculate the numerical value of $\bar{x}$, and substitute this value into the expressions $\bar{x} - 1.96\,\sigma/\sqrt{n}$ and $\bar{x} + 1.96\,\sigma/\sqrt{n}$ along with the known numerical values of σ and n. In a specific numerical case this will produce 2 numbers, such as 1.7 and 4.9, for the limits. But here we run into a snag. We cannot assert that the *probability* that μ lies between 1.7 and 4.9 is 0.95. The quantities 1.7, 4.9, and μ are all *constants*. In fact, either μ lies between these numbers or it does not, and it is not sensible to assign a probability of 0.95 to the statement that it is between them.

The difficulty here arises at the point of substitution of the observed numerical value for $\bar{x}$. The random variation in $\bar{x}$ is variation from random sample to random sample, and it is this variation to which the probability refers. However, if we denote the observed numerical value of $\bar{x}$ by $\bar{x}_0$, we know that this value arose from a process that could have produced infinitely many different values of $\bar{x}$ and hence infinitely many different intervals of the form $\bar{x} \pm 1.96\,\sigma/\sqrt{n}$. The probability statement says that 95 per cent of these intervals would actually include or trap μ between their limits. Since we have only one random sample, one value x_0 of x and one interval $x_0 \pm 1.96\,\sigma/\sqrt{n}$, we say we are 95 per cent *confident* that μ lies between these limits. We might write

$$C\left(\bar{x}_0 - 1.96\,\frac{\sigma}{\sqrt{n}} < \mu < \bar{x}_0 + 1.96\,\frac{\sigma}{\sqrt{n}}\right) = 0.95.$$

This indicates that the *confidence* is 0.95, not the *probability*. The interval $\bar{x}_0 \pm 1.96\,\sigma/\sqrt{n}$ is called a 95 per cent confidence interval for μ and is the interval estimator we have been seeking. The figure 0.95 is ordinarily called the *confidence level*.

To help fix these ideas we shall give the results of a sampling experiment in which a number of random samples are drawn from the same normally distributed population and a number of 95 per cent confidence intervals calculated. These results are artificial but are constructed to resemble hemoglobin levels of white males over the age of 11 as reported by Milam and Muench (35). We suppose that the hemoglobin levels are normally distributed with $\sigma = 1.209$ grams per 100 ml. Table 7-1 gives 95 per cent level confidence intervals for 20 different random samples of size $n = 4$ from such a population.

In Fig. 7-2 we have plotted each of the intervals as a bar above the axis of measurement. Notice that the position of the interval changes from random sample to random sample. There is, in fact, a whole infinite population of these intervals, of which we have graphed only 20. In a given practical situation we will have only one interval, but it is well to retain a mental picture

Table 7-1

NINETY-FIVE PER CENT CONFIDENCE INTERVALS FOR MEAN HEMOGLOBIN LEVEL
FOR WHITE MALES OVER THE AGE OF 11 YEARS FOR 20 RANDOM SAMPLES
OF SIZE 4—ARTIFICIAL DATA

Serial Number of Sample	Sample Mean	Lower Confidence Limit	Upper Confidence Limit
1	14.41	13.23	15.60
2	14.53	13.34	15.71
3	13.83	12.65	15.02
4	14.88	13.70	16.07
5	12.91	11.73	14.10
6	14.91	13.72	16.09
7	14.08	12.89	15.26
8	14.18	12.99	15.36
9	14.45	13.27	15.64
10	16.20	15.01	17.38
11	14.52	13.33	15.70
12	15.11	13.93	16.30
13	13.95	12.77	15.14
14	15.27	14.08	16.45
15	13.89	12.71	15.08
16	14.54	13.36	15.73
17	15.03	13.85	16.22
18	13.79	12.61	14.98
19	13.67	12.49	14.86
20	15.04	13.86	16.23

of Fig. 7-2 and remember that all the other intervals are there as well. They correspond to other random samples we might have selected.

An important feature of a confidence interval is that we expect a certain proportion of intervals to miss the true parameter value. In fact, if 95 per cent of all the intervals $\bar{x} \pm 1.96 \, \sigma/\sqrt{n}$ include μ between their limits, then 5 per cent will miss μ. If this is too large a proportion of misses, we can increase the confidence level, say, to 99 per cent by using the interval $x \pm 2.576 \cdot \sigma/\sqrt{n}$. The ± 2.576 comes from Table A-4 and gives 3-place accuracy for the values from the standard normal distribution between which the central 99 per cent of the total area lies. A 99 per cent interval corresponding to the first sample of Table 7-1 would have limits 12.85 and 15.97 instead of 13.23 and 15.60. In general, increasing the level of confidence increases the length of the interval.

If we wanted to be absolutely certain that μ lies between the limits, we would have to construct a 100 per cent interval. This would use values between which 100 per cent of the standard normal area lies. But since the

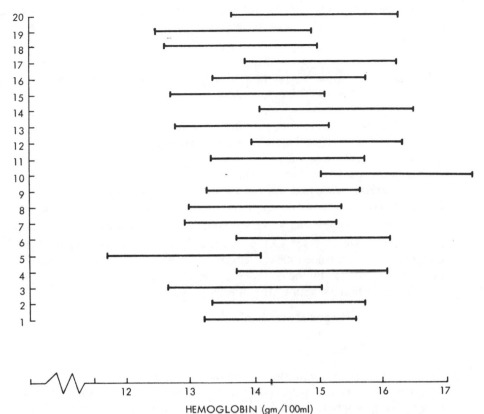

Figure 7-2   **Ninety-five per cent confidence intervals for mean hemoglobin level for white males over the age of 11 years for 20 repeated random samples of size 4—artificial data.**

normal distribution has tails of infinite length, this would produce an interval from minus infinity to plus infinity. Of course, you would not get much argument if you asserted that the mean hemoglobin level of some population is between minus and plus infinity, but a more useless result is hard to imagine. You would not have needed to do any sampling at all to make such a statement.

Since 100 per cent confidence intervals are useless, the only alternative is an interval with some chance of missing μ. For the hemoglobin values the sampling was actually from a population with $\mu = 14.25$. Notice that samples 5 and 10 produced intervals that missed the true value. In general, we expect a long-run percentage of 5 per cent of intervals at the 95 per cent confidence level to miss the parameter value.

We have seen that increasing the confidence level increases the length of the interval. Let us examine the effect of increasing the sample size.

Suppose the sample mean of the first hemoglobin sample had been calculated from a random sample of size 50 instead of size 4. The limits would then have been

$$14.41 \pm 1.96\left(\frac{1.209}{\sqrt{50}}\right) = 14.41 \pm 0.34 \quad \text{or} \quad 14.07 \quad \text{and} \quad 14.75.$$

These limits are narrower, i.e., the interval is shorter, than the corresponding interval of 13.23 to 15.60 based on 4 observations.

In general, short confidence intervals are more desirable than long ones, since they express a more limited range of uncertainty concerning the location of the parameter value and thus give a tighter interval estimate. The length of the interval is determined by the value that is added to and subtracted from $\bar{x}$ to produce the limits. This quantity is $1.96\ \sigma/\sqrt{n}$ for a 95 per cent interval and $2.576\ \sigma/\sqrt{n}$ for a 99 per cent interval. There are 3 parts of this quantity which influence interval length: σ, n, and the factor, such as 1.96 or 2.576, determined by the confidence level.

To decrease the length of the interval, 3 possibilities exist:

1. Reduce the confidence level.
2. Reduce σ.
3. Increase n.

The first alternative produces narrower intervals at the cost of decreased confidence that the true parameter will be trapped by the interval. This may be entirely appropriate in some cases. For example, 99 per cent may be an unnecessarily high level for some applications. Most investigators are unwilling to use a level of confidence lower than 90 per cent, feeling that a risk greater than 10 per cent of missing the parameter value is unreasonably large. Reducing σ may be a problem, since this means a change in the underlying population. In some instances it may be appropriate to restrict the sampled population of individuals. In the hemoglobin example we might look only at a single group with very similar nutritional status, age, and race composition, and from a single geographic area. This would, however, also have the effect of limiting the population to which our inferences apply and might necessitate separate studies in other groups. In some applications it is possible to adopt a more precise technique of measurement, thereby reducing σ by reducing the error of measurement. An increase in sample size is the third possibility for reducing interval length. Here the cost is partly economic because of the direct increase in cost of collecting information. However, the resulting increased time length of the study may also be a factor. For example, if we are studying a new treatment for a relatively rare disease, the time required to accumulate a longer case series might result in delayed availability of a truly superior treatment. Notice, too, that the interval length decreases with the square root of n rather than with n itself. For example, to cut the interval length in half, the sample size must be 4 times as large.

The moral to this story is that while we have several ways to influence interval length, each of these is subject to practical problems and must be carefully considered for any real situation. Notice that the question of interval length is a problem of study design rather than data analysis. Changes in σ and n can only be made prior to analysis. Whereas decisions about confidence level can be made after the data have been inspected, most statisticians encourage the investigator to think about this in advance to avoid the bias of forcing the interval to agree with preconception rather than with data.

Although the majority of confidence intervals in practice are constructed at either the 95 per cent or 99 per cent levels, a notational device will be convenient in providing a generalization. We shall use the notation z_p for the $100p$ percentile of the standard normal distribution. In other words, $z_{0.95}$ is the 95th percentile, $z_{0.975}$ the 97.5th percentile, etc. The subscript is a probability; it is multiplied by 100 when one is speaking of a percentile. In this notation the quantity 1.96, which arose in the 95 per cent confidence interval, is $z_{0.975}$. The 0.975 comes from the fact that the original event $-1.96 < z < 1.96$ allowed z to fall outside the limits ± 1.96 with probability 0.025 in either direction. That is, the balance between a probability of 1.00 and 0.95 was divided equally between the 2 tails of the distribution. In general, if we want confidence level $1 - \alpha$ then we must use the quantity $z_{1-(\alpha/2)}$ in forming the interval. To check this, for a confidence level of $1 - \alpha = 0.99$ we have $\alpha = 0.01$. Then $z_{1-(\alpha/2)} = z_{0.995} = 2.576$, agreeing with our earlier result. The general form of a $1 - \alpha$ level confidence interval is then $\bar{x} \pm z_{1-(\alpha/2)}\sigma/\sqrt{n}$.

From the central limit theorem (Sec. 6-4-7) the distribution of $\bar{x}$ will be approximately normal with mean μ and variance σ^2/n for samples even from a nonnormal distribution, provided n is sufficiently large. It follows that the confidence intervals of this section still apply fairly well if the underlying population is nonnormal. Of course, the nature and degree of nonnormality influence the size of sample required to overcome it, but experience and sampling experiments suggest that unless the underlying population is very markedly skewed, a sample size of 20 to 25 is large enough to insure an adequate approximation.

EXAMPLE

Find a 90 per cent confidence interval for the mean increase in heart rate in a sample of trained athletes subjected to a standard work load. Suppose that the standard deviation of the population is known from earlier studies to be 15 beats per minute, that 25 athletes participate in the current study, and that the sample mean heart rate increase is 85 beats per minute.

The limits will be

$$\bar{x}_0 \pm z_{0.95} \frac{\sigma}{\sqrt{n}} = 85 \pm (1.645)\left(\frac{15}{5}\right) = 85 \pm 4.9, \quad \text{or} \quad 80.1 \quad \text{and} \quad 89.9.$$

The results of this section can be used to determine the sample size needed to estimate μ with a given level of confidence to within a certain degree of precision. Returning to the hemoglobin example, suppose that we wish to estimate the population mean to within 0.5 gram per 100 ml with 95 per cent confidence. In other words, we want to be 95 per cent confident that $\bar{x}$ is within one-half unit of μ. Since $\bar{x}$ is the midpoint of the confidence interval and we must move up and down a distance of 1.96 $\sigma/\sqrt{n}$ units from $\bar{x}$ to reach the upper and lower confidence limits, it follows that we want 1.96 $\sigma/\sqrt{n} = 0.5$. Since $\sigma = 1.209$ we have $2.37/\sqrt{n} = 0.5$ or $\sqrt{n} = 4.74$, whence $n = 22.5$. Since n must be a whole number, it is conventional to use the next larger value, 23 in this case, in order to guarantee at least the specified precision with at least the specified level of confidence.

In general, if we want $\bar{x}$ to be within d units of μ with confidence $1 - \alpha$, then we set

$$z_{1-(\alpha/2)} \cdot \frac{\sigma}{\sqrt{n}} = d,$$

whence

$$n = \frac{z_{1-(\alpha/2)}^2 \sigma^2}{d^2}.$$

This result makes good intuitive sense, since it says that the sample size varies directly with the variance and inversely with the allowable deviation of $\bar{x}$ from μ. Furthermore, it shows that sample size varies directly with the confidence level, since the larger the value of $1 - \alpha$, the larger the value of $z_{1-(\alpha/2)}$.

7-2-3 CONFIDENCE INTERVAL FOR THE MEAN OF A NORMAL DISTRIBUTION WITH UNKNOWN VARIANCE

In the preceding section no confidence interval for μ, the mean of a normal distribution, could be calculated without having available a numerical value for σ^2 or σ. It is reasonable that if σ^2 is unknown the data in the random sample should be used to estimate its value. The sample variance s^2 is the natural candidate for this estimation. Furthermore, since we started constructing the earlier interval from the quantity $z = (\bar{x} - \mu)/(\sigma/\sqrt{n})$, we might now think of substituting s, the sample standard deviation for σ and using the quantity $(\bar{x} - \mu)/(s/\sqrt{n})$. This quantity is clearly a random variable because of its dependence upon $\bar{x}$ and s, but its distribution is not the standard normal distribution. In fact, it follows a distribution known as Student's t distribution or more briefly, the t distribution.

The t distribution is important in statistical inference and was first discovered in 1908 by W. S. Gosset (1876–1937) (22), a brewer with Messrs. Guinness in Dublin. He was responsible for introducing statistical techniques to the brewing industry, left his brewery post briefly to study statistics at University College of the University of London with Prof. Karl Pearson, later directed the statistics department of the brewery, and was sent in 1935 to head the new Guinness brewery in London. His purely statistical publications were published under the pseudonym "Student," perhaps to avoid making Guinness's competitors aware of the fact that breweries might profit (literally) from statistical methods. Gosset is considered by many to be the founder of the field of exact statistical inference. Until his 1908 paper (22), it was common to treat $(\bar{x} - \mu)/(s/\sqrt{n})$ as a standard normal variable.

Again it is reasonable that for large samples s should be quite close to σ with high probability. If n is large $(\bar{x} - \mu)/(s/\sqrt{n})$ will approximately follow the standard normal distribution. We would expect in general that the replacement of the constant σ by the random variable s would produce a quantity with increased variability, and this is also true.

The t distribution depends on a single parameter, called the number of degrees of freedom. Table A-5 gives percentiles of the t distribution for various degrees of freedom. Similar to the preceding section, we will use the notation t_p for the $100p$ percentile of the t distribution. The distribution is symmetric about zero, so the percentiles for $p < 0.5$ can be found by affixing a minus sign to the corresponding percentile for $p > 0.5$, and the lower percentiles are thus omitted from Table A-5. This correspondence is such that, for example, $t_{0.20} = -t_{0.80}$, $t_{0.025} = -t_{0.975}$, or, more generally, $t_p = -t_{1-p}$.

The value of the "degrees of freedom" parameter is determined by the problem under consideration. For the quantity $t = (\bar{x} - \mu)/(s/\sqrt{n})$ it is equal to $n - 1$. As we have noted, when n increases we expect the t distribution approximately to agree with the standard normal. Recalling that the 97.5 percentile of the standard normal distribution is 1.96, it is instructive to examine the values of this percentile of the t distribution as the number of degrees of freedom (d.f.) increases. Using Table A-5 for d.f. $= 1$, we find that $t_{0.975} = 12.706$; for d.f. $= 10$, $t_{0.975} = 2.2281$; for d.f. $= 20$, $t_{0.975} = 2.0860$. The entry for an infinite number of degrees of freedom is 1.9600, the standard normal value, a situation equivalent to knowing the value of σ. Examination of other percentiles shows a similar effect. Since in the present problem the number of degrees of freedom is one less than the sample size, this convergence to the standard normal values conforms to our earlier statement that for large n, $(\bar{x} - \mu)/(s/\sqrt{n})$ is approximately normal. However, it will become clear as we consider different problems that other random variables also follow the t distribution. For these variables, although there may not be such a close relationship between d.f. and sample size, the convergence to the normal percentiles with increasing degrees of freedom still holds.

Returning to the problem of constructing a $1 - \alpha$ level confidence interval for μ, the mean of a normal distribution with unknown variance, we have

$$P\left(-t_{1-(\alpha/2)} < \frac{\bar{x} - \mu}{s/\sqrt{n}} < t_{1-(\alpha/2)}\right) = 1 - \alpha.$$

As in the preceding section we seek to recast the above event to produce an equivalent event of the form $c < \mu < d$ which will also have probability $1 - \alpha$. A similar sequence of algebraic steps is used to produce an event of this form. The 3 terms in the chain inequality are multiplied by $s/\sqrt{n}$, reduced by $\bar{x}$, and multiplied by -1. The terms of the resulting chain inequality are then written in reverse order to produce

$$P\left(\bar{x} - t_{1-(\alpha/2)} \frac{s}{\sqrt{n}} < \mu < \bar{x} + t_{1-(\alpha/2)} \frac{s}{\sqrt{n}}\right) = 1 - \alpha.$$

The factor $t_{1-(\alpha/2)}$ is the $100(1 - \alpha/2)$ percentile of the t distribution with $n - 1$ degrees of freedom. Notice that again it is the end points of the interval that are random variables—in fact, they now depend upon $\bar{x}$ and s, both of which vary from random sample to random sample. Thus, as illustrated by Fig. 7-2, there are many different confidence intervals corresponding to the many different random samples of size n that can be selected. In fact, there is an entire infinite population of such intervals.

An important difference exists here, however. If we examine the length of the interval in the case of known population variance considered in the preceding section, we find that it equals, subtracting the lower limit from the upper,

$$\left(\bar{x} + z_{1-(\alpha/2)} \frac{\sigma}{\sqrt{n}}\right) - \left(\bar{x} - z_{1-(\alpha/2)} \frac{\sigma}{\sqrt{n}}\right) = 2z_{1-(\alpha/2)} \frac{\sigma}{\sqrt{n}},$$

a constant not varying for different random samples. In the case of unknown variance, however, the interval length is

$$\left(\bar{x} + t_{1-(\alpha/2)}\frac{s}{\sqrt{n}}\right) - \left(\bar{x} - t_{1-(\alpha/2)}\frac{s}{\sqrt{n}}\right) = 2t_{1-(\alpha/2)}\frac{s}{\sqrt{n}},$$

a random variable depending upon s. This implies that different random samples will produce intervals that differ not only in their midpoint, which is $\bar{x}$ whether or not σ is known, but also in length. The discussion given in Sec. 7-2-2 about factors influencing interval length is still relevant and applicable, but now we must speak in terms of average or mean interval length.

To find a specific confidence interval, we select a random sample of size n from the population and calculate specific numerical values of the sample mean and standard deviation, say $\bar{x}_0$ and s_0. But now $\bar{x}_0$, s_0, and μ are all fixed, thus fixing all 3 terms in the chain inequality. Again we cannot say that μ lies between the resulting limits with probability $1 - \alpha$. As before, we write

$$C\left(\bar{x}_0 - t_{1-(\alpha/2)}\frac{s_0}{\sqrt{n}} < \mu < \bar{x}_0 + t_{1-(\alpha/2)}\frac{s_0}{\sqrt{n}}\right) = 1 - \alpha.$$

We mean that we have produced an interval from a random sampling process which generates infinitely many intervals, a proportion $1 - \alpha$ of which include μ between their limits. We say that our confidence is $1 - \alpha$ that μ lies between the fixed numerical limits. Again this confidence interval, though technically based on the normal distribution, holds well for moderate-sized samples from nonnormal populations. Extreme skewness or asymmetry is probably the most serious kind of nonnormality, but if the data have been screened for obvious extreme values that could only be the result of contamination such as clerical error, faulty instrumentation, or similar factors, then the assumption of normality should probably be regarded as essentially nominal for reasonable sample size.

EXAMPLE

Using specimens obtained from 10 individuals, determinations of per cent calcium content of sound teeth gave the following results: 36.39, 36.19, 34.20, 35.15, 35.47, 35.22, 36.11, 35.63, 36.63, 35.59. Find 95 and 99 per cent confidence intervals for the mean percentage of calcium in the population.

We find: $\sum_{i=1}^{10} x_i = 356.58$; $\sum_{i=1}^{10} x_i^2 = 12{,}719.5140$; $\bar{x}_0 = 35.6580$;

$s_0^2 = 0.5094$; $s_0 = 0.7137$; $\dfrac{s_0}{\sqrt{n}} = 0.2257$; number of degrees of freedom

$= n - 1 = 9$; $t_{0.975} = 2.262$; $t_{0.975}\dfrac{s_0}{\sqrt{n}} = 0.5105$; $t_{0.995} = 3.250$; $t_{0.995}\dfrac{s_0}{\sqrt{n}}$

$= 0.7335$.

95 per cent interval for μ: $\bar{x}_0 \pm t_{0.975}\dfrac{s_0}{\sqrt{n}} = 35.66 \pm 0.51$[1]

lower 95 per cent limit: 35.15; upper 95 per cent limit: 36.17.

99 per cent interval for μ: $\bar{x}_0 \pm t_{0.995}\dfrac{s_0}{\sqrt{n}} = 35.66 \pm 0.73$

lower 99 per cent limit: 34.93: upper 99 per cent limit 36.39.

7-2-4 CONFIDENCE INTERVALS FOR THE VARIANCE AND STANDARD DEVIATION OF A NORMAL DISTRIBUTION

In Sec. 7-1 we stated that s^2 was the minimum variance unbiased estimator of σ^2, the variance of a normal distribution. It is curious but true that s is not an unbiased estimator of σ and therefore cannot be minimum variance unbiased. However, the bias gets small quite rapidly as n increases. It seems natural to base an interval estimate of σ^2 on s^2.

In constructing such an interval it is necessary to introduce the chi-square distribution. Percentiles of this distribution are given in Table A-6. Like the t distribution, the chi-square distribution depends upon a single parameter called the number of degrees of freedom. Again this parameter takes positive, whole number values. Unlike either the normal or t distribution, the chi-square distribution is positively skewed and has a tail extending infinitely only to the right. It is the distribution of a continuous random variable, and all the area under the density corresponds to positive values of the random variable. That is, if v follows the chi-square distribution with f degrees of freedom, then $P(v < 0) = 0$. The mean of the distribution is f, the number of degrees of freedom. The $100u$ percentiles of the chi-square distribution will be denoted by χ_u^2, using the Greek symbol for the letter chi. The random variable $(n - 1)s^2/\sigma^2$ follows the chi-square distribution

---

[1] Notice that in this and other numerical work in this text we have carried more decimal places in intermediate stages of computation than we intend to report in the final results. Rounding off is delayed until the final step to avoid the cumulative effect of frequent intermediate roundings.

with $n - 1$ degrees of freedom. The fact that $(n - 1)s^2/\sigma^2$ takes only non-negative values is consistent with the property noted above that the chi-square density has no area over negative values of the random variable. Furthermore, since $(n - 1)s^2/\sigma^2$ is a continuous random variable, it takes the value zero (or any other single value for that matter) with probability zero. For computational purposes we notice that

$$(n - 1)s^2 = \sum_{i=1}^{n}(x_i - \bar{x})^2 = \frac{n\sum_{i=1}^{n}x_i^2 - \left(\sum_{i=1}^{n}x_i\right)^2}{n},$$

where $x_1, x_2, \ldots, x_n$ are the observations in the random sample.

To form a $1 - \alpha$ level confidence interval for σ^2, notice that an area $\alpha/2$ lies to the left of $\chi^2_{\alpha/2}$ and the same area lies to the right of $\chi^2_{1-(\alpha/2)}$. Thus

$$P\left[\chi^2_{\alpha/2} < \frac{(n - 1)s^2}{\sigma^2} < \chi^2_{1-(\alpha/2)}\right] = 1 - \alpha.$$

We want an equivalent event of the form $g < \sigma^2 < h$. Dividing the 3 terms of the chain inequality by $(n - 1)s^2$, we obtain the equivalent event:

$$\frac{\chi^2_{\alpha/2}}{(n - 1)s^2} < \frac{1}{\sigma^2} < \frac{\chi^2_{1-(\alpha/2)}}{(n - 1)s^2},$$

recalling that the sense of an inequality is not affected by division by a positive quantity. Now, taking reciprocals of the 3 terms and using property 4 of Sec. 6-4-3, we have the equivalent event:

$$\frac{(n - 1)s^2}{\chi^2_{\alpha/2}} > \sigma^2 > \frac{(n - 1)s^2}{\chi^2_{1-(\alpha/2)}},$$

or, reversing the order of the terms,

$$\frac{(n - 1)s^2}{\chi^2_{1-(\alpha/2)}} < \sigma^2 < \frac{(n - 1)s^2}{\chi^2_{\alpha/2}}.$$

Thus,

$$P\left[\frac{(n - 1)s^2}{\chi^2_{1-(\alpha/2)}} < \sigma^2 < \frac{(n - 1)s^2}{\chi^2_{\alpha/2}}\right] = 1 - \alpha.$$

We again have a class of random intervals, since s^2 is a random variable.

Substituting a specific numerical value s_0^2 from a single sample, we write

$$C\left[\frac{(n-1)s_0^2}{\chi_{1-(\alpha/2)}^2} < \sigma^2 < \frac{(n-1)s_0^2}{\chi_{\alpha/2}^2}\right] = 1 - \alpha.$$

The intervals calculated in this section and in the preceding 2 sections are not the only possible $1 - \alpha$ level intervals for their respective parameters. We have elected in each of these 3 cases to start with an event that chops off an area $\alpha/2$ from each tail of the distribution. For a 95 per cent confidence interval this involves excluding a probability of 0.025 in each tail. Why not exclude probability 0.01 from one tail and probability 0.04 from the other? That is, in the case of the interval in Sec. 7-2-2, why not start with the event $z_{0.01} < z < z_{0.96}$? Obviously, many events have the desired probability of 0.95. The answer to the question is that for both the interval of Sec. 7-2-2 based on the standard normal distribution and the interval of Sec. 7-2-3 based on the t distribution, the symmetry of the distribution implies that intervals of minimum length are produced by cutting off equal areas from the 2 tails. Since we have agreed that short intervals are more desirable in general than long ones, this is a strong rationale for the recommendations we have made.

The chi-square distribution, however, is asymmetric, and, in fact, slightly shorter intervals at the $1 - \alpha$ level can be found by cutting off unequal areas from the 2 tails. For moderate size n the saving in length is small, and since the form of the resulting interval is more complicated, we do not recommend the use of this nicety.

The event

$$\frac{(n-1)s^2}{\chi_{1-(\alpha/2)}^2} < \sigma^2 < \frac{(n-1)s^2}{\chi_{\alpha/2}^2},$$

as we have seen, has probability $1 - \alpha$. The square roots of the 3 terms will be in the same relationship, and we can write

$$\sqrt{\frac{(n-1)s^2}{\chi_{1-(\alpha/2)}^2}} < \sigma < \sqrt{\frac{(n-1)s^2}{\chi_{\alpha/2}^2}}.$$

This event is equivalent to the original event. Random samples for which the first chain inequality is true also guarantee the truth of the second inequality, and vice versa. Thus the interval for σ is also at the $1 - \alpha$ level. Again, to calculate a specific interval we substitute s_0^2 for s^2. In practice the procedure is very simple. After calculating the confidence limits for σ^2, we merely extract their square roots to find the limits for σ.

EXAMPLE

The data on the per cent calcium content considered in the example of Sec. 7-2-3 will be used to produce 95 and 99 per cent confidence intervals for the variance and standard deviation of the population.

$$s_0^2 = 0.5094; \quad (n-1)s_0^2 = 4.5846; \quad s_0 = 0.7137; \quad \text{d.f.} = 9;$$

$$\chi_{0.025}^2 = 2.70; \quad \chi_{0.975}^2 = 19.02; \quad \chi_{0.005}^2 = 1.73; \quad \chi_{0.995}^2 = 23.59.$$

95 per cent interval for σ^2:

lower 95 per cent limit: $\dfrac{4.5846}{19.02} = 0.241.$

upper 95 per cent limit: $\dfrac{4.5846}{2.70} = 1.698.$

99 per cent interval for σ^2:

lower 99 per cent limit: $\dfrac{4.5846}{23.59} = 0.194.$

upper 99 per cent limit: $\dfrac{4.5846}{1.73} = 2.650.$

95 per cent interval for σ:

lower 95 per cent limit: $\sqrt{0.241} = 0.491.$

upper 95 per cent limit: $\sqrt{1.698} = 1.303.$

99 per cent interval for σ:

lower 99 per cent limit: $\sqrt{0.194} = 0.440.$

upper 99 per cent limit: $\sqrt{2.650} = 1.628.$

Nonnormality can affect the intervals on the variance and standard deviation. The type of nonnormality that is most serious involves the kurtosis of the distribution, the relation of the height or peakedness of the center of a distribution to the length of its tails. If the relationship between the center and the tails of the distribution is not like that of the normal distribution, our nominal confidence level of $1 - \alpha$ can be seriously in error. In Chap. 8 we shall introduce a specific procedure to test for this characteristic. At this stage we can say that extreme or contaminating values in the data can distort these intervals. It is important that the data be scrutinized carefully to make sure that clerical errors or other kinds of contaminating factors have not produced clearly aberrant values. On the other hand, extreme values do not always indicate contamination but can be evidence that we are dealing with a skewed distribution. For this reason caution must be used in making the decision to exclude any aberrant values. Unfortunately, here we are in an area where hard and fast rules do not apply—common sense and detailed study of the data are the best guides.

7-2-5 CONFIDENCE INTERVALS FOR THE DIFFERENCE BETWEEN THE MEANS OF TWO NORMAL DISTRIBUTIONS

The comparison of 2 sets of observations is a very common problem in biology. The agronomist comparing yields from a new barley hybrid with those from an old one; the pharmacologist comparing the hypotensive effects of 2 agents, one containing a chlorine and the other a fluorine atom at a certain point on the molecule; the surgeon comparing late effects of radical and simple mastectomy in breast cancer; the health educator comparing the effects of strong and weak fear-arousing messages in inducing smokers to decrease their cigarette consumption; the epidemiologist comparing serum lipid levels in white and black subjects; the gastroenterologist comparing 2 diets for treating peptic ulcer; the health officer comparing diphtheria immunization levels among children in 2 sections of a large city—all of these investigators are comparing 2 populations. Very often the data from such investigations will be roughly normally distributed, and the comparison of interest can be given as the difference between the means of the 2 populations. Even if normality does not hold exactly, an extension of the central limit theorem (Sec. 6-4-7) guarantees that the methods to be discussed in this section, though nominally assuming normality, hold approximately for moderate-sized samples from populations that are quite nonnormal.

7-2-5-1 Difference between Means—Known Population Variances. It is instructive first to consider the comparison of the means of 2 normal distributions whose variances are known. We assume that random samples of size n_1 and n_2, respectively, are independently drawn from the populations. The assumption of independence is important. It means that the results of sampling one population in no way influence the outcome of sampling the other or alter the probability of any outcomes from the other. The assumption would not be expected to hold, for example, in comparing readings taken on the same subjects before and after treatment from some agent. It is nearly always true in practice that there is some relationship, however slight, between readings taken at different times on the same subject. Such data can be handled, but not by the procedures of this section—in Sec. 7-2-5-4 we will discuss this situation in detail. In terms of probability, the assumption of independent sampling means that the probability of any sets of outcomes from the 2 samples factors into the product of the probability of specified outcomes from the first sample times the corresponding probability from the second sample. Knowledge of the outcomes from any part of the first sample must not influence the probability of any outcomes from the second.

We must introduce some additional notation. Suppose that the populations have means μ_1 and μ_2, respectively, and variances σ_1^2 and σ_2^2. We suppose that the variances are known, but of course the means are not. Let x_{ij} denote the observation of the jth unit selected from the ith population. Then $i = 1, 2$, and $j = 1, 2, \ldots, n_i$. Let $\bar{x}_1$ and $\bar{x}_2$ denote the sample means. We saw in Sec. 6-4-7 that $\bar{x}_1$ is normally distributed with mean μ_1 and variance σ_1^2/n_1 and $\bar{x}_2$ is normal with mean μ_2 and variance σ_2^2/n_2, since $\bar{x}_1$ and $\bar{x}_2$ are sample means from normally distributed populations.

In estimating $\mu_1 - \mu_2$ it is natural to use $\bar{x}_1 - \bar{x}_2$, the corresponding difference in sample means. In fact, the sampling distribution of $\bar{x}_1 - \bar{x}_2$ is normal with mean $\mu_1 - \mu_2$ and variance $(\sigma_1^2/n_1) + (\sigma_2^2/n_2)$. It is not surprising that this distribution has mean $\mu_1 - \mu_2$, but the expression for the variance may seem strange at first. Our first inclination might be to think that the correct variance is the difference rather than the sum of the variances of $\bar{x}_1$ and $\bar{x}_2$. This cannot be right, because there is nothing to prevent σ_2^2/n_2 from being larger than σ_1^2/n_1, producing a negative value for the variance of $\bar{x}_1 - \bar{x}_2$, a distinct embarrassment since negative variances just do not happen. In fact, because of independent sampling there is no constraint on the relationship between $\bar{x}_1$ and $\bar{x}_2$, and a large value of one mean may be accompanied by a small value of the other, producing a very large difference. Thus, the difference in means should be more variable than the individual means, and it is. A more complete argument demonstrating that the variance of $\bar{x}_1 - \bar{x}_2$ must indeed be $(\sigma_1^2/n_1) + (\sigma_2^2/n_2)$ involves mathematics beyond the scope of this text.

Standardizing the normal variable $\bar{x}_1 - \bar{x}_2$, we find

$$z = \frac{(\bar{x}_1 - \bar{x}_2) - (\mu_1 - \mu_2)}{\sqrt{\dfrac{\sigma_1^2}{n_1} + \dfrac{\sigma_2^2}{n_2}}}$$

and

$$P\left[-z_{1-(\alpha/2)} < \frac{(\bar{x}_1 - \bar{x}_2) - (\mu_1 - \mu_2)}{\sqrt{\dfrac{\sigma_1^2}{n_1} + \dfrac{\sigma_2^2}{n_2}}} < z_{1-(\alpha/2)}\right] = 1 - \alpha.$$

We want an equivalent event with $\mu_1 - \mu_2$ as the middle term. The sequence of algebraic steps to produce that result is: multiplication by $\sqrt{\sigma_1^2/n_1 + \sigma_2^2/n_2}$, subtraction of $\bar{x}_1 - \bar{x}_2$, multiplication by -1, reversal of the order of terms.

The result is

$$P\left[(\bar{x}_1 - \bar{x}_2) - z_{1-(\alpha/2)}\sqrt{\frac{\sigma_1^2}{n_1} + \frac{\sigma_2^2}{n_2}} < \mu_1 - \mu_2 < (\bar{x}_1 - \bar{x}_2)\right.$$

$$\left. + z_{1-(\alpha/2)}\sqrt{\frac{\sigma_1^2}{n_1} + \frac{\sigma_2^2}{n_2}}\right] = 1 - \alpha.$$

Sample values are again substituted for $\bar{x}_1$ and $\bar{x}_2$ to produce specific numerical confidence limits. All our earlier remarks about confidence versus probability still hold and will continue to hold throughout the rest of this chapter.

EXAMPLE

To illustrate the method of this section we again make use of artificial data based upon the results of Milam and Muench (35) on hemoglobin levels. Suppose we can assume that the levels in white males over age 11 are normally distributed with variance 1.462 and that those for white males age 11 and under are normal with variance 0.867. Independent random samples of 10 older and 20 younger males are selected and produce sample means of 14.47 mg/100 ml and 12.64 mg/100 ml, respectively. We find $\bar{x}_1 - \bar{x}_2 = 1.83$,

$$\frac{\sigma_1^2}{n_1} = \frac{1.462}{10} = 0.1462, \qquad \frac{\sigma_2^2}{n_2} = \frac{0.867}{20} = 0.0434,$$

$$\frac{\sigma_1^2}{n_1} + \frac{\sigma_2^2}{n_2} = 0.1896, \qquad \sqrt{\frac{\sigma_1^2}{n_1} + \frac{\sigma_2^2}{n_1}} = 0.4354.$$

For a 95 per cent interval for $\mu_1 - \mu_2$, $z_{1-(\alpha/2)} = 1.960$ and

$$z_{1-(\alpha/2)}\sqrt{\frac{\sigma_1^2}{n_1} + \frac{\sigma_2^2}{n_2}} = (1.960)(0.4354) = 0.8534.$$

The lower 95 per cent limit is $1.83 - 0.85 = 0.98$.

The upper 95 per cent limit is $1.83 + 0.85 = 2.68$.

It is important to notice that zero is not included between these limits. If the 2 populations had the same mean, i.e., if $\mu_1 = \mu_2$, then $\mu_1 - \mu_2 = 0$ and we would expect to find this value among those we would believe at the 95 per cent confidence level. The fact that it is not included in the interval indicates that our populations probably differ in mean hemoglobin level, and we think that they may differ by as little as 0.98 mg/100 ml or as much as 2.68 mg/100 ml.

The following several sections consider procedures to be employed when the population variances are unknown. In preparation, consider the case in which the 2 population variances are equal, and suppose that the common value is denoted by σ^2. Then the standard normal variable on which our interval is based becomes

$$z = \frac{(\bar{x}_1 - \bar{x}_2) - (\mu_1 - \mu_2)}{\sqrt{\dfrac{\sigma^2}{n_1} + \dfrac{\sigma^2}{n_2}}} = \frac{(\bar{x}_1 - \bar{x}_2) - (\mu_1 - \mu_2)}{\sigma\sqrt{\dfrac{1}{n_1} + \dfrac{1}{n_2}}}.$$

7-2-5-2 Difference between Means—Unknown but Equal Population Variances. The assumption of equal population variances is, as you will see, a pervasive and important assumption in many of the techniques that we shall consider. The assumption, perhaps too simply worded to suit some jargon-hunting individuals, is often labeled with the status-elevating terminology—the assumption of homoscedasticity. Because our field is not as richly endowed with jargon as some others, you will perhaps allow us the indulgence of occasionally using the term. Of course, it must follow that populations with unequal variances are heteroscedastic.

The equal variance assumption is not as restrictive as it appears at first. It holds or nearly holds in many practical situations, and data violating the assumption can frequently be transformed to a new scale of measurement which makes variances nearly equal. It seems to be true that means are more unstable and more subject to alteration by a variety of factors than are variances. This is obviously not a mathematical theorem, but a statement of experience, although there are some fairly convincing mathematical analogs that tend to give support to this impression. In biological data it is frequently true that means and variances tend to vary together, larger means being accompanied by larger variances, although in terms of relative magnitude of change the property just noted for variances to be more stable still holds. Fortunately, situations in which means and variances change together tend to be admirably suited to transformations; shifting to measurements that are logarithms or square roots of the original measurements, if appropriate to the particular application, often makes variances approximately equal. Such transformations may often be regarded as simply replacing the scale template on a measuring instrument by a new template with numerical values in different absolute positions but with different values still in the same order relationship.

We assume independent (again a critically important assumption) random samples of sizes n_1 and n_2 from normally distributed populations with means μ_1 and μ_2 and common variance σ^2. The sample means are still denoted by $\bar{x}_1$ and $\bar{x}_2$, the sample variances by s_1^2 and s_2^2, and x_{ij} is the jth observation from the ith population with $i = 1, 2$ and $j = 1, 2, \ldots, n_i$.

How should we estimate the common variance? Since both s_1^2 and s_2^2 are estimating the same quantity σ^2, averaging or pooling them in some way seems appropriate. The sample sizes may be different, however, and this should be taken into account in the averaging process by calculating a weighted mean in the sense of Sec. 3-3-5-4. Using the quantities $n_1 - 1$ and $n_2 - 1$ produces an unbiased estimator having minimum variance among all estimators that are weighted means of s_1^2 and s_2^2. That is, $n_1 - 1$ and $n_2 - 1$ are in a sense the "best" weights. The resulting estimator is denoted by s_p^2, to denote the fact that it is a pooled estimator, and we have

$$s_p^2 = \frac{(n_1 - 1)s_1^2 + (n_2 - 1)s_2^2}{(n_1 - 1) + (n_2 - 1)} = \frac{(n_1 - 1)s_1^2 + (n_2 - 1)s_2^2}{(n_1 + n_2 - 2)}.$$

If we are calculating s_p^2 from the raw data rather than from s_1^2 and s_2^2, a better formula is available. Notice that

$$(n_1 - 1)s_1^2 = \sum_{j=1}^{n_1}(x_{1j} - \bar{x}_1)^2, \qquad (n_2 - 1)s_2^2 = \sum_{j=1}^{n_2}(x_{2j} - \bar{x}_2)^2,$$

and

$$s_p^2 = \frac{\sum_{j=1}^{n_1}(x_{1j} - \bar{x}_1)^2 + \sum_{j=1}^{n_2}(x_{2j} - \bar{x}_2)^2}{n_1 + n_2 - 2}.$$

But

$$\sum_{j=1}^{n_1}(x_{1j} - \bar{x}_1)^2 = \frac{n_1 \sum_{j=1}^{n_1} x_{1j}^2 - \left(\sum_{j=1}^{n_1} x_{1j}\right)^2}{n_1}$$

and

$$\sum_{j=1}^{n_2}(x_{2j} - \bar{x}_2)^2 = \frac{n_2 \sum_{j=1}^{n_2} x_{2j}^2 - \left(\sum_{j=1}^{n_2} x_{2j}\right)^2}{n_2}.$$

Finally,

$$s_p^2 = \frac{1}{n_1 + n_2 - 2}\left[\frac{n_1 \sum_{j=1}^{n_1} x_{1j}^2 - \left(\sum_{j=1}^{n_1} x_{1j}\right)^2}{n_1} + \frac{n_2 \sum_{j=1}^{n_2} x_{2j}^2 - \left(\sum_{j=1}^{n_2} x_{2j}\right)^2}{n_2}\right].$$

Although this expression looks a bit ghastly, it has advantages both of speed and precision for computation on a desk calculator. For technical reasons this may not be true on an electronic computer. If in doubt, consult your friendly statistician or programmer or both.

In Sec. 7-2-3 we replace σ by s in the formula for z to produce a t-distributed variable. Here we might consider replacing σ by s_p in the expression

$$z = \frac{(\bar{x}_1 - \bar{x}_2) - (\mu_1 - \mu_2)}{\sigma\sqrt{\dfrac{1}{n_1} + \dfrac{1}{n_2}}},$$

which we came up with at the end of the preceding section. If this is done the resulting quantity follows the t distribution with $n_1 + n_2 - 2$ degrees of freedom. That is,

$$t = \frac{(\bar{x}_1 - \bar{x}_2) - (\mu_1 - \mu_2)}{s_p\sqrt{\dfrac{1}{n_1} + \dfrac{1}{n_2}}}$$

has, under the assumptions we have made (independent random samples from normal populations with equal variances), the t distribution with $n_1 + n_2 - 2$ degrees of freedom.

Thus, for a $1 - \alpha$ level confidence interval,

$$P\left[-t_{1-(\alpha/2)} < \frac{(\bar{x}_1 - \bar{x}_2) - (\mu_1 - \mu_2)}{s_p\sqrt{\dfrac{1}{n_1} + \dfrac{1}{n_2}}} < t_{1-(\alpha/2)} \right] = 1 - \alpha.$$

Repeating a sequence of algebraic manipulations that by now should be horribly familiar, we find

$$P\left[(\bar{x}_1 - \bar{x}_2) - t_{1-(\alpha/2)}s_p\sqrt{\frac{1}{n_1} + \frac{1}{n_2}} < \mu_1 - \mu_2 \right.$$

$$\left. < (\bar{x}_1 - \bar{x}_2) + t_{1-(\alpha/2)}s_p\sqrt{\frac{1}{n_1} + \frac{1}{n_2}} \right] = 1 - \alpha.$$

This gives the required interval.

EXAMPLE

The following data represent resting systolic blood pressures of a group of children having one hypertensive parent (group 1) and a group of children both of whose parents have normal blood pressure (group 2).

Calculate a 95 per cent confidence interval for the difference between the means of the populations from which the observations are selected, and use the interval to determine if the 2 population means are likely to be different.

Group 1:   100, 102, 96, 106, 110, 110, 120, 112, 112, 90.

Group 2:   104, 88, 100, 98, 102, 92, 96, 100, 96, 96.

We assume that the 2 sets of data are independent random samples from normally distributed populations having common variance and find:

$$n_1 = n_2 = 10; \quad \sum_{j=1}^{n_1} x_{1j} = 1{,}058; \quad \bar{x}_1 = 105.8; \quad \sum_{j=1}^{n_1} x_{1j}^2 = 112{,}644;$$

$$n_1 \sum_{j=1}^{n_1} x_{1j}^2 - \left(\sum_{j=1}^{n_1} x_{1j} \right)^2 = 7{,}076; \quad \sum_{j=1}^{n_2} x_{2j} = 972; \quad \bar{x}_2 = 97.2;$$

$$\sum_{j=1}^{n_2} x_{2j}^2 = 94{,}680; \quad n_2 \sum_{j=1}^{n_2} x_{2j}^2 - \left(\sum_{j=1}^{n_2} x_{2j} \right)^2 = 2{,}016;$$

$$s_p^2 = \frac{\left[\dfrac{n_1 \sum_{j=1}^{n_1} x_{1j}^2 - \left(\sum_{j=1}^{n_1} x_{1j} \right)^2}{n_1} \right] + \left[\dfrac{n_2 \sum_{j=1}^{n_2} x_{2j}^2 - \left(\sum_{j=1}^{n_2} x_{2j} \right)^2}{n_2} \right]}{n_1 + n_2 - 2}$$

$$= \frac{707.6 + 201.6}{18} = 50.5111; \quad s_p = 7.1071;$$

$$\sqrt{\frac{1}{n_1} + \frac{1}{n_2}} = \sqrt{0.2} = 0.4472; \quad t_{0.975} = 2.101;$$

$$t_{0.975} s_p \sqrt{\frac{1}{n_1} + \frac{1}{n_2}} = (2.101)(7.1071)(0.4472) = 6.6776;$$

$$\bar{x}_1 - \bar{x}_2 = 8.6.$$

lower 95 per cent limit for $\mu_1 - \mu_2$: $8.6 - 6.7 = 1.9.$

upper 95 per cent limit for $\mu_1 - \mu_2$: $8.6 + 6.7 = 15.3.$

Since the value zero is not included between the limits, we are confident that the population means are different. We estimate the difference to be between approximately 2 and 15 mm Hg.

The additional assumption of equal variances makes consequences of violations of the assumptions in this interval rather more complicated. First, as before, the assumption of normality is not in itself particularly important or restrictive. If the departure from normality is general in the sense that the 2 populations have a similar shape, then rather severe departures can be tolerated if the sample sizes are at least moderate. Extreme skewness in one direction in one population and in the other direction in the second population can cause difficulties.

Violation of the assumption of equal variances can be serious, but only if the sample sizes are unequal. For equal or nearly equal-sized samples the assumption can essentially be ignored. For unequal-sized samples, if the larger sample comes from the population with the larger variance the calculated confidence interval is conservative in the sense that the true confidence level is larger than the nominal level associated with the interval. The most serious case is that in which the population with larger variance is sampled at a lower rate. This tends to give the calculated interval a lower confidence level than advertised. In doubtful cases the approximate interval of the next section is probably the method of choice. Unfortunately, most of the traditional procedures designed to investigate the truth of the assumption of equal variances are sensitive to departures from normality. As before, the assumption of independent sampling must be satisfied.

7-2-5-3 Difference between Means—Unknown and Unequal Population Variances. Development of the theory underlying the confidence interval estimate of the difference between the means of 2 normal populations has been marked by continuing controversy dating at least from the 1930's. There is general agreement that no interval, exact in the sense that the other intervals of this chapter are exact, can be computed when the population variances are unknown and unequal. Either the basic concepts and definitions of probability and sampling distributions as we have presented them must be fundamentally changed, or an approximate procedure must be used. We recommend the latter alternative.

Welch (58) has suggested an approximate procedure that treats the quantity

$$\frac{(\bar{x}_1 - \bar{x}_2) - (\mu_1 - \mu_2)}{\sqrt{\dfrac{s_1^2}{n_1} + \dfrac{s_2^2}{n_2}}}$$

as a *t*-distributed variable with *f* "degrees of freedom," yielding confidence limits of the form

$$(\bar{x}_1 - \bar{x}_2) \pm t_{1-(\alpha/2)}\sqrt{\frac{s_1^2}{n_1} + \frac{s_2^2}{n_2}}$$

for an interval of confidence level $1 - \alpha$. In this case, however, the quantity f is not a positive whole number. In fact,

$$f = \frac{\left(\dfrac{s_1^2}{n_1} + \dfrac{s_2^2}{n_2}\right)^2}{\dfrac{\left(\dfrac{s_1^2}{n_1}\right)^2}{n_1 + 1} + \dfrac{\left(\dfrac{s_2^2}{n_2}\right)^2}{n_2 + 1}} - 2.$$

The required value of $t_{1-(\alpha/2)}$ can either be found by interpolating in Table A-5 or by taking the whole number closest to f given in the table. The latter approach is ordinarily sufficient.

EXAMPLE

The following data represent total histidine excretions in mg from 24-hour urine samples for 5 men and 10 women on unrestricted diets.

Men: 229, 236, 435, 172, 432.

Women: 197, 224, 115, 74, 138, 135, 107, 204, 200, 138.

Compute a 95 per cent confidence interval for the difference between the means of the 2 populations from which the observations are selected.
Letting population 1 be males and population 2 be females, we have

$$n_1 = 5; \quad \sum_{j=1}^{n_1} x_{1j} = 1{,}504; \quad \bar{x}_1 = 300.8; \quad \sum_{j=1}^{n_1} x_{1j}^2 = 513{,}570;$$

$$s_1^2 = 15{,}291.7; \quad n_2 = 10; \quad \sum_{j=1}^{n_2} x_{2j} = 1{,}532; \quad \bar{x}_2 = 153.2;$$

$$\sum_{j=1}^{n_2} x_{2j}^2 = 257{,}064; \quad s_2^2 = 2{,}484.62.$$

We assume that the 2 populations are normally distributed, but notice a fairly large discrepancy between the sample variances. Furthermore, the larger variance is associated with the smaller sample—the sample of males, in this case. This makes it unwise to use the conventional confidence interval of Sec. 7-2-5-2, which assumes equal variances. We apply the Welch approximation instead and find

$$f = \frac{\left(\dfrac{s_1^2}{n_1} + \dfrac{s_2^2}{n_2}\right)^2}{\dfrac{\left(\dfrac{s_1^2}{n_1}\right)^2}{n_1 + 1} + \dfrac{\left(\dfrac{s_2^2}{n_2}\right)^2}{n_2 + 1}} - 2 = \frac{(3{,}058.340 + 248.462)^2}{\dfrac{9{,}353{,}444}{6} + \dfrac{61{,}733}{11}} - 2$$

$$= \frac{10{,}934{,}939}{1{,}558{,}907 + 5{,}612} - 2 = 6.99 - 2 = 4.99$$

or approximately 5. Thus the approximate confidence interval will be based upon the t distribution with 5 degrees of freedom rather than the $n_1 + n_2 - 2 = 13$ degrees of freedom associated with the corresponding interval for Sec. 7-2-5-2.

$$\bar{x}_1 - \bar{x}_2 = 147.6; \qquad t_{0.975} = 2.571;$$

$$\sqrt{\frac{s_1^2}{n_1} + \frac{s_2^2}{n_2}} = \sqrt{3{,}306.802} = 57.5048.$$

lower 95 per cent limit for $\mu_1 - \mu_2$

$$147.6 - (2.571)(57.5048) = 147.6 - 147.8 = -0.2.$$

upper 95 per cent limit for $\mu_1 - \mu_2$

$$147.6 + 147.8 = 295.4.$$

Since zero is included between the limits, we cannot conclude that a difference between means has been established by this study, operating at the 95 per cent confidence level. However, the extreme length of the confidence interval, partly due to the very small sample sizes, indicates that we do not have a very sensitive study.

Another approximate procedure is available to us for large samples. With reference to the situation discussed in Sec. 7-2-5-1 when the population variances were known, there seems to be little loss when n_1 and n_2 are large in substituting sample variances for the corresponding population variances and carrying out the calculations as in that section. That is, we use

$$\frac{(\bar{x}_1 - \bar{x}_2) - (\mu_1 - \mu_2)}{\sqrt{\dfrac{s_1^2}{n_1} + \dfrac{s_2^2}{n_2}}}$$

as an approximate standard normal variable, producing limits

$$(\bar{x}_1 - \bar{x}_2) - z_{1-(\alpha/2)}\sqrt{\frac{s_1^2}{n_1} + \frac{s_2^2}{n_2}} \quad \text{and} \quad (\bar{x}_1 - \bar{x}_2) + z_{1-(\alpha/2)}\sqrt{\frac{s_1^2}{n_1} + \frac{s_2^2}{n_2}}.$$

Recall that in Sec. 7-2-3 a similar situation held; we found that the percentiles of the t distribution were close to those from the standard normal when the sample size got large, indicating that the replacement of σ by s, though very important for small samples, was of little effect for large n. Here again the influence of moderate nonnormality is not large. The assumption of independence of sampling between the 2 populations is very important, however. If it does not hold, then methods like those to be described in Sec. 7-2-5-4 may be appropriate.

7-2-5-4 Pairing—Confidence Interval for the Mean Difference. Many biological studies are designed to produce observations in pairs. For example, a clinical trial of a blood glucose-reducing drug may involve examining the glucose level before and after treatment in each subject studied. This device of using a subject as his own control is useful in a variety of circumstances and has the advantage of controlling many extraneous sources of variation in the data. Pairs of identical twins are sometimes used to remove the effect of hereditary differences in making various comparisons. In agricultural field trials, 2 varieties of grain will be compared by planting both varieties in each of a series of blocks of land in order that comparisons may be made within blocks or fields, thereby controlling for meteorologic factors and soil fertility differences.

In each of these instances we wish to compare observations from 2 populations—before versus after treatment readings, observations from members of twin pairs assigned to one group versus their twins assigned to the other group, and observations for grain of the first variety versus grain of the second. A confidence interval for the difference between the means of the 2 populations will often be a useful device for making such comparisons.

However, the methods described so far for constructing such intervals do not apply to paired data. In such cases the 2 samples are not independent. Readings taken before and after treatment on the same patient, observations on the 2 members of the same pair of twins, and yields from 2 varieties of grain planted in the same block cannot be assumed to be unrelated. In fact, the decision to use paired data and the advantages such a design presents are directly related to the strength of this relationship. Twin pairs are used precisely because they are so similar with respect to many characteristics, the effects of which can then be eliminated in making the comparison of interest. Data collected in pairs require special forms of analysis and can, in certain circumstances, produce very refined inferences if such analyses are used.

Pairing implies that the sample sizes of the 2 groups will be equal. If we let x_{ij} denote the observation in the jth pair from group i, then $i = 1, 2$ and $j = 1, 2, \ldots, n$, where n is the number of pairs. Again denoting the sample means by $\bar{x}_1$ and $\bar{x}_2$ and letting $d_j = x_{1j} - x_{2j}$, the difference between the observations on the jth pair, we find that an interesting fact emerges— namely, $\bar{d} = \bar{x}_1 - \bar{x}_2$, where $\bar{d}$ denotes the mean difference $\left(\sum\limits_{j=1}^{n} d_j \right) / n$. This property follows directly from the properties of summation signs discussed in Chap. 3 and states that the mean difference equals the difference between means.

This suggests an approach to our problem. Instead of considering that we have 2 populations yielding observations x_{1j} and x_{2j}, we think of a single population whose units are pairs. The corresponding population of observations consists of differences d_j between the x values measured on the members of each pair. The result $\bar{d} = \bar{x}_1 - \bar{x}_2$ is reflected in the populations and $\mu_d = \mu_1 - \mu_2$, where μ_d is the mean of the population of differences. A confidence interval for μ_d will thus be an interval for $\mu_1 - \mu_2$. This approach eliminates the assumption of independence by replacing the 2 samples by a single sample. Notice that the question of equal population variances is also eliminated, since there is now only one population.

Our situation then is as follows. Suppose that $d_1, d_2, \ldots, d_n$ is a random sample from a normally distributed population with mean μ_d and variance σ_d^2. We seek a $1 - \alpha$ level confidence interval for μ_d. This puts us back into the situation covered in Secs. 7-2-2 and 7-2-3. If σ_d^2 is known, the interval for μ_d or $\mu_1 - \mu_2$ is $\bar{d} \pm z_{1-(\alpha/2)}(\sigma_d/\sqrt{n})$. If σ_d is unknown, the required interval is $\bar{d} \pm t_{1-(\alpha/2)}(s_d/\sqrt{n})$, where s_d is the standard deviation of the sample of differences and $t_{1-(\alpha/2)}$ is the $100[1 - (\alpha/2)]$ percentile of the t distribution with $n - 1$ degrees of freedom. All the remarks and discussion of the earlier sections apply here to the population and sample of differences.

EXAMPLE

A pharmaceutical company is engaged in preliminary investigation of a new drug which may have serum cholesterol-lowering properties. A small study is designed using 6 prison inmates. Serum cholesterol determinations in milligrams per 100 milliliters are made before and after treatment on each prisoner. Find a 95 per cent confidence interval for the mean difference between the before and after treatment readings. Does the interval provide evidence of a treatment effect?

Prisoner number	1	2	3	4	5	6
Cholesterol level before treatment (x_{1j})	217	252	229	200	209	213
Cholesterol level after treatment (x_{2j})	209	241	230	208	206	211
Difference $(d_j = x_{1j} - x_{2j})$	8	11	−1	−8	3	2

$$n = 6; \qquad \sum_{j=1}^{n} d_j = 15 \qquad \bar{d} = 2.5; \qquad \sum_{j=1}^{n} d_j^2 = 263;$$

$$n \sum_{j=1}^{n} d_j^2 - \left(\sum_{j=1}^{n} d_j \right)^2 = 1353;$$

$$s_d^2 = 45.1; \qquad \sqrt{n} = 2.4495; \qquad s_d = 6.7157; \qquad \frac{s_d}{\sqrt{n}} = 2.7417.$$

For 5 degrees of freedom,

$$t_{0.975} = 2.571; \qquad t_{0.975} \frac{s_d}{\sqrt{n}} = 7.0489.$$

lower 95 per cent limit for μ_d: $2.5 - 7.0 = -4.5$.

upper 95 per cent limit for μ_d: $2.5 + 7.0 = 9.5$.

Although the sample mean difference is 2.5 mg/100 ml, indicating a slight average effect, 2 of the 6 men show an adverse response to treatment, i.e., a lower level before than after treatment. The 95 per cent confidence interval includes zero, and thus we cannot conclude from these data that the treatment necessarily has any cholesterol-lowering action in the population.

7-2-6 CONFIDENCE INTERVAL FOR THE RATIO OF THE VARIANCES OF TWO NORMAL POPULATIONS

It is sometimes important to compare the variances of 2 populations. A change in the variability of potency of a drug, though perhaps less important than a change in average potency, might result in an increased number of lots having either uselessly low or dangerously high potency. In some chronic diseases investigators feel that the transition from health to disease is marked first by an increased lability or variability in some indicator variable.

As a consequence, one might expect that if children having a hypertensive parent are at increased risk of developing hypertension, then their casual blood pressure levels might be more variable than those of control children whose parents are normotensive.

Suppose we have independent random samples of size n_1 and n_2 from 2 normally distributed populations with means μ_1 and μ_2 and variances σ_1^2 and σ_2^2. Let x_{ij} be the observation on the jth unit selected from the ith population: $i = 1, 2; j = 1, 2, \ldots, n_i$. Let s_1^2 and s_2^2 be the sample variances.

The sampling distribution of the quantity $(s_1^2/\sigma_1^2)/(s_2^2/\sigma_2^2)$ is known and is called the F distribution. The distribution was named by Professor G. W. Snedecor in honor of Sir Ronald A. Fisher, one of the true giants of statistics and discoverer of many of the most important parts of current statistical theory and method. Fisher, who died in 1963, was an active contributor to this field almost to the end of his long and productive life. The F distribution depends upon 2 parameters called the numerator and denominator degrees of freedom. These parameters, as in the case of the t and χ^2 distributions, take positive whole number values and in our present application take specific values, $n_1 - 1$ and $n_2 - 1$, respectively. Recall that in Secs. 7-2-3 and 7-2-4 we used t and χ^2 variables having $n - 1$ degrees of freedom. The situation here is analogous in that s_1^2/σ_1^2, the numerator or upstairs quantity in our random variable, is based on a sample size of n_1 corresponding to $n_1 - 1$ degrees of freedom and s_2^2/σ_2^2, the denominator or downstairs quantity, on a sample size of n_2, i.e., $n_2 - 1$ degrees of freedom.

Table A-7 gives percentiles of the F distribution. It is important that the numerator and denominator degrees of freedom not be confused. Scrutiny of the table reveals that incorrect values result from interchanging them. The F distribution is positively skewed, the degree of skewness depending jointly upon the numerator and denominator degrees of freedom.

To find a $1 - \alpha$ level confidence interval for the ratio σ_1^2/σ_2^2, we start, as usual, with a basic probability statement:[1]

$$P\left(F_{\alpha/2} < \frac{s_1^2/\sigma_1^2}{s_2^2/\sigma_2^2} < F_{1-(\alpha/2)}\right) = 1 - \alpha.$$

But

$$\frac{s_1^2/\sigma_1^2}{s_2^2/\sigma_2^2} = \frac{s_1^2}{s_2^2} \cdot \frac{\sigma_2^2}{\sigma_1^2},$$

and dividing the 3 terms in the inequality by s_1^2/s_2^2, we have the equivalent event

$$\frac{F_{\alpha/2}}{s_1^2/s_2^2} < \frac{\sigma_2^2}{\sigma_1^2} < \frac{F_{1-(\alpha/2)}}{s_1^2/s_2^2}.$$

[1]The relation $F_{\alpha/2}(f_1, f_2) = \dfrac{1}{F_{1-(\alpha/2)}(f_2, f_1)}$ gives lower percentiles of the F distribution. Note the reversal of degrees of freedom.

Taking reciprocals, noting that the 3 terms are positive, and recalling Property 4 of Sec. 6-4-3, we have the equivalent event

$$\frac{s_1^2/s_2^2}{F_{\alpha/2}} > \frac{\sigma_1^2}{\sigma_2^2} > \frac{s_1^2/s_2^2}{F_{1-(\alpha/2)}}.$$

Reversing the order, we find

$$P\left(\frac{s_1^2/s_2^2}{F_{1-(\alpha/2)}} < \frac{\sigma_1^2}{\sigma_2^2} < \frac{s_1^2/s_2^2}{F_{\alpha/2}}\right) = 1 - \alpha.$$

Values of the sample variances are substituted for s_1^2/s_2^2 to produce the required interval.

EXAMPLE

We will use the data of the example in Sec. 7-2-5-2 on resting systolic blood pressures of children having one hypertensive parent and children with both parents normotensive to calculate a 99 per cent confidence interval for the ratio of population variances. We find

$$s_1^2 = \frac{7076}{90} = 78.622; \qquad s_2^2 = \frac{2016}{90} = 22.400.$$

The relevant F distribution has 9 degrees of freedom in both numerator and denominator.

$$F_{0.005} = 0.153; \qquad F_{0.995} = 6.54; \qquad \frac{s_1^2}{s_2^2} = 3.510.$$

lower 99 per cent limit for σ_1^2/σ_2^2: $\dfrac{3.510}{6.54} = 0.537.$

upper 99 per cent limit for σ_1^2/σ_2^2: $\dfrac{3.510}{0.153} = 22.9.$

Since the value $\sigma_1^2/\sigma_2^2 = 1$, corresponding to $\sigma_1^2 = \sigma_2^2$, is included between the limits, there is no reason, on the basis of these data and at this level of confidence, to conclude that the population variances are different. Notice, however, that we would not expect an experiment with only

10 children in each group to be very sensitive to population differences, even if those differences were fairly large.

The assumption of normality is important and restricts the applicability of inferential procedures on variances. As in Sec. 7-2-4 the type of non-normality that is most serious is in the kurtosis characteristic, which can be seriously upset by extreme or contaminating values. We will present a procedure for testing for normal kurtosis in Chap. 8.

7-2-7 CONFIDENCE INTERVAL FOR p, THE PARAMETER OF THE BINOMIAL DISTRIBUTION

The exact $1 - \alpha$ level confidence interval for p, the binomial parameter, has a complicated analytical form involving mathematics beyond the level of this book. However, a graphic procedure adequate for most ordinary purposes is available. Furthermore, for reasonable sample sizes there is an interval based upon the normal approximation to the binomial distribution.

Using the notation of Sec. 6-1, we assume that x is the random variable from $B(x; n, p)$; i.e., x is the total number of successes in n independent trials where the probability of success in a single trial is p. The relative frequency of successes in the sample is x/n, which we will denote by $\hat{p}$, the natural estimator of p.

Table A-8 gives 2 graphs that can be used to construct 95 per cent and 99 per cent confidence intervals for p. Each graph shows values of $\hat{p}$ along the top and bottom and values of p along the left and right sides. The bottom scale gives values of $\hat{p}$ between 0.00 and 0.50, whereas the top scale gives the values from 0.50 to 1.00. It is important to remember that when $\hat{p}$ takes one of the values shown on the bottom scale, the confidence limits for p are read from the left-hand scale; when $\hat{p}$ takes a value from the top scale, the limits are read from the right-hand scale. Two curves are given for each value of n: one concave upward and one concave downward. The confidence limits for p are located as 2 points, one corresponding to each of the curves for the given value of n. It will be necessary to interpolate visually if the exact values of $\hat{p}$, p, or n are not given in the graphs.

EXAMPLE

An experimental flock of 200 chickens is inoculated with an organism suspected of causing a set of clinical signs observed in several commerical flocks in recent months. Within 14 days after inoculation, 137 of the birds have exhibited the characteristic signs. Suppose that the chickens are housed and handled in such a way that cross infection is not a problem and that independence can be assumed. Find 95 per cent and 99 per cent

confidence limits on p, the probability that an individual inoculated chicken will show the signs within 14 days.
Here

$$n = 200, \quad x = 137, \quad \text{and} \quad \hat{p} = \frac{x}{n} = 0.685.$$

We use the upper and right-hand scales of the 2 graphs and read:

lower 95 per cent limit for p: 0.61.

upper 95 per cent limit for p: 0.75.

lower 99 per cent limit for p: 0.59.

upper 99 per cent limit for p: 0.77.

The confidence intervals for p based on the normal approximation to the binomial distribution use the fact that if n is sufficiently large, $\hat{p}$ is approximately normally distributed with mean p and variance $p(1 - p)/n = pq/n$, as discussed in Sec. 6-1. The continuity correction is ordinarily ignored. Since p is unknown, $\hat{p}(1 - \hat{p})$ is substituted for $p(1 - p)$ in the expression for the variance.
We use the fact that

$$\frac{\hat{p} - p}{\sqrt{\dfrac{\hat{p}(1 - \hat{p})}{n}}}$$

is approximately a standard normal variable. Then we have

$$P\left[-z_{1-(\alpha/2)} < \frac{\hat{p} - p}{\sqrt{\dfrac{\hat{p}(1 - \hat{p})}{n}}} < z_{1-(\alpha/2)}\right] \doteq 1 - \alpha,$$

using the symbol "$\doteq$" for "is approximately equal to." Manipulations similar to others in this chapter enable us to conclude

$$P\left[\hat{p} - z_{1-(\alpha/2)}\sqrt{\frac{\hat{p}(1 - \hat{p})}{n}} < p < \hat{p} + z_{1-(\alpha/2)}\sqrt{\frac{\hat{p}(1 - \hat{p})}{n}}\right] \doteq 1 - \alpha,$$

which produces the interval we seek.

EXAMPLE

Use the normal approximation to calculate 95 per cent and 99 per cent limits for p in the experiment with chickens described in the preceding example.

$$\hat{p} = 0.685; \qquad \hat{p}(1 - \hat{p}) = 0.2158; \qquad \frac{\hat{p}(1 - \hat{p})}{n} = 0.001079;$$

$$\sqrt{\frac{\hat{p}(1 - \hat{p})}{n}} = 0.03285; \qquad z_{0.975} = 1.960; \qquad z_{0.995} = 2.576;$$

$$z_{0.975}\sqrt{\frac{\hat{p}(1 - \hat{p})}{n}} = 0.06439; \qquad z_{0.995}\sqrt{\frac{\hat{p}(1 - \hat{p})}{n}} = 0.08462.$$

lower 95 per cent limit for p: $0.685 - 0.064 = 0.621$.

upper 95 per cent limit for p: $0.685 + 0.064 = 0.749$.

lower 99 per cent limit for p: $0.685 - 0.085 = 0.600$

upper 99 per cent limit for p: $0.685 + 0.085 = 0.770$.

Notice the close agreement with the values obtained by reading the graphs of Table A-8.

7-2-8   CONFIDENCE INTERVAL FOR $p_1 - p_2$, THE DIFFERENCE BETWEEN THE PARAMETERS OF TWO BINOMIAL DISTRIBUTIONS

Just as it is often of interest to compare the means of 2 normal distributions, so when we are dealing with dichotomous variables we often wish to compare the probability parameters of 2 binomial distributions. In comparing 2 modes of treatment in a clinical trial, for example, we may record for each patient whether or not he shows improvement as assessed by the disappearance of a particular sign or symptom after treatment. Patients treated by either of 2 regimes are then compared to see whether the improvement rates (probabilities of improvement) differ, and, if so, by how much. Incidence rates for coronary artery disease (probabilities of developing disease within some time period) in 2 populations or subpopulations may be compared by examining samples of subjects from the populations. Many examples of this sort can be cited, and certainly the technique to be described here is as useful and pervasive for dichotomous variables as are the techniques described in Sec. 7-2-5 for continuous variables.

We assume independent observations x_1 and x_2 of the total number of successes in n_1 and n_2 independent trials from the binomial distributions

$B(x_1; n_1, p_1)$ and $B(x_2; n_2, p_2)$. The independence of x_1 and x_2 as well as the independence of the sequences of trials within each population are important assumptions. Let $\hat{p}_1 = x_1/n_1$ and $\hat{p}_2 = x_2/n_2$, the sample proportions of successes.

Now if n_1 and n_2 are reasonably large, the quantity $\hat{p}_1 - \hat{p}_2$ will be approximately normally distributed with mean $p_1 - p_2$ and variance

$$\frac{p_1(1 - p_1)}{n_1} + \frac{p_2(1 - p_2)}{n_2}.$$

This result is analogous to that of Sec. 7-2-5, since here again the population mean is the difference between the means of the individual populations, and the population variance is the sum of the individual population variances. As in the preceding section, the substitution of

$$\frac{\hat{p}_1(1 - \hat{p}_1)}{n_1} + \frac{\hat{p}_2(1 - \hat{p}_2)}{n_2} \quad \text{for} \quad \frac{p_1(1 - p_1)}{n_1} + \frac{p_2(1 - p_2)}{n_2}$$

does not seriously affect the approximation if n_1 and n_2 are reasonably large. Thus

$$P\left[-z_{1-(\alpha/2)} < \frac{(\hat{p}_1 - \hat{p}_2) - (p_1 - p_2)}{\sqrt{\dfrac{\hat{p}_1(1 - \hat{p}_1)}{n_1} + \dfrac{\hat{p}_2(1 - \hat{p}_2)}{n_2}}} < z_{1-(\alpha/2)}\right] \doteq 1 - \alpha.$$

A familiar sequence of algebraic manipulations produces

$$P\left[(\hat{p}_1 - \hat{p}_2) - z_{1-(\alpha/2)}\sqrt{\frac{\hat{p}_1(1 - \hat{p}_1)}{n_1} + \frac{\hat{p}_2(1 - \hat{p}_2)}{n_2}} < p_1 - p_2\right.$$
$$\left. < (\hat{p}_1 - \hat{p}_2) + z_{1-(\alpha/2)}\sqrt{\frac{\hat{p}_1(1 - \hat{p}_1)}{n_1} + \frac{\hat{p}_2(1 - \hat{p}_2)}{n_2}}\right] \doteq 1 - \alpha.$$

EXAMPLE

A second flock of 150 chickens is inoculated with an organism isolated during a different outbreak of the disease described in the example of Sec. 7-2-7. This time, within 14 days of inoculation 98 birds are clinically ill. Find a 95 per cent confidence interval for the difference in the population proportions of birds ill within 14 days of inoculation by the 2 isolations.

data from 1st isolation (see Sec. 7-2-7):

$$n_1 = 200; \qquad \hat{p}_1 = 0.685.$$

data from 2nd isolation:

$$n_2 = 150; \qquad \hat{p}_2 = \frac{98}{150} = 0.653.$$

$$\frac{\hat{p}_1(1 - \hat{p}_1)}{n_1} = 0.001079; \qquad \frac{\hat{p}_2(1 - \hat{p}_2)}{n_2} = 0.001511;$$

$$\sqrt{\frac{\hat{p}_1(1 - \hat{p}_1)}{n_1} + \frac{\hat{p}_2(1 - \hat{p}_2)}{n_2}} = \sqrt{0.002590} = 0.05089; \qquad z_{0.975} = 1.960;$$

$$z_{0.975}\sqrt{\frac{\hat{p}_1(1 - \hat{p}_1)}{n_1} + \frac{\hat{p}_2(1 - \hat{p}_2)}{n_2}} = 0.09974 \doteq 0.100; \quad \hat{p}_1 - \hat{p}_2 = 0.032.$$

lower 95 per cent limit for $p_1 - p_2$: $0.032 - 0.100 = -0.068.$

upper 95 per cent limit for $p_1 - p_2$: $0.032 + 0.100 = 0.132.$

Since the interval includes zero, we cannot conclude from the data that there is a difference in virulence between the 2 isolations as measured by 14-day incidence of clinical signs.

7-2-9   CONFIDENCE INTERVAL FOR μ, THE PARAMETER OF THE POISSON DISTRIBUTION

The Poisson distribution, as you will recall from Sec. 6-2, is the distribution of a random variable arising as a count. The distribution involves the single parameter μ, and both the mean and variance of the distribution are equal to μ.

Table A-9 provides confidence limits for μ. The table is largely self-explanatory, but we should note that the true confidence level may be larger than the level indicated in the table. This is due to the discontinuity of the distribution and to the manner in which the limits were calculated; the error is largest for small values of μ. In this sense, then, the tabulated limits are conservative.

EXAMPLE

A radioactive sample emits a count of 9 in a standard counting period. Find 95 per cent confidence limits for μ, the mean of the population of counts from the source over counting periods of the same length. From Table A-9 we read:

lower 95 per cent limit for μ: 4.12.

upper 95 per cent limit for μ: 17.08.

7-3 Tolerance Limits

Frequently an investigator wants to determine limits between which most of the observations in a population will be found. In previous sections we have developed techniques to estimate a particular *parameter*. For example, if the data are drawn from a normal distribution, we know that 95 per cent confidence limits for the mean μ are $\bar{x} \pm 1.96 \, \sigma/\sqrt{n}$. Consider, however, the problem of finding two limits which with a given level of belief include between them a certain proportion of all observations in the population. For instance, a medical investigator may be studying a variable useful in predicting disease. If the investigator can discern the range for the "normal" population, then the value for an individual subject can be compared with the limits to determine whether he is like or unlike the population of normals. Such limits are termed tolerance limits.

If the underlying population follows the normal distribution and a random sample is drawn from this population, then these data may be used to find tolerance limits as follows.

Limits that cover at least a proportion Π of the population with level of confidence $1 - \alpha$ are of the form $\bar{x} \pm ks$, where s is the standard deviation of the sample and k is a value obtained from Table A-18. Generally the tolerance limits are wider than corresponding confidence limits. This is consistent with the distinction between them. Confidence limits are used to estimate a single value, a parameter, of a population, whereas tolerance limits are used to estimate a high proportion of the individual values in a population.

As an example, consider the study of radioisotope renograms by Pircher, et al. (45). As one aspect of that study the range of expected normal renogram values for the maximum of counts per minute was needed. The necessary data for the right kidney are $\bar{x} = 58.81$, $s = 15.83$, and $n = 9$. Computations of the 95 per cent tolerance limits designed to include 90 per cent of the population are $58.81 \pm (2.967)(15.83)$, yielding $(11.84, 105.78)$.

In Sec. 12-6 tolerance limits are described for situations when we cannot assume an underlying normal distribution.

EXERCISES

7-1 As a budding dermatologist you are investigating a certain skin cancer. Twenty-five rats have this cancer induced in them and are treated with a new drug. You count the number of hours until the cancer is gone with the following results: $\bar{x}$ is 322 hours and $s = 101$ hours. Assuming normality,

a. Compute a 90 per cent confidence interval for μ.
b. Suppose $\sigma = 100$ hours; now compute a 90 per cent confidence interval for μ.
c. Give a careful, detailed statement of the exact meaning of the final, computed intervals in parts a and b. Be sure to distinguish between the concepts of confidence and probability.
d. What size sample do we need to be 95 per cent confident that we are estimating μ within 5 hours if $\sigma = 100$ hours?

7-2 Ages at death for a random sample of 9 individuals who died of tuberculosis yield a mean of 49 years and a standard deviation of 5 years. Assuming normality,

a. Give the values of unbiased point estimates of μ and σ^2.
b. Find a 95 per cent confidence interval for μ.

7-3 Show that the sample mean and median have the same variance for samples of size 2.

7-4 Show that a symmetric confidence interval $\bar{x} \perp z(\sigma/\sqrt{n})$ for μ when σ is known has minimum length for fixed confidence level.

7-5 State whether true or false. When false, discuss what is wrong:

a. An unbiased statistic is the result of an unbiased sample.
b. Use of a random variable will yield a random sample.
c. The pooled variance is an example of a weighted mean.

7-6 For more than how many degrees of freedom will all percentiles from $t_{0.80}$ to $t_{0.995}$ be within 1 per cent of the standard normal value?

7-7 As the physician in charge of new drug experimentation in a pharmaceutical company, you are to plan a study concerned with a diuretic agent. Twenty edematous patients are available. Design a study, including statistical procedures, to:

a. Estimate the effect of the diuretic on urine sodium concentrations.
b. Compare this diuretic with a placebo.

7-8 As part of the study of Exercise 7-7, 11 edematous patients were studied. By a random sampling technique 6 patients were given the diuretic agent and and 5 a placebo. Urine sodium concentrations (mEq/L) were measured 24 hours after admission of the agent or placebo. The results were:

    Agent: 20.4,   92.5,   61.3,   44.2.   11.1,   23.7.
    Placebo: 1.2,   6.9,   33.7,   20.4,   17.2.

Assuming normality,

a. Compute a 95 per cent confidence interval for the mean under the diuretic agent.
b. Compute a 99 per cent confidence interval for the ratio of the variances

7-9 In a dental clinic a study investigating the ability of cements to hold single crowns was initiated. A new cement was to be compared with the one previously in use. Casts of teeth were used to test the holding strength of cements by recording the number of foot-pounds pressure required to pull the crown off the tooth. Fifty casts for the new cement and 50 for the old were studied with results:

Old: $\bar{x}_1 = 39$ ft lb, $s_1 = 4.5$ ft lb
New: $\bar{x}_2 = 42$ ft lb, $s_2 = 5.2$ ft lb

Stating all assumptions,

a. Find a 90 per cent confidence interval for the difference in means, $\mu_1 - \mu_2$.
b. Find 95 per cent confidence intervals for σ_1^2 and σ_1.

7-10 Using the properties of summation signs presented in Sec. 3-2, show in the notation of Sec. 7-2-5-4 that $\bar{d}$, the mean difference in a paired sample, equals $\bar{x}_1 - \bar{x}_2$, the difference between the means.

7-11 In a pediatric clinic a study was carried out to see how effective aspirin was in reducing temperature. Twelve 4-year-old girls suffering from influenza had their temperatures taken immediately before and one hour after administration of aspirin. The results are as follows:

Patient	Before	After
1	102.4	99.6
2	103.2	100.1
3	101.9	100.2
4	103.0	101.1
5	101.2	99.8
6	100.7	100.2
7	102.5	101.0
8	103.1	100.1
9	102.8	100.7
10	102.3	101.1
11	101.9	101.3
12	101.4	100.2

a. Find a 90 per cent confidence interval for the mean difference, assuming normality.
b. Is the paired confidence interval more like a one or two-population interval? Discuss.
c. What 2 values needed for a two-population confidence interval for the difference of means are altered when a paired confidence interval is used?

7-12 Two large university hospitals are cooperating in a study of hypertension. At hospital A 20 patients have a mean systolic blood pressure of $\bar{x} = 167.5$ mm with $s = 27$ mm. Hospital B has 24 patients with mean $\bar{x} = 164.2$ mm with $s = 6$ mm. Compute a 90 per cent confidence interval for the difference in means, assuming normality. Do not assume that the variances are equal.

7-13 A survey indicated that of 200 health officers queried, 92 favored increased salaries for public health dentists in their community.

    a. Find a 99 per cent confidence interval for the true proportion favoring the issue.

    b. If it is assumed that no one in the population changed his mind when the issue came up at budget time, do the dentists have a chance to buy those extras they have wanted? Discuss.

7-14 As research physician of the cancer institute at your medical school, you set up a study to see if a promising new drug prolongs life in patients with terminal cancer of the throat. Two hundred patients are randomly assigned to a drug or placebo group, 100 to each. Later it is found that:

    Drug: $\bar{x} = 4.6$ months, $s = 2.5$ months.
    Placebo: $\bar{x} = 3.4$ months, $s = 3.1$ months.

Assuming normality,

    a. Find a 95 per cent confidence interval for the difference in means, assuming the variances are equal.

    b. Find a 90 per cent confidence interval for the ratio of standard deviations.

7-15 In the study discussed in Exercise 7-14, how large a sample would be required to estimate the mean prolongation of life in the drug group to the nearest $\frac{1}{2}$ month with 95 per cent confidence if $\sigma = 2.0$ months?

7-16 In the manufacture of a particular type of pipette, it is necessary to maintain a very tight tolerance on the number of air bubbles in the glass. A sampling of pipettes showed an average of 3 bubbles per pipette. Find a 95 per cent confidence interval for μ, the mean of the population of air bubbles in such pipettes.

8   *Hypothesis Testing*

An investigator is often interested in comparing his data against some numerical standard. This standard may be a stated numerical value of a population parameter or a value of the difference or ratio of parameters from 2 populations. For example, in a taxonomic study the lengths of larvae for samples from 2 insect populations might be compared in an attempt to determine whether the populations had the same mean. That is, the investigator might wonder whether the difference between the population means could be zero. He would then examine the data to see whether they were consistent or inconsistent with his hypothesis. Presumably he would take one of 2 possible actions, based upon a decision that the data did or did not support the statement of zero difference. The actions might simply be different approaches to his continuing line of research into insect classification. The important thing to us from a statistical point of view is that he is using data from a sample or samples to support or deny a statement about the population or populations. We say he is testing a statistical hypothesis.

Hypothesis testing can be regarded as an example of a decision process, in which data are assembled in a particular way to produce a quantity that leads to a choice between 2 decisions. Each decision then leads to an action. Because data arise from a sampling process, there is some risk that an incorrect decision will be made and some loss attendant to the resulting incorrect action. In this context, hypothesis testing is an example of a more general study known as decision theory. Many situations in life require a choice between 2 decisions, whether or not the choice is actually made on the basis of data from a sample. The decisions to purchase or not to purchase a particular piece of laboratory equipment, to translate or not to translate a certain research paper from a foreign journal, to market or not to market a particular drug, to operate or not to operate upon a particular surgical candidate, or to initiate or withhold a mass screening program for disease detection among the employees of a large industry are all examples in which a decision must be made between 2 alternatives. Each decision carries a risk of potential loss, ranging from the loss of money spent to purchase a piece of equipment later found unnecessary to the possible loss of life involved in an incorrect decision to release or withold a drug.

Perhaps the most important thing to remember at the outset is that although great quantities of excellent and expensive data may be amassed and although great effort and cost may be involved in computer reduction and analysis of the data, the final decision may of necessity be a simple yes or no. It may be possible to defer the decision by collecting more data, but often a black or white action must be based upon data that are some shade of gray.

The classical approach to hypothesis testing is only one example of a decision process. For an elementary treatment of the general subject of decision theory, see Chernoff and Moses (6).

8-1 Basic Concepts and Definitions

8-1-1 AN EXAMPLE

We will use an artificial example to illustrate the basic ideas of hypothesis testing. Such an example has the advantage that the essential statistical features can be clarified in the absence of the inevitable and interesting complexities that may obscure these features in a more realistic set of data.

Suppose we again find ourselves involved with a Mississippi riverboat gambler like the one described in Sec. 4-2. He proposes a game with a coin produced from his vest pocket. Suspecting that something might be wrong with the coin, we obtain his consent to an initial investigation. In addition to allowing us to inspect the coin superficially, he will let us toss it 10 times and decide either to quit and not enter the serious game which is to follow or to go ahead. The actions available to us are thus to withdraw or to participate in further gambling, and our decision will be based upon the outcome of the initial 10 tosses of the coin.

We set up the statistical hypothesis that the coin is fair and will either accept or reject that hypothesis depending upon the results of the 10 tosses. If we let x be the number of heads in 10 tosses and if we assume that the individual tosses are independent—not an unreasonable assumption, particularly if we do the tossing—then x is binomially distributed. If the hypothesis is true, then p, the binomial parameter here indicating the probability of a head on a single trial, is 0.5.

Before we proceed to toss the coin, it is probably a good idea to set up a detailed plan for reaching a decision. Table A-2 gives the binomial probabilities for the case we are testing: $n = 10, p = 0.5$. We notice that if the coin is fair, 5 heads are more likely than any other single outcome—hardly a surprising result. However, this result only occurs with probability 0.2461, and thus if we insist that exactly 5 heads result in order that we continue playing, we have probability $1 - 0.2461 = 0.7539$ of avoiding a fair game and unfairly accusing our opponent of dishonesty.[1]

However, values of x far enough from 5 will arouse our suspicion that the coin is biased. The values $x = 0$ and $x = 10$ each occur with probability 0.0010 when $p = 0.5$. As a result, if either of these values occurs, we should probably reject the hypothesis and decline to continue. Notice carefully what we are doing in this case. We do not say that the values 0 and 10 could not occur if the coin were true, but rather that a true coin produces each of these results only once in 1,000 repetitions of such an experiment. Furthermore, since these outcomes agree with our notion of the behavior of a biased coin, we feel that, should one of them occur, a value of p other than 0.5 is considerably more likely.

So far we agree to accept the hypothesis if $x = 5$ and to reject it if $x = 0$ or $x = 10$, but what about the other possible outcomes? Suppose we adopt a decision rule that says reject the hypothesis if $x = 0, 1, 9$ or 10? Let us examine the probability of rejecting a true hypothesis under such a rule. If $p = 0.5$, then $P(x = 0$ or 1 or 9 or $10) = 0.0010 + 0.0098 + 0.0098 + 0.0010 = 0.0216$. If the coin is fair, this rule gives probability 0.0216 of rejection.

This probability, the probability of rejecting the tested hypothesis when it is true, is called the *level of significance* of the test. The rule itself, which specifies that we examine x, the number of heads, reject the hypothesis if $x = 0, 1, 9,$ or 10 and accept it otherwise, is called a *test of the hypothesis*. The x values 0, 1, 9, 10 are collectively called the *critical region* or *rejection region* and the remaining values, 2 through 8, are called the *acceptance region*. The quantity x on which the decision of acceptance or rejection is based is called the *test statistic*.

Other decision rules will have other critical regions and other significance levels. For example, if we reject the hypothesis whenever x takes one of the values 0, 1, 2, 8, 9, 10, the level of significance is 0.1094.

We might wish to reject the hypothesis for values of x that are extreme only in one direction. If for example, the game that is to follow the initial experiment consists of our winning one unit when the coin comes up heads and losing one unit when it comes up tails, then we would definitely want to continue playing if there is evidence that p is greater than 0.5. In such a case we would want to include only small values of x in the critical region, since these would tend to indicate a coin unfavorable to us. In that case the

---

[1] We assume that the initial experiment is a perfectly adequate model of the serious game that is to follow. This means, among other things, that the same coin will be used and that it will be tossed in the same way.

values 0, 1, 2 might form the critical region, and the level of significance would become 0.0547. A test whose critical region consists of values of the test statistic extreme in only one direction is called a *one-sided test* or a *one-tailed test*. If the critical region consists of extreme values in both directions, we have a *two-sided test* or *two-tailed test*.

Another aspect of our decision rule requires careful investigation. If the coin is biased, we hope to detect the fact and reject the hypothesis that it is fair. A biased coin might have any particular value of p other than 0.5. Suppose, in fact, that $p = 0.2$. The decision rule we adopted first—reject the hypothesis that $p = 0.5$ if $x = 0$, 1, 9, 10 and accept it otherwise— should be examined in this light. If $p = 0.2$, $P(x = 0,$ or 1 or 9 or 10) $= 0.1074 + 0.2684 + 0.0000 + 0.0000 = 0.3758$, again using Table A-2. This is the probability of rejecting the tested hypothesis when it is false and is called *the power of the test.* Notice that this is the probability of correctly rejecting the hypothesis and that we want this probability to be as large as possible. The statement $p = 0.5$ is called *the tested hypothesis*[2] and the statement $p = 0.2$ *the alternative hypothesis.*

We must again remember that the true value of the parameter p is unknown to us and will remain so. We study the performance of a hypothesis testing procedure by examining probabilities of the individual decisions for various values of p.

The power of the test certainly depends upon the true value of p, i.e., upon the alternative hypothesis. Intuitively, if $p \doteq 0.1$ we would expect a greater probability of rejecting the tested hypothesis, since 0.1 is farther from 0.5 than is 0.2. From Table A-2 we find that if $p = 0.1$, then $P(x = 0$ or 1 or 9 or 10) $= 0.3487 + 0.3874 + 0.0000 + 0.0000 = 0.7361$, a value greater than 0.3758, the rejection probability when $p = 0.2$, supporting our intuition. The probability of accepting the tested hypothesis is 1 minus the probability of rejecting it, since we must either accept or reject the hypothesis. The events "accept" and "reject" are thus complementary events and $P(\text{accept}) + P(\text{reject}) = 1$. From this it follows that if the tested hypothesis is false, $P(\text{accept}) = 1 - P(\text{reject}) = 1 - \text{power}$. In the present case if $p = 0.2$ we have $P(\text{accept}) = 1 - 0.3758 = 0.6242$ and if $p = 0.1$, $P(\text{accept}) = 1 - 0.7361 = 0.2639$. This again agrees with intuition, since the farther the true value of p from the tested value, the smaller the probability of accepting the tested hypothesis.

Summarizing the hypothesis testing problem, we see that it is possible to make 2 distinct types of error:

Type I error: Reject the tested hypothesis when it is true.
Type II error: Accept the tested hypothesis when it is false.

---

[2] Many statisticians use the term *null hypothesis* instead of *tested hypothesis*. We prefer to reserve the former term for hypotheses involving some element of nullity, i.e., hypotheses stating that some quantity is zero.

In the terminology of hypothesis testing, and using the symbols α and β for the 2 error probabilities:

$$P(\text{type I error}) = \alpha = \text{level of significance};$$

$$P(\text{type II error}) = \beta = 1 - \text{power}.$$

Some authors call a type I error an α error and a type II error a β error. Notice that power $= 1 - \beta$.

Of course, we would like both α and β to be as small as possible. We might reduce the size of α by including fewer values of x in the critical region. For example, if we use the critical region $x = 0$, 10 in testing the hypothesis that $p = 0.5$, α is reduced to 0.002. Notice what happens to β. If $p = 0.2$, power $= 0.1074 + 0.0000 = 0.1074$ and $\beta = 1 - 0.1074 = 0.8926$, whereas if $p = 0.1$, power $= 0.3487 + 0.0000 = 0.3487$ and $\beta = 1 - 0.3487 = 0.6513$. Thus, β has increased to 0.8926 and 0.6513 from values of 0.6242 and 0.2639, respectively, when the critical region consisted of $x = 0, 1, 9, 10$. We have reduced α, the probability of a type I error, but β, the probability of a type II error, has increased. It is true, in general, that for a fixed sample size, α and β vary inversely. This result is again in agreement with intuition, since we reduce α by including fewer values of x in the critical region, thereby reducing the probability of rejecting the hypothesis even if it is false. Table 8-1 summarizes our numerical results and includes an additional decision rule and alternative hypothesis for comparison. Notice that the specification of a critical region completely determines a decision rule in that it specifies exactly which values x lead to rejection, the remaining values leading to acceptance.

Table 8-1

PROBABILITIES ASSOCIATED WITH VARIOUS CRITICAL REGIONS FOR TESTING THE HYPOTHESIS $p = 0.5$ FOR $n = 10$

Values of x in Critical Region	Level of Significance α	Alternative Hypothesis	Power	Type II Error Probability β
0, 1, 2, 8, 9, 10	0.1094	$p = 0.3$	0.3844	0.6156
0, 1, 2, 8, 9, 10	0.1094	$p = 0.2$	0.6779	0.3221
0, 1, 2, 8, 9, 10	0.1094	$p = 0.1$	0.9298	0.0702
0, 1, 9, 10	0.0216	$p = 0.3$	0.1495	0.8505
0, 1, 9, 10	0.0216	$p = 0.2$	0.3758	0.6242
0, 1, 9, 10	0.0216	$p = 0.1$	0.7361	0.2639
0, 10	0.0020	$p = 0.3$	0.0283	0.9717
0, 10	0.0020	$p = 0.2$	0.1074	0.8926
0, 10	0.0020	$p = 0.1$	0.3487	0.6513

If we wish to reduce β for fixed α, increasing the sample size will do the job. Intuitively, we feel that if the gambler will let us toss the coin a greater number of times, we shall be better able to detect bias. In the language of hypothesis testing, this says that the power is increased, or equivalently the type II error probability β is decreased. In Table 8-2 we show the effect of increasing the sample size from 10 to 25. The discreteness of the binomial distribution prevents us from matching exactly the earlier levels of significance, but we have chosen critical regions that produce α values as close as possible to those of Table 8-1. The numerical values in Table 8-2 were obtained from an expanded version of Table A-2.

Table 8-2

PROBABILITIES ASSOCIATED WITH VARIOUS CRITICAL REGIONS FOR
TESTING THE HYPOTHESIS $p = 0.5$ FOR $n = 25$

Values of x in Critical Region	Level of Significance α	Alternative Hypothesis	Power	Type II Error Probability β
0–8, 17–25*	0.1078	$p = 0.3$	0.6769	0.3231
0–8, 17–25	0.1078	$p = 0.2$	0.9532	0.0468
0–8, 17–25	0.1078	$p = 0.1$	0.9995	0.0005
0–6, 19–25	0.0146	$p = 0.3$	0.3407	0.6593
0–6, 19–25	0.0146	$p = 0.2$	0.7800	0.2200
0–6, 19–25	0.0146	$p = 0.1$	0.9905	0.0095
0–4, 21–25	0.0009	$p = 0.3$	0.0905	0.9095
0–4, 21–25	0.0009	$p = 0.2$	0.4207	0.5793
0–4, 21–25	0.0009	$p = 0.1$	0.9020	0.0980

* The description of critical region, 0–8, 17–25, indicates that all values of x between 0 and 8 inclusive and 17 and 25 inclusive lead to rejection of the tested hypothesis. The other descriptions in this column are interpreted similarly.

Study of Table 8-2 reveals several important properties of hypothesis testing. Again, as the level of significance decreases, the power against a fixed alternative hypothesis decreases and the type II error probability increases. For a fixed level of significance, as the alternative hypothesis deviates by a greater amount from the tested hypothesis, the power increases and the type II error probability decreases. Comparing Tables 8-1 and 8-2, we see that for comparable levels of significance and a fixed alternative hypothesis the power increases and type II error probability decreases when n is increased. For example, for $n = 10$, $\alpha = 0.1094$ and alternative hypothesis $p = 0.3$, power $= 0.3844$, $\beta = 0.6156$; whereas for $n = 25$, $\alpha = 0.1078$, $p = 0.3$ we find power $= 0.6769$, $\beta = 0.3231$.

The usual practice in hypothesis testing is to consider carefully the nature of the type I and type II errors and on the basis of this consideration to choose an appropriate level of significance. In making the choice, a balance between the 2 error probabilities must be made; recall that the smaller the level of significance, the larger the type II error probability.

In the present problem a type I error is the error of rejecting the hypothesis $p = 0.5$ when, in fact, it is true. In other words, in making such an error we conclude that the coin is biased when in fact it is fair. In this case we will probably decide at the very least to avoid the game proposed to us by the gambler, when in fact it is a fair game. At the worst we might publicly and erroneously accuse the gambler of dishonesty, press legal charges against him, or decide to take matters into our own hands. The consequences to us might range from eventual private or public humiliation to the loss of an expensive suit for defamation of character, or even grave bodily harm. A type II error is made in the present case when we accept the coin to be true and it is in fact biased. The consequence here is possible financial loss. We enter a game that is rigged to the gambler's advantage.

Decisions about the seriousness of the 2 kinds of error and the degree of control to be placed upon each are, of course, largely subjective. It is common in practice to choose a level of significance α equal to 0.05 or 0.01, although other values are used occasionally. In the present example the discreteness of the test statistic prevents us from choosing these exact values. In tests based upon a continuously distributed test statistic this will not be a problem.

In recognition of the fact that different individuals may choose different levels of significance for the same testing situation, it is sometimes useful to quote the level of significance at which the observed value of the test statistic would just have led to rejection of the tested hypothesis. This level is called the P value of the test. In our example suppose that the coin is tossed 10 times and we find $x = 2$; i.e., 2 heads occur. The critical region just barely large enough to include the value 2 consists of the x values 0, 1, 2, 8, 9, 10. As we have seen, the level of significance corresponding to this region is 0.1094. We say $P = 0.1094$. Anyone inspecting a report of our investigation can examine this P value and decide whether it is smaller or larger than the value of α he would have chosen for this test and thus whether he would reject or accept the hypothesis, respectively.

In some instances it is possible to choose a value of α and to specify an alternative hypothesis to which we attach some importance. We may then fix the power $1 - \beta$ against this alternative, thus also determining β, the type II error probability. Once α and β have been fixed, it is ordinarily possible to choose a value of n, the sample size, that will guarantee at least this degree of control over the 2 kinds of error. This, however, often becomes a sufficiently technical operation that the advice of a professional statistician should be sought. In Sec. 8-2-1 we shall carry out such a procedure in a simple case.

We cannot let this opportunity for a bit of sermonizing pass unnoticed. The choice of sample size for fixed α and β is a matter of study design that must be completed before the collection of data begins. We have seen that statistics plays a role at this stage. The appropriate framework for testing a hypothesis entirely precedes data collection. The level of significance, test statistic, critical region, and power may all be set up before a single observation is taken. This is clearly the most important point at which statistical concepts and procedures can operate. A badly designed study cannot be made good after all the data are in. Yet too many investigators expect the statistician to perform some kind of alchemy over a set of data from a study that he has had no part in designing. To say that this limits the statistician's usefulness is a gross understatement. The time for statistics is very early in the evolution of a study, whether the statistical role is to be played by a professional statistician or by an investigator with a grasp of statistical principles. Just take it from us that this role is there and becomes a speaking part in Act 1, Scene 1. The only real issue is whether or not the role is to be adequately filled or allowed to drift until the entire production is endangered.

8-1-2 DEFINITIONS

Here we shall collect the definitions important in hypothesis testing.

Statistical hypothesis. A statement about the population or populations being sampled, or occasionally a statement about the sampling procedure.

Tested hypothesis. A statistical hypothesis the truth of which is to be investigated by sampling; sometimes called the null hypothesis, or, when there is no danger of ambiguity, simply the hypothesis.

Test statistic. A random variable related to the tested hypothesis.

Critical region. A set of values of the test statistic leading to rejection of the tested hypothesis; sometimes called the rejection region.

Significant value of test statistic. A value of the test statistic falling into the critical region; generally when a sample produces a value of the test statistic in the critical region, the results are said to be statistically significant.

Acceptance region. A set of values of the test statistic leading to acceptance of the tested hypothesis, i.e., the values of the test statistic not included in the critical region.

Test of hypothesis. A procedure whereby the truth or falseness of the tested hypothesis is investigated by examining a value of the test statistic computed from a sample and then deciding to reject or accept the tested hypothesis according to whether the value falls into the critical region or acceptance region, respectively.

Type I error. The error of rejecting the tested hypothesis when in fact it is true; sometimes called the α error or error of the first kind.

Type II error. The error of accepting the tested hypothesis when in fact it is false; sometimes called the β error or error of the second kind.

Level of significance. The probability of a type I error; denoted by α.

Alternative hypothesis. A statistical hypothesis that disagrees with the tested hypothesis.

Power. The probability of rejecting the tested hypothesis when it is false, i.e., when an alternative hypothesis is true; denoted by $1 - \beta$, where β is the probability of a type II error.

Two-sided test. A test for which the critical region consists of values of the test statistic extreme in 2 directions; sometimes called a two-tailed test.

One-sided test. A test for which the critical region consists of values of the test statistic extreme in only one direction; sometimes called a one-tailed test.

P value. The level of significance at which the observed value of the test statistic would just be significant, i.e., would just fall into the critical region.

8-2 *Tests for the Parameters of a Normal Distribution*

8-2-1 THE NORMAL MEAN WHEN THE VARIANCE IS KNOWN

Here we are concerned with testing the hypothesis that μ, the mean of a normal distribution, takes the specific numerical value μ_0 when σ^2, the population variance, is known.[3] If $\bar{x}$ is the sample mean based upon a random

---

[3] Notice that μ_0 denotes any specific numerical value of μ. The subscript zero merely indicates that μ_0 is a number such as 10.6, 129, 0.73, etc. Do not be confused into thinking that μ_0 is necessarily equal to zero. After all, x_1 certainly is not necessarily equal to one.

sample of size n, then $z = (\bar{x} - \mu_0)/(\sigma/\sqrt{n})$ follows the standard normal distribution when the hypothesis is true. Furthermore, since μ_0, σ, and n are numerically known, z can be used as a test statistic.

Denoting the tested hypothesis $\mu = \mu_0$ by H, we see that if H is true we would expect $\bar{x}$ to be distributed about the central value μ_0. On the other hand, if H should be false, then we would expect $\bar{x}$ to fall farther from μ_0. But if $\bar{x}$ is far from μ_0, then $\bar{x} - \mu_0$ will be large in numerical value; i.e., $\bar{x} - \mu_0$ will be a large positive number if $\bar{x}$ is above μ_0 or a large negative number if $\bar{x}$ is below μ_0. This implies in turn that $z = (\bar{x} - \mu_0)/(\sigma/\sqrt{n})$ will be numerically large and that the critical region for testing $\mu = \mu_0$ should consist of large positive and large negative values of z.

But how large is large? Suppose we fix the level of significance α. Then the critical region should consist of all those values of the test statistic z so extreme that they would only occur with probability α when the hypothesis is true. Since z is a standard normal variable when H is true, we can choose as critical region values of z above $z_{1-(\alpha/2)}$ or below $z_{\alpha/2}$; these quantities $z_{1-(\alpha/2)}$ and $z_{\alpha/2}$ chop off area $\alpha/2$ from the upper and lower tails of the distribution. Because the values $z_{1-(\alpha/2)}$ and $z_{\alpha/2}$ separate the critical region from the acceptance region, they are sometimes called *critical values*. Figure 8-1 shows the situation graphically. Notice in the figure that the critical

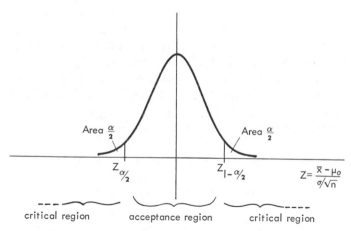

Figure 8-1 **Critical and acceptance regions for testing** $\mu = \mu_0$ **in a normal distribution when** σ^2 **is known.**

region here is two-sided and consists of values of z in both the upper and lower tails of the distribution. Consulting Table A-4, we find that if $\alpha = 0.05$ the critical values are ± 1.96 and if $\alpha = 0.01$ they are ± 2.576.

EXAMPLE

Suppose chest circumference of presumably normal newborn baby girls is normally distributed with $\mu = 13.0$ inches and $\sigma = 0.7$ inches. A group of 25 newborn girls from a population group living in a remote region and thought perhaps to constitute a genetic isolate are studied and found to have an average chest circumference of 12.6 inches. Is this evidence that the group of 25 come from a population with parameter values different from the values $\mu = 13.0$ and $\sigma = 0.7$?

Recasting the problem in our terminology, we shall assume that the 25 observations are a random sample from a normally distributed population. If the population has the same standard deviation as the population of presumably normal infants, then $\sigma = 0.7$. We may test the hypothesis $H:\mu = \mu_0 = 13.0$. Suppose we agree to use a level of significance $\alpha = 0.05$. Then the critical values of the test statistic $z = (\bar{x} - \mu_0)/(\sigma/\sqrt{n})$ are $z_{0.025} = -1.96$ and $z_{0.975} = 1.96$. But for the sample we have found $\bar{x} = 12.6$ and thus

$$z = \frac{12.6 - 13.0}{0.7/\sqrt{25}} = \frac{-0.4}{0.14} = -2.9.$$

Since this value is less than -1.96, it falls into the critical region and we reject the hypothesis. Our conclusion is that the 25 infants cannot (at the 0.05 level of significance) be regarded to be a sample from the same population as the population of presumed normals. In particular, we find the sample mean to be significantly different from the standard value, 13.0 inches. The P value is 0.0037, since this is the area under the normal curve outside ± 2.9, and P is the level of significance at which the observed value, 2.9, would just be significant.

Notice that a 95 per cent confidence interval for the mean of the population from which the sample of 25 is drawn would be $\bar{x} \pm 1.96\,(\sigma/\sqrt{n}) = 12.6 \pm (1.96)(0.14) = 12.6 \pm 0.2744 = 12.3$ and 12.9. The hypothetical value 13.0 does not fall within this interval.

To determine the power of the test of $H:\mu = \mu_0$ against the alternative hypothesis $H_a:\mu = \mu_1$ we must find the probability of rejecting H when in fact H_a is true. But we reject H whenever

$$\frac{\bar{x} - \mu_0}{\sigma/\sqrt{n}} \le z_{\alpha/2} \quad \text{or} \quad \frac{\bar{x} - \mu_0}{\sigma/\sqrt{n}} \ge z_{1-(\alpha/2)}.$$

When H_a is true, then μ_1 is the true population mean and the quantity $(\bar{x} - \mu_0)/(\sigma/\sqrt{n})$ will not follow the standard normal distribution. In this case $(\bar{x} - \mu_1)/(\sigma/\sqrt{n})$ will be standard normal.

EXAMPLE

In the preceding example, in which we were testing the hypothesis that mean chest circumference of a population of newborn girls is 13.0 inches when σ is 0.7 inches, assuming that chest circumference is normally distributed, find the power of the test against the alternative hypothesis that $\mu = \mu_1 = 12.8$, for random samples of size 25 and $\alpha = 0.05$.

The critical region here consists of values:

$$\frac{\bar{x} - 13.0}{0.14} \leq -1.96 \quad \text{or} \quad \frac{\bar{x} - 13.0}{0.14} \geq 1.96.$$

But multiplying both inequalities through by 0.14 and adding 13.0 we find that the critical region can be expressed as

$$\bar{x} \leq 13.0 - (1.96)(0.14) = 12.73 \quad \text{or} \quad \bar{x} \geq 13.0 + (1.96)(0.14) = 13.27.^4$$

Now we must find the probability of this event given that the alternative hypothesis is true, i.e., $\mu = 12.8$; this probability will be the power. If $\mu = 12.8$, then $\bar{x}$ is normal with mean 12.8 and standard deviation $\sigma/\sqrt{n} = 0.7/\sqrt{25} = 0.14$, and $z = (\bar{x} - 12.8)/0.14$ will be a standard normal variable. Therefore, we find that

$$\text{Power} = P(\text{rejecting hypothesis } \mu = 13.0 \mid \mu = 12.8)$$

$$= P(\bar{x} \leq 12.73 \quad \text{or} \quad \bar{x} \geq 13.27 \mid \mu = 12.8)$$

$$= P\left(z \leq \frac{12.73 - 12.8}{0.14} \quad \text{or} \quad z \geq \frac{13.27 - 12.8}{0.14}\right)$$

$$= P(z \leq -0.50 \quad \text{or} \quad z \geq 3.36)$$

$$= P(z \leq -0.50) + P(z \geq 3.36),$$

using the notation for conditional probability presented in Sec. 4-10 and noting that the 2 events connected by "or" are mutually exclusive. But from Table A-4 we find that $P(z \leq -0.50) = 0.3085$ and $P(z \geq 3.36) = 0.0004$, whence power $= 0.3089$. $\beta = P(\text{type II error}) = 1 - \text{power} = 0.6911$.

---

[4] Notice that we have converted the critical values of z to critical values of $\bar{x}$. This shows that $\bar{x}$ could have been used as test statistic for this problem in place of $z = (\bar{x} - \mu_0)/(\sigma/\sqrt{n})$. It is quite common in hypothesis testing to be able to find 2 or more equivalent test statistics for a given hypothesis.

Extending this example, we examine the effect on the power of increasing the sample size from 25 to 100. We shall continue to consider $H_a: \mu = 12.8$ and $\alpha = 0.05$. The critical region now changes to $(\bar{x} - 13.0)/(0.7/\sqrt{100}) \leq -1.96$ or $(\bar{x} - 13.0)/(0.7/\sqrt{100}) \geq 1.96$. If we multiply these inequalities by $0.7/\sqrt{100} = 0.07$ and add 13.0 we obtain the critical region in terms of $\bar{x}$:

$$\bar{x} \leq 12.86 \quad \text{or} \quad \bar{x} \geq 13.14.$$

But

$$\text{Power} = P(\bar{x} \leq 12.86 \quad \text{or} \quad \bar{x} \geq 13.14 \,|\, \mu = 12.8)$$

$$= P(\bar{x} \leq 12.86 \,|\, \mu = 12.8) + P(\bar{x} \geq 13.14 \,|\, \mu = 12.8)$$

$$= P\left(z \leq \frac{12.86 - 12.8}{0.07}\right) + P\left(z \geq \frac{13.14 - 12.8}{0.07}\right)$$

$$= P(z \leq 0.86) + P(z \geq 4.86) = 0.8051 + 0.0000 = 0.8051.$$

$$P(\text{type II error}) = 1 - \text{power} = 0.1949.$$

Thus, increasing the sample size from 25 to 100 has increased the power from 0.3089 to 0.8051. Furthermore, the contribution of the upper tail of the critical region corresponding to $(\bar{x} - 13.0)/0.07 \geq 1.96$ has, to 4 decimal places, become negligible. This reflects the fact that if the true value of μ is 12.8, the probability of rejecting $H: \mu = 13.0$ due to an observed value of the test statistic lying above the upper critical value when $n = 100$ is small—in fact to 4 decimal places is zero. A further increase in sample size will produce an even smaller contribution to the power from the upper tail, and thus to 4-place accuracy for larger samples the upper tail can be ignored when one is considering $H_a: \mu = 12.8$.

Figures 8-2 and 8-3 illustrate the numerical results. In each figure the shaded area corresponds to the power. The curves are the distributions of $\bar{x}$ under H and H_a and the critical regions are given in terms of $\bar{x}$. Study of the figures reveals why the power increases as n increases. When $n = 100$ the standard error of $\bar{x}$ (standard deviation of the sampling distribution of $\bar{x}$) is only half as large as when $n = 25$. This produces a better separation between the distributions of $\bar{x}$ under H and H_a. Clearly we are better able to distinguish between the curves in Fig. 8-3 than in Fig. 8-2. The test is also better able to distinguish between them, and this is reflected by larger power. Notice that only one of the curves can actually be the distribution of $\bar{x}$, depending upon whether in fact H

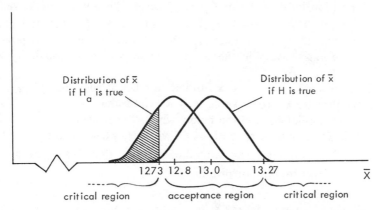

Figure 8-2   **Power of the test of $H:\mu = 13.0$ against $H_a:\mu = 12.8$ when $\sigma = 0.7$, $\alpha = 0.05$, $n = 25$.**

or H_a is true. It is possible or even likely that neither of the distributions is exactly the distribution of $\bar{x}$.

To complete our consideration of this example, suppose that we wish to perform a test of $H:\mu = 13.0$ at the 5 per cent level of significance, but want to guarantee that the test has power 0.90 against the alternative $H_a:\mu = 12.8$. How large a random sample should we take?

From the earlier parts of this example we see that n will be larger than 100, since $n = 100$ only produces power $= 0.8051$ against H_a. Furthermore, this implies that the contribution to power of the upper tail of the critical region can be ignored since even when $n = 100$ that

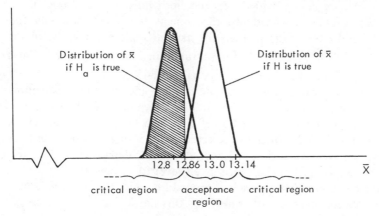

Figure 8-3   **Power of the test of $H:\mu = 13.0$ against $H_a:\mu = 12.8$ when $\sigma = 0.7$, $\alpha = 0.05$, $n = 100$.**

contribution is, to 4 decimal places, zero. The lower critical value is $(\bar{x} - 13.0)/(0.7/\sqrt{n}) = -1.96$ or $\bar{x} = 13.0 - 1.96(0.7/\sqrt{n})$. If $\bar{x}$ is at or below this value we will reject H. But in order to produce a power of 0.90, we must have $\bar{x}$ in the critical region with this probability when $\mu = 12.8$.

If $\mu = 12.8$ we want to find a number u such that $\bar{x} \leq u$ with probability 0.90. From Table A-4 we find that the value from the standard normal distribution below which 90 per cent of the area lies is 1.2816. Thus for the distribution of $\bar{x}$ when H_a is true we find $u = 12.8 + 1.2816(0.7/\sqrt{n})$. But to fix α at 0.05 we saw that this value must be $13.0 - 1.96(0.7/\sqrt{n})$. Thus we have

$$12.8 + 1.2816\left(\frac{0.7}{\sqrt{n}}\right) = 13.0 - 1.96\left(\frac{0.7}{\sqrt{n}}\right)$$

or

$$\left(\frac{0.7}{\sqrt{n}}\right)(1.2816 + 1.96) = 13.0 - 12.8,$$

whence

$$\sqrt{n} = \frac{(0.7)(3.2416)}{0.2} = 11.3456$$

and

$$n = 128.7 \quad \text{or} \quad 129.$$

One-sided tests of $H:\mu = \mu_0$ are formed in a straightforward manner. If it seems appropriate, the entire level of significance can be included in one tail instead of being divided between both tails. For example, if $\alpha = 0.05$ and we want a one-sided test that rejects the hypothesis only for small values of the test statistic, then the critical region should be of the form $(\bar{x} - \mu_0)/(\sigma/\sqrt{n}) \leq -1.645$. For a one-tailed upper tail test the critical region is of the form $(\bar{x} - \mu_0)/(\sigma/\sqrt{n}) \geq 1.645$.

A word of caution is in order here. One-sided tests have sometimes been used to obtain significance that was not available from a corresponding two-sided procedure. Notice that if $\alpha = 0.05$ the critical values for a two-sided test are ± 1.96, whereas the critical value for a one-sided test using the upper-tailed critical region is 1.645. A smaller departure of $\bar{x}$ from μ_0 can thus be declared significant at a given level for a one-sided procedure than for its corresponding two-sided procedure, and this is true under quite general circumstances, not merely for this particular test. In case of doubt it is probably better to perform a two-sided procedure.

In some situations there will be a qualitative difference between departures from the hypothetical value in one direction and departures in the opposite direction. For example, if we are testing for bacteriologic contamination of whole milk, we may have a standard that specifies a value for the average bacterial count. An observed count higher than this value indicates contamination, but a value lower than the standard does not. In this case if we test the hypothesis that a shipment of milk conforms to the standard, it may well be appropriate to reject the hypothesis only for observed values sufficiently larger than the standard, i.e., to perform a one-sided test. Again if the *P* value of a test is quoted and if it is made clear whether that *P* value refers to a one-sided or two-sided procedure, then the reader of a publication presenting results of the investigation can decide whether the results are *for him* significant.

Another general characteristic of significance tests should be mentioned. Notice that the decision to reject the tested hypothesis is a stronger decision than the decision to accept. The error to which we are subject when we reject a tested hypothesis is a type I error, and the probability of this kind of error is controlled at α by adjusting the size of the critical region.[5] When we accept the hypothesis, we are subject to a type II error. The probability of making this kind of error, however, depends upon the alternative hypothesis. We are more likely to make a type II error for an alternative value close to the tested value than for one farther from it.

For this reason, when we accept a hypothesis we do not really mean that we think the hypothesis is literally or mathematically true. Some statisticians prefer the phrase "unable to reject" in place of "accept," because when we accept the hypothesis we are saying that the data do not contradict it—not that the hypothesis is mathematically true. If we consider, for instance, the depressingly large number of values that μ might take, it does not seem very very likely that $\mu = \mu_0$ down to the last millimicron or even far closer than that. The hypothesis test should probably be regarded as a guide to whether μ is sufficiently far from μ_0 to be of importance. A study design taking into account type II error probability or power against an important alternative hypothesis can help in this context.

8-2-2 THE RELATIONSHIP BETWEEN HYPOTHESIS
TESTING AND CONFIDENCE INTERVAL
ESTIMATION

There is ordinarily a close relationship between a test of a hypothesis concerning a parameter or parameters and the corresponding confidence interval. To illustrate this relationship we will examine the $1 - \alpha$ level confidence interval for μ, the mean of a normal distribution with known

[5] Mathematical statisticians often call the level of significance the *size of the test*.

variance σ^2 and sample size n, which we considered in Sec. 7-2-2. That interval had limits $\bar{x} - z_{1-(\alpha/2)}(\sigma/\sqrt{n})$ and $\bar{x} + z_{1-(\alpha/2)}(\sigma/\sqrt{n})$.

Now consider a particular numerical value, say μ_0, between the limits; that is,

$$\bar{x} - z_{1-(\alpha/2)} \frac{\sigma}{\sqrt{n}} < \mu_0 < \bar{x} + z_{1-(\alpha/2)} \frac{\sigma}{\sqrt{n}}.$$

By the opposite sequence of algebraic steps to that used in Sec. 7-2-2 to derive the interval, i.e., reversing the order of terms in this inequality and using $>$ signs instead of $<$ signs, subtracting $\bar{x}$, multiplying by -1, and dividing by $\sigma/\sqrt{n}$, we have the equivalent chain inequality

$$-z_{1-(\alpha/2)} < \frac{\bar{x} - \mu_0}{\sigma/\sqrt{n}} < z_{1-(\alpha/2)},$$

or, recalling that $z_{\alpha/2} = -z_{1-(\alpha/2)}$,

$$z_{\alpha/2} < \frac{\bar{x} - \mu_0}{\sigma/\sqrt{n}} < z_{1-(\alpha/2)}.$$

But from the work of Sec. 8-2-1 this is the form of the acceptance region for testing $H:\mu = \mu_0$ in a two-sided test at α level of significance, for known σ^2 and sample size n. This says that a value μ_0 lying within the a $1 - \alpha$ confidence interval corresponds to a value of the test statistic $(\bar{x} - \mu_0)/(\sigma/\sqrt{n})$ that would lead to acceptance of the hypothesis. If we start with a value of μ_0 lying outside the confidence interval, we find that the corresponding value of the test statistic will fall into the rejection region.

This is an important finding. We have shown that a $1 - \alpha$ level confidence interval for μ contains between its limits all values μ_0 of μ for which $H:\mu = \mu_0$ would have been accepted in a two-sided α level test. Furthermore, all values μ_0 lying outside the interval correspond to hypotheses $\mu = \mu_0$, which would have been rejected. Thus the confidence interval simultaneously provides a decision for *all hypotheses* of the form $\mu = \mu_0$, i.e., for *any* value of μ_0.

Similar results hold for other tests and intervals and in particular for all the hypotheses to be discussed in the remainder of Secs. 8-2 and 8-3. For this reason these tests will be presented in a condensed catalog form. Consequences of violations of assumptions in the tests are similar to those discussed in Chap. 7 for the corresponding confidence intervals. Notation is as defined in Chap. 7.

8-2-3 THE NORMAL MEAN WHEN THE VARIANCE IS UNKNOWN

Tested hypothesis $H:\mu = \mu_0$.

Assumptions: random sample of size n from a normally distributed population.

Test statistic:

$$t = \frac{\bar{x} - \mu_0}{s/\sqrt{n}}.$$

Distribution of test statistic when H is true: Student's t distribution with $n - 1$ degrees of freedom.

Two-sided critical region at level of significance α:

$$t \leq t_{\alpha/2} \quad \text{or} \quad t \geq t_{1-(\alpha/2)}.$$

Lower one-sided, α-level critical region: $t \leq t_\alpha$.

Upper one-sided, α-level critical region: $t \geq t_{1-\alpha}$.

8-2-4 THE NORMAL VARIANCE AND STANDARD DEVIATION

Tested hypothesis $H:\sigma^2 = \sigma_0^2$.

Assumptions: random sample of size n from a normally distributed population.

Test statistic:

$$\chi^2 = \frac{(n - 1)s^2}{\sigma_0^2}.$$

Distribution of test statistic when H is true: chi-square distribution with $n - 1$ degrees of freedom.

Two-sided critical region at level of significance α:

$$\chi^2 \leq \chi_{\alpha/2}^2 \quad \text{or} \quad \chi^2 \geq \chi_{1-(\alpha/2)}^2.$$

Lower one-sided, α-level critical region: $\chi^2 \leq \chi_\alpha^2$.

Upper one-sided, α-level critical region: $\chi^2 \geq \chi_{1-\alpha}^2$.

The hypothesis on the standard deviation $H:\sigma = \sigma_0$ can be immediately converted to the hypothesis $\sigma^2 = \sigma_0^2$ by squaring σ_0. We suggest that the test be carried out in this form.

8-3   Tests for the Parameters
of Two Normal Distributions

In this section we will consider sampling from 2 underlying normally distributed populations with means μ_1 and μ_2 and variances σ_1^2 and σ_2^2, respectively.

8-3-1   DIFFERENCE BETWEEN MEANS
WHEN VARIANCES ARE KNOWN

We shall consider 2 separate hypotheses concerning $\mu_1 - \mu_2$. The first will specify that the difference takes some specific value Δ_0; the second will, because of the frequency of interest in whether the populations have the same mean, specify that $\mu_1 - \mu_2 = 0$, although this is only a special case of the first hypothesis, with $\Delta_0 = 0$.

Tested hypothesis $H: \mu_1 - \mu_2 = \Delta_0$.

Assumptions: independent, random samples of size n_1 and n_2 from 2 normally distributed populations.

Test statistic:

$$z = \frac{(\bar{x}_1 - \bar{x}_2) - \Delta_0}{\sqrt{\dfrac{\sigma_1^2}{n_1} + \dfrac{\sigma_2^2}{n_2}}}.$$

Distribution of test statistic when H is true: standard normal distribution.

Two-sided critical region at level of significance α.

$$z \leq z_{\alpha/2} \quad \text{or} \quad z \geq z_{1-(\alpha/2)}.$$

Lower one-sided α-level critical region: $z \leq z_\alpha$.

Upper one-sided α-level critical region: $z \geq z_{1-\alpha}$.

Tested hypothesis $H: \mu_1 - \mu_2 = 0$.

Assumptions: independent, random samples of size n_1 and n_2 from 2 normally distributed populations.

Test statistic:

$$z = \frac{\bar{x}_1 - \bar{x}_2}{\sqrt{\dfrac{\sigma_1^2}{n_1} + \dfrac{\sigma_2^2}{n_2}}}.$$

Distribution of test statistic when H is true: standard normal distribution.

Two-sided critical region at level of significance α:

$$z \leq z_{\alpha/2} \quad \text{or} \quad z \geq z_{1-(\alpha/2)}.$$

Lower one-sided α-level critical region: $z \leq z_{\alpha}$.

Upper one-sided α-level critical region: $z \geq z_{1-\alpha}$.

The relationship between the two-sided tests of $H: \mu_1 - \mu_2 = 0$ and the corresponding confidence interval of Sec. 7-2-5 is again direct. If the value zero is included between the confidence limits, then the corresponding test produces a decision to accept; otherwise we reject. Again confidence level $1 - \alpha$ corresponds to significance level α.

8-3-2   DIFFERENCE BETWEEN MEANS WHEN VARIANCES ARE UNKNOWN BUT EQUAL

Again, both $H: \mu_1 - \mu_2 = \Delta_0$ and $H: \mu_1 - \mu_2 = 0$ will be discussed individually. The notation is that of Sec. 7-2-5-2. Consequences of violations of assumptions are as discussed in that section.

Tested hypothesis $H: \mu_1 - \mu_2 = \Delta_0$.

Assumptions: independent, random samples of size n_1 and n_2 from 2 normally distributed populations with equal variances, i.e., $\sigma_1^2 = \sigma_2^2 = \sigma^2$.

Test statistic:

$$t = \frac{(\bar{x}_1 - \bar{x}_2) - \Delta_0}{s_p \sqrt{\dfrac{1}{n_1} + \dfrac{1}{n_2}}}.$$

Distribution of test statistic when H is true: Student's t distribution with $n_1 + n_2 - 2$ degrees of freedom.

Two-sided critical region at level of significance α:

$$t \leq t_{\alpha/2} \quad \text{or} \quad t \geq t_{1-(\alpha/2)}.$$

Lower one-sided α-level critical region: $t \leq t_{\alpha}$.

Upper one-sided α-level critical region: $t \geq t_{1-\alpha}$.

Tested hypothesis $H: \mu_1 - \mu_2 = 0$.

Assumptions: independent, random samples of size n_1 and n_2 from 2 normally distributed populations with equal variances, i.e., $\sigma_1^2 = \sigma_2^2 = \sigma^2$.

Test statistic:

$$t = \frac{\bar{x}_1 - \bar{x}_2}{s_p \sqrt{\dfrac{1}{n_1} + \dfrac{1}{n_2}}}.$$

Distribution of test statistic when H is true: Student's t distribution with $n_1 + n_2 - 2$ degrees of freedom.

Two-sided critical region at level of significance α:

$$t \leq t_{\alpha/2} \quad \text{or} \quad t \geq t_{1-(\alpha/2)}.$$

Lower one-sided α-level critical region: $\quad t \leq t_\alpha$.

Upper one-sided α-level critical region: $\quad t \geq t_{1-\alpha}$.

8-3-3   Difference between Means When Variances Are Unknown And Unequal

We again suggest the Welch approximation of Sec. 7-2-5-3. Notation, assumptions, interpretation, and indications for using this procedure in preference to that of the preceding section, where variances are assumed equal, are as in the earlier discussion of this approximation. The 2 hypotheses $\mu_1 - \mu_2 = \Delta_0$ and $\mu_1 - \mu_2 = 0$ are both discussed.

Tested hypothesis $H: \mu_1 - \mu_2 = \Delta_0$.

Assumptions: independent, random samples of size n_1 and n_2 from 2 normally distributed populations with unknown and presumably unequal variances.

Test statistic:

$$t = \frac{(\bar{x}_1 - \bar{x}_2) - \Delta_0}{\sqrt{\dfrac{s_1^2}{n_1} + \dfrac{s_2^2}{n_2}}}.$$

Distribution of test statistic when H is true: Approximately Student's t distribution with

$$\frac{\left(\dfrac{s_1^2}{n_1} + \dfrac{s_2^2}{n_2}\right)^2}{\dfrac{\left(\dfrac{s_1^2}{n_1}\right)^2}{n_1 + 1} + \dfrac{\left(\dfrac{s_2^2}{n_2}\right)^2}{n_2 + 1}} - 2 \quad \text{degrees of freedom.}$$

Two-sided critical region at level of significance α:

$$t \leq t_{\alpha/2} \quad \text{or} \quad t \geq t_{1-(\alpha/2)}.$$

Lower one-sided α-level critical region: $\quad t \leq t_\alpha$.

Upper one-sided α-level critical region: $\quad t \geq t_{1-\alpha}$.

Tested hypothesis $H : \mu_1 - \mu_2 = 0$.

Assumptions: independent, random samples of size n_1 and n_2 from 2 normally distributed populations with unknown and presumably unequal variances.

Test statistic:

$$t = \frac{\bar{x}_1 - \bar{x}_2}{\sqrt{\dfrac{s_1^2}{n_1} + \dfrac{s_2^2}{n_2}}}$$

Distribution of test statistic when H is true: Approximately Student's t distribution with

$$\frac{\left(\dfrac{s_1^2}{n_1} + \dfrac{s_2^2}{n_2}\right)^2}{\dfrac{\left(\dfrac{s_1^2}{n_1}\right)^2}{n_1 + 1} + \dfrac{\left(\dfrac{s_2^2}{n_2}\right)^2}{n_2 + 1}} - 2 \quad \text{degrees of freedom.}$$

Two-sided critical region at level of significance α:

$$t \leq t_{\alpha/2} \quad \text{or} \quad t \geq t_{1-(\alpha/2)}.$$

Lower one-sided α-level critical region: $\quad t \leq t_\alpha$.

Upper one-sided α-level critical region: $\quad t \geq t_{1-\alpha}$.

8-3-4   PAIRING—THE MEAN DIFFERENCE

The discussion in Sec. 7-2-5-4 concerning pairing applies to hypothesis testing as well as to interval estimation. The notation of that section will be used here, and we shall discuss the 2 hypotheses $\mu_d = \Delta_0$ and $\mu_d = 0$.

Tested hypothesis $H : \mu_d = \Delta_0$.

Assumptions: a random sample of n paired differences from a normally distributed population of differences.

Test statistic:

$$t = \frac{\bar{d} - \Delta_0}{\frac{s_{\bar{d}}}{\sqrt{n}}}.$$

Distribution of test statistic when H is true: Student's t distribution with $n - 1$ degrees of freedom.

Two-sided critical region at level of significance α:

$$t \leq t_{\alpha/2} \quad \text{or} \quad t \geq t_{1-(\alpha/2)}.$$

Lower one-sided α-level critical region: $t \leq t_{\alpha}$.

Upper one-sided α-level critical region: $t \geq t_{1-\alpha}$.

Tested hypothesis $H: \mu_d = 0$.

Assumptions: a random sample of n paired differences from a normally distributed population of differences.

Test statistic:

$$t = \frac{\bar{d}}{\frac{s_{\bar{d}}}{\sqrt{n}}}.$$

Distribution of test statistic when H is true: Student's t distribution with $n - 1$ degrees of freedom.

Two-sided critical region at level of significance α:

$$t \leq t_{\alpha/2} \quad \text{or} \quad t \geq t_{1-(\alpha/2)}.$$

Lower one-sided α-level critical region: $t \leq t_{\alpha}$.

Upper one-sided α-level critical region: $t \geq t_{1-\alpha}$.

8-3-5 RATIO OF VARIANCES

The notation, assumptions, and consequences of violations of assumptions are as presented in Sec. 7-2-6. We shall consider both the hypotheses

$$\frac{\sigma_1^2}{\sigma_2^2} = R_0 \quad \text{and} \quad \frac{\sigma_1^2}{\sigma_2^2} = 1, \quad \text{i.e., } \sigma_1^2 = \sigma_2^2.$$

Tested hypothesis:

$$H: \frac{\sigma_1^2}{\sigma_2^2} = R_0.$$

Assumptions: independent, random samples of size n_1 and n_2 from 2 normally distributed populations.

Test statistic:

$$F = \frac{s_1^2}{R_0 s_2^2}.$$

Distribution of test statistic when H is true: F distribution with $n_1 - 1$ numerator degrees of freedom and $n_2 - 1$ denominator degrees of freedom.

Two-sided critical region at level of significance α:

$$F \leq F_{\alpha/2} \quad \text{or} \quad F \geq F_{1-(\alpha/2)}.$$

Lower one-sided α-level critical region: $F \leq F_\alpha$.

Upper one-sided α-level critical region: $F \geq F_{1-\alpha}$.

Tested hypothesis:

$$H: \frac{\sigma_1^2}{\sigma_2^2} = 1.$$

Assumptions: independent, random samples of size n_1 and n_2 from 2 normally distributed populations.

Test statistic:

$$F = \frac{s_1^2}{s_2^2}.$$

Distribution of test statistic when H is true: F distribution with $n_1 - 1$ numerator degrees of freedom and $n_2 - 1$ denominator degrees of freedom.

Two-sided critical region at level of significance α:

$$F \leq F_{\alpha/2} \quad \text{or} \quad F \geq F_{1-(\alpha/2)}.$$

Lower one-sided α-level critical region: $F \leq F_\alpha$.

Upper one-sided α-level critical region: $F \geq F_{1-\alpha}$.

8-4   Tests on Binomial Distributions

8-4-1   Tests of p, the Parameter of a Binomial Distribution

Here we shall discuss only tests based upon the normal approximation to the binomial distribution. The notation of Sec. 7-2-7 is used. We let $\hat{p} = x/n$ = sample proportion of successes, and p_0 is the hypothetical value of p to be tested.

Tested hypothesis $H:p = p_0$.

Assumptions: Sampling is binomial. That is, we have a sequence of independent observations, each resulting in a "success" or "failure," so that the probability of success on a single trial is p.

Test statistic:

$$z = \frac{\hat{p} - p_0}{\sqrt{\dfrac{\hat{p}(1 - \hat{p})}{n}}}.$$

Distribution of test statistic when H is true: approximately standard normal, the approximation improving as n increases.

Two-sided critical region at level of significance α:

$$z \leq z_{\alpha/2} \quad \text{or} \quad z \geq z_{1-\alpha/2}.$$

Lower one-sided α-level critical region: $z \leq z_{\alpha}$.

Upper one-sided α-level critical region: $z \geq z_{1-\alpha}$.

Some statisticians prefer to use the quantity

$$z = \frac{\hat{p} - p_0}{\sqrt{\dfrac{p_0(1 - p_0)}{n}}}.$$

as test statistic in this case, reasoning that if H is true, then $p_0(1 - p_0)/n$ is the variance of $\hat{p}$. This quantity would then be approximately standard normal if H is true. Because of the importance of the connection between confidence intervals and tests, we prefer the test using $\hat{p}(1 - \hat{p})/n$ as the approximate variance. No value of p_0 is available for use in the corresponding confidence interval; the interval contains all values p_0 for which the corresponding hypothesis would be accepted.

8-4-2   TESTS ON THE DIFFERENCE BETWEEN THE PARAMETERS OF TWO BINOMIAL DISTRIBUTIONS

As in Sec. 7-2-8 we let $\hat{p}_1 = x_1/n_1$ and $\hat{p}_2 = x_2/n_2$, the respective proportions of successes in the 2 samples, while p_1 and p_2 are the corresponding population parameters. We consider the hypotheses $p_1 - p_2 = \Delta_0$ and $p_1 - p_2 = 0$.

Tested hypothesis $H: p_1 - p_2 = \Delta_0$.

Assumptions: independent sampling from 2 binomial distributions; that is, n_1 trials from a binomial distribution with parameter p_1 and n_2 trials from a second binomial distribution with parameter p_2, the trials from the first distribution being independent of those from the second.

Test statistic:

$$z = \frac{(\hat{p}_1 - \hat{p}_2) - \Delta_0}{\sqrt{\dfrac{\hat{p}_1(1 - \hat{p}_1)}{n_1} + \dfrac{\hat{p}_2(1 - \hat{p}_2)}{n_2}}}.$$

Distribution of test statistic when H is true: approximately standard normal, the approximation improving as n_1 and n_2 increase.

Two-sided critical region at level of significance α:

$$z \le z_{\alpha/2} \quad \text{or} \quad z \ge z_{1-(\alpha/2)}.$$

Lower one-sided α-level critical region: $z \le z_\alpha$.

Upper one-sided α-level critical region: $z \ge z_{1-\alpha}$.

Tested hypothesis $H: p_1 - p_2 = 0$.

Assumptions: independent sampling from 2 binomial distributions; that is, n_1 trials from a binomial distribution with parameter p_1 and n_2 trials from a second binomial distribution with parameter p_2, the trials from the first distribution being independent of those from the second.

Test statistic:

$$z = \frac{\hat{p}_1 - \hat{p}_2}{\sqrt{\dfrac{\hat{p}_1(1 - \hat{p}_1)}{n_1} + \dfrac{\hat{p}_2(1 - \hat{p}_2)}{n_2}}}.$$

Distribution of test statistic when H is true: approximately standard normal, the approximation improving as n_1 and n_2 increase.

Two-sided critical region at level of significance α:

$$z \leq z_{\alpha/2} \quad \text{or} \quad z \geq z_{1-(\alpha/2)}.$$

Lower one-sided α-level critical region: $z \leq z_\alpha$.

Upper one-sided α-level critical region: $z \geq z_{1-\alpha}$.

There is a rationale for using a different test statistic for the hypothesis $H: p_1 - p_2 = 0$. If the hypothesis is true, then $p_1 = p_2$ and denoting the common value of these 2 parameters by p, we consider how we might best estimate this quantity. A reasonable choice would be the overall proportion of successes in the 2 samples combined. Since there are $x_1 + x_2$ successes out of a total of $n_1 + n_2$ trials, the overall proportion is $\hat{p} = (x_1 + x_2)/(n_1 + n_2)$. This quantity might replace $\hat{p}_1$ and $\hat{p}_2$ in the denominator of the test statistic, producing

$$z' = \frac{\hat{p}_1 - \hat{p}_2}{\sqrt{\dfrac{\hat{p}(1 - \hat{p})}{n_1} + \dfrac{\hat{p}(1 - \hat{p})}{n_2}}}.$$

This statistic is again approximately standard normal if H is true.

In practice the difference between z and z' is not particularly important. It can be shown that if $n_1 = n_2$ then $|z| \geq |z'|$. On the other hand, there is a direct and simple connection between z' and the chi-square statistic for a 2×2 contingency table to be discussed in Sec. 9-4. The analogy with the confidence interval for $p_1 - p_2$ and the direct relation between the test based upon z and that for $H: p_1 - p_2 = \Delta_0$ tends to favor z, but the argument that when $p_1 = p_2$ the common estimator $\hat{p}$ is natural is also impressive. As far as we are concerned, you "pays your money and you takes your choice." If problems like the difference between z and z' are among the largest ones facing you in the treatment of your data, you are indeed fortunate. Be our guest. Use whichever of the 2 test statistics you prefer.

8-5   Tests for Normality of the Underlying Distribution

The assumption that a set of data come from a normally distributed population occurs repeatedly in statistical inference. The consequences of violating the assumption vary (as we have seen in Chaps. 7 and 8) from relatively mild in the case of inferences on means to relatively severe for inferences on variances. A collection of techniques for investigating the probable truth or falsity of this assumption is therefore of interest. The available procedures can be divided into 2 categories—general purpose and special purpose techniques. The general techniques retain some sensitivity to nonnormality of many types, but may be relatively insensitive to departures of a particular type. The special techniques, on the other hand, look only at nonnormality of a specific type and show little or no sensitivity to other kinds of non-normality. The special purpose techniques will be discussed in this section, and one general purpose procedure will be presented in Sec. 9-3.

8-5-1 SKEWNESS AND TESTING FOR IT

The normal distribution has the property of balance or symmetry about its mean value. If we look at the height of the normal curve at a distance of c units above the mean, we find that this height is identical to the height c units below the mean for any value of c. To put this another way, if we draw a straight line connecting the mean with the peak or maximum value of the curve and the fold along this line we find that the 2 sides match exactly. We say that the normal curve is symmetric with axis of symmetry $x = \mu$, the line along which we fold the curve. The normal curve is not the only example of a symmetric distribution. The 4 curves shown in Fig. 8-4 are all symmetric, with axis of symmetry indicated in each case by a dotted line.

A distribution that is not symmetric is said to be asymmetric or *skewed*. Skewed distributions are divided into 2 categories: positively skewed or

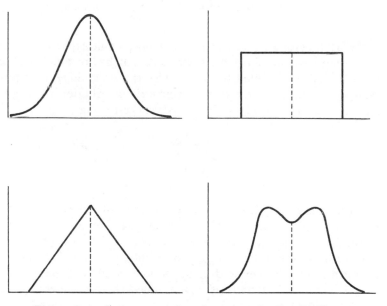

Figure 8-4   Some examples of symmetric distributions.

skewed to the right, and negatively skewed or skewed to the left. A positively skewed distribution has a longer tail to the right or in a positive direction from the mean while the reverse is true for a negatively skewed distribution. Figure 8-5 shows 2 skewed distributions.

As a measure of skewness in a set of observations $x_1, x_2, \ldots, x_n$, the quantity

$$m_3 = \frac{\sum_{i=1}^{n} (x_i - \bar{x})^3}{n}$$

is often used. A similar quantity can be defined as a measure of skewness in the population. The argument that favors using such a measure takes

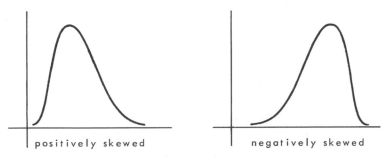

positively skewed   negatively skewed

Figure 8-5   Examples of skewed distributions.

note of the fact that for a symmetric distribution we would expect each observation at a certain distance above the mean to be matched by another observation at the same distance below the mean. That is, we should expect a positive deviation $x_i - \bar{x}$ to be matched in a symmetric distribution by a negative deviation $x_i - \bar{x}$. This would at first glance suggest using

$$\frac{\sum_{i=1}^{n} (x_i - \bar{x})}{n}$$

as a skewness measure. We saw, however, in Sec. 3-4-3 that this quantity is always equal to zero for any values of x_i, even for a skewed distribution, and would not be a satisfactory measure of skewness. The quantity m_3 will, however, be zero for any symmetric distribution and will ordinarily be positive for a positively skewed distribution and negative if the distribution is negatively skewed. The units of m_3 are, however, rather awkward, being cubes of the units of x.

In order to make m_3 unit-free and furthermore to permit the comparison of skewness for distributions having different degrees of dispersion or variability, m_3 is frequently divided by

$$\left(\sqrt{\frac{\sum_{i=1}^{n} (x_i - \bar{x})^2}{n}} \right)^3$$

to produce the standardized measure of skewness,

$$a_3 = \frac{m_3}{m_2^{3/2}},$$

where

$$m_2 = \frac{\sum_{i=1}^{n} (x_i - \bar{x})^2}{n}.$$

Note that

$$m_2 = \frac{(n - 1)s^2}{n}.$$

If a_3 is to be computed from grouped data, the quantities $\sum_{i=1}^{n} (x_i - \bar{x})^2$ and $\sum_{i=1}^{n} (x_i - \bar{x})^3$ are replaced by $\sum_{i=1}^{c} f_i(x_i - \bar{x})^2$ and $\sum_{i=1}^{c} f_i(x_i - \bar{x})^3$, respectively, where the x_i are class midpoints, the f_i are class frequencies, c is the number of classes, and $\bar{x}$ is the grouped mean. If you need to refresh your memory concerning grouped data, take a look at Sec. 3-7 again.

Since the normal distribution is symmetric rather than skewed, one specific way in which a distribution may be nonnormal is by being positively or negatively skewed. A test sensitive to skewness would then investigate that specific kind of normality.

Table A-10 shows the approximate 95th and 99th percentiles of the distribution of a_3 for random samples from a normally distributed population. The sample must have $n \geq 25$ because of the nature of the mathematical methods used in deriving the percentiles. Since the distribution of a_3 for samples from a normal distribution is symmetric about zero, it follows that the 1st and 5th percentiles are the negatives of the 99th and 95th percentiles, respectively. That is, $(a_3)_{0.01} = -(a_3)_{0.99}$ and $(a_3)_{0.05} = -(a_3)_{0.95}$, where $(a_3)_\alpha$ denotes the 100α percentile of the distribution of a_3 for random samples from a normal distribution.

An upper one-sided skewness test of the hypothesis, H: the observations are a random sample from a normally distributed population, using a_3 as test statistic and level of significance 0.05 or 0.01 can be carried out quite simply. The quantity a_3 is calculated and compared with the appropriate critical value from Table A-10, $(a_3)_{0.95}$ being used for a 5 per cent level test and $(a_3)_{0.99}$ for a test at the 1 per cent level. If a_3 is greater than or equal to the tabulated critical value, then H is rejected; otherwise it is accepted. Lower one-sided skewness tests at the 5 per cent or 1 per cent level are available by affixing minus signs to the tabulated upper percentiles and using the resulting negative quantities as critical values. For two-sided tests, Table A-10 can be used for 2 per cent and 10 per cent level procedures. The critical region for a two-sided 2 per cent level skewness test consists of values of a_3 such that $a_3 \leq (a_3)_{0.01}$ or $a_3 \geq (a_3)_{0.99}$. The 10 per cent critical region is $a_3 \leq (a_3)_{0.05}$ or $a_3 \geq (a_3)_{0.95}$. Remember that these tests are designed to investigate nonnormality only in terms of skewness and are thus special purpose procedures.

EXAMPLE

Test the 195 heights presented in Table 6-3 for normality, using the skewness test.

We shall perform a two-sided 10 per cent level test and must calculate a_3, using the procedures for grouped data. We found in Sec. 6-4-5 that the grouped mean and standard deviation of these heights were 69.47 inches and 2.8164 inches, respectively.

Notice that

$$m_2 = \frac{\sum_{i=1}^{c} f_i(x_i - \bar{x})^2}{n} = s^2\left(\frac{n-1}{n}\right) = (2.8164)^2\left(\frac{194}{195}\right)$$

and

$$m_2^{3/2} = (2.8164)^3(\tfrac{194}{195})^{3/2} = (22.3400)(0.99487)^{3/2}$$

$$= (22.3400)(\sqrt{0.99487})^3 = (22.3400)(0.99743)^3$$

$$= (22.3400)(0.99231) = 22.1682.$$

Now

$$\sum_{i=1}^{c} f_i(x_i - \bar{x})^3 = 2(62.5 - 69.47)^3 + 16(64.5 - 69.47)^3 + \cdots$$

$$+ 11(74.5 - 69.47)^3 + 1(76.5 - 69.47)^3$$

$$= -677.218 - 1964.215 + \cdots + 1399.899 + 347.429$$

$$= -586.497,$$

and

$$m_3 = \frac{\sum_{i=1}^{c} f_i(x_i - \bar{x})^3}{n} = \frac{-586.497}{195} = -3.0077.$$

$$a_3 = \frac{m_3}{m_2^{3/2}} = \frac{-3.0076}{22.1682} = -0.1357.$$

From Table A-10 the critical values for $n = 200$ are ± 0.280. We might interpolate on n to find more exact values, but since our value of a_3 is well within the acceptance region for both of the closest tabulated values, i.e., $n = 175$ and $n = 200$, there is no point to such an interpolation in this case. We accept the hypothesis of normality based on the skewness test. With respect to the P value of the procedure, we can say $P > 0.10$.

8-5-2   KURTOSIS AND TESTING FOR IT

Another characteristic of distributions that is of particular relevance to inferences on variances and standard deviations is *kurtosis*. This property has been alluded to in Secs. 7-2-4 and 7-2-6, but we will now discuss it in detail. Roughly speaking, the kurtosis of a distribution describes its degree of peakedness or, more exactly, its peakedness relative to the length and size of its tails. A distribution that is relatively flat and has short tails is of low kurtosis and is said to be *platykurtic*. A distribution with a sharp peak and long tapering tails is of high kurtosis and is termed *leptokurtic*. The normal distribution is used for reference or comparison, is thus considered to be of intermediate kurtosis, and is said to be *mesokurtic*. Figure 8-6 shows 3 symmetric distributions with different grades of kurtosis.

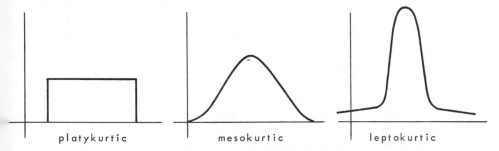

platykurtic   mesokurtic   leptokurtic

Figure 8-6   Symmetric distributions of varying kurtosis.

The quantity

$$m_4 = \frac{\sum\limits_{i=1}^{n}(x_i - \bar{x})^4}{n}$$

is sometimes used as a measure of kurtosis. As in the case of the skewness measure, this quantity is frequently made unit-free by dividing by

$$m_2^2 = \left[\frac{\sum\limits_{i=1}^{n}(x_i - \bar{x})^2}{n}\right]^2$$

to produce the quantity $a_4 = m_4/m_2^2$. The population parameter analogous to a_4 takes the value 3 for the normal distribution, higher values for lepto-kurtic distributions, and lower for platykurtic distibutions. Unfortunately, no adequate approximation to the distribution of a_4 for random samples from a normal distribution is available for $n < 50$.

R. C. Geary (21) proposed an alternative measure that does not have this defect. His measure is

$$g = \frac{\sum\limits_{i=1}^{n}|x_i - \bar{x}|}{s\sqrt{n(n-1)}},$$

where $|x_i - \bar{x}|$ is the absolute value of $x_i - \bar{x}$, obtained by discarding the sign and considering the difference to be positive. The analogous parameter for a normally distributed population takes the value 0.7979; smaller values of g indicate leptokurtosis, and larger values show platykurtosis— just the reverse of a_4.

Table A-11 gives 6 percentiles for the distribution of g for random samples from a normally distributed population, i.e., when the hypothesis, H: the observations are a random sample from a normally distributed population, is true. These percentiles permit upper or lower one-sided kurtosis tests of H at the 1 per cent, 5 per cent, or 10 per cent levels of significance and two-sided tests at the 2 per cent, 10 per cent, or 20 per cent levels. For example, the critical region for an upper 5 per cent level one-sided test is $g \geq g_{0.95}$, while the critical region for a two-sided 10 per cent level test is $g \leq g_{0.05}$ or $g \geq g_{0.95}$.

The heavy dependence of inferential procedures for variances and standard deviations upon the assumption of normality, particularly with respect to kurtosis, makes this an important and useful test.

EXAMPLE

Test the 195 heights of Table 6-3 for nonnormal kurtosis.

Since the data are grouped, we will use the grouped version of the formula for g—namely,

$$g = \frac{\sum_{i=1}^{c} f_i |x_i - \bar{x}|}{s\sqrt{n(n-1)}}.$$

We use the grouped mean and standard deviation of these data, calculated in Sec. 6-4-5 to be $\bar{x} = 69.47$ inches and $s = 2.8164$ inches. Now

$$\sum_{i=1}^{c} f_i |x_i - \bar{x}| = 2|62.5 - 69.47| + 16|64.5 - 69.47| + \cdots$$

$$+ 11|74.5 - 69.47| + 2|76.5 - 69.47|$$

$$= 2(6.97) + 16(4.97) + \cdots + 11(5.03) + 1(7.03) = 459.09$$

$$\sqrt{n(n-1)} = \sqrt{(195)(194)} = 194.50;$$

$$g = \frac{459.09}{(2.8164)(194.50)} = 0.8381.$$

If we perform a two-sided 10 per cent level test, then the critical region for $n = 201$ is $g \leq 0.7738$ or $g \geq 0.8220$, and we reject the hypothesis. Again interpolation is unnecessary, since the same decision would be reached at the next lower sample size, $n = 101$. We conclude that there is evidence of nonnormal kurtosis in the direction of platykurtosis. The P value is $P < 0.02$, indicating that we would even have rejected the hypothesis (two-sided test) at the 2 per cent level of significance.

E X E R C I S E S

In these exercises, give the P value for each hypothesis you test.

8-1   A certain breed of rat shows a mean weight gain of 65 grams during the first 3 months of life. Sixteen of these rats were fed a new diet from birth until age 3 months. These 16 had a mean $\bar{x} = 60.75$ and a standard deviation $s = 3.84$.

a. Is there reason to believe, at the 5 per cent level of significance, that the new diet causes a change in the average amount of weight gained?
b. State the assumptions.
c. Suppose that the population variance is known to be $\sigma^2 = 10.00$ and is assumed to be the same for the new diet; recompute your test statistic and again test the hypothesis $H: \mu = 65$.

8-2 As a geneticist interested in human populations, you have been studying growth patterns in United States males since 1900. A monograph written in 1902 states that the mean height of adult United States males is 67.5 inches with a standard deviation of 3.2 inches. Wishing to compare these values with the current measures, you draw a random sample of 100 adult males and find: $\bar{x} = 69.2$ inches and $s = 4.1$ inches.

a. Test the hypothesis that the current mean height has not changed from the 1902 value, i.e., $H: \mu = 67.5$, using as standard deviation $\sigma = 3.2$.
b. Repeat this test, but use the current sample standard deviation of $s = 4.1$ instead.
c. Test the hypothesis that the current variance is the same as that of 1902, i.e., $H: \sigma^2 = (3.2)^2$.

8-3 A particular drug company claims that its headache remedy stops the average headache in 14 minutes. Being skeptical, you randomly select 25 headache patients from the outpatient clinic, asking them to take one of these pills when they have a headache and record the length of time (in minutes) until it disappears. The results of your study are

$$\bar{x} = 19 \text{ minutes}, \qquad s = 7 \text{ minutes.}$$

Do the results of your study reinforce or disagree with the drug company's claim?

8-4 As a microbiologist you want to obtain microscope slides of uniform thickness. One company claims that its slides have a very small variance, in fact, $\sigma^2 = 0.0121$ micron^2. Using a sensitive micrometer, you randomly sample 50 slides with a resultant sample variance of $s^2 = 0.0213$ micron^2. Is the company's quoted value consistent with your data? State your assumptions.

8-5 As one part of a study of the ability of dental cements to hold single crowns, 2 cements were tested on 50 different tooth casts each. The amount of force in foot pounds required to pull each cemented crown from the casting was reported:

Cement 1: $\bar{x} = 45$ ft lb; $s = 6.2$ ft lb; $n = 50$.
Cement 2: $\bar{x} = 42$ ft lb; $s = 5.8$ ft lb; $n = 50$.

a. Test the hypothesis $H: \mu_1 = \mu_2$, assuming $\sigma_1^2 = \sigma_2^2$.
b. What other assumptions are made?
c. Perform the test, using $\sigma_1 = 4.1$ and $\sigma_2 = 3.4$.

8-6 Twenty-three yearling sheep fed pingue (a toxin-producing rubber-weed), were studied by Aanes (1). Thirteen died, while the remaining 10 survived.

Their weights (in pounds) at the time of introduction of the pingue into their diets were

Died	Survived
46, 55, 61, 75, 64, 75, 71,	57, 62, 67, 45, 73,
59, 64, 67, 60, 63, 66	74, 57, 66, 64, 74

a. Test the hypothesis $H:\mu_1 = \mu_2$, using the assumption that $\sigma_1^2 = \sigma_2^2$.
b. Repeat your test, but without assuming $\sigma_1^2 = \sigma_2^2$.
c. What assumptions were made in parts a and b?
d. Discuss the basis for choosing between the tests in parts a and b. Which test would you use on these data in practice?

8-7 Suppose one wishes to see whether aspirin and a buffered product are equally effective in alleviating symptoms accompanying influenza. Length of time (in minutes) from taking the drug to patient's saying he feels improved is recorded with results as follows:

Aspirin	Buffered Product
$\bar{x} = 15.2$	$\bar{x} = 13.4$
$s = 8.7$	$s = 6.9$
$n = 10$	$n = 20$

a. State and test the appropriate hypothesis.
b. List all the assumptions used.

8-8 Suppose 2 cholesterol determination techniques are thought to be equally variable. Samples of sizes $n_1 = 25$ and $n_2 = 61$ yield standard deviations of $s_1 = 8.4$ units and $s_2 = 11.5$ units.

a. Test the hypothesis $H:\sigma_1^2 = \sigma_2^2$.
b. List all assumptions.

8-9 Standing and supine systolic blood pressures are taken on 12 subjects with results as follows:

Subject	Standing	Supine
1	132	136
2	146	145
3	135	140
4	141	147
5	139	142
6	162	160
7	128	137
8	137	136
9	145	149
10	151	158
11	131	130
12	143	150

a. Using the appropriate statistic, test the hypothesis that the means for standing and supine pressure are the same.
b. Should the test statistic have 22, 11, or some other number of degrees of freedom? Defend your answer.
c. State your assumptions.

8-10 In a dental study suppose it is hypothesized that 90 per cent of all 4-year-old children give no evidence of dental caries. In a study of 100 children, 82 per cent gave no such evidence. On the basis of these data, would you accept the quoted hypothetical value of 90 per cent?

8-11 Use the normal approximation to the binomial distribution to calculate approximate significance levels for the 3 critical regions of Table 8-1.

8-12 Using the normal approximation to the binomial distribution, check all probabilities in Table 8-2. (Recall the rule stated in Chap. 6 that np and nq should both be greater than 5 for the approximation to be adequate.)

8-13 Two communities were sampled to learn their positions concerning fluoridation prior to campaigns' being launched. The results of these surveys (p being proportion favorable) were
Community 1: $n = 110$, $p = 0.52$.
Community 2: $n = 75$, $p = 0.55$.
Do you feel that it is unlikely that the 2 communities have equal proportions of their populations favoring fluoridation?

8-14 Consider the following frequency distribution of 52 heart rates.

Class Boundaries	Frequency
54–62	0
62–70	6
70–78	6
78–86	16
86–94	14
94–102	4
102–110	4
110–118	2
118–126	0

Are these data consistent with an underlying normal distribution with respect to

a. Skewness?
b. Kurtosis?

9

Chi-square Tests
for
Frequency Data

9-1 General Background and the Basic Chi-square Statistic

We have seen that observational data frequently take the form of counts or frequencies. Distributions like the binomial and Poisson are distributions of random variables taking nonnegative whole-number values, arising as counts of various kinds of occurrences. Whenever we classify data from a sample into a frequency table (Sec. 3-7-1), we are reducing the observations to counts of the number of occurrences in each class. In this way, even observations from an underlying continuous distribution can be presented as counts or frequencies. An example is the set of 195 heights in Table 6-3.

Denoting the set of observed frequencies by $O_1, O_2, \ldots, O_k$, we find that it is frequently of interest to compare these frequencies with a set of expected frequencies $E_1, E_2, \ldots, E_k$. The expected frequencies arise from a hypothetical model under study. But do not let the words "hypothetical model" intimidate you. They might, for example, refer to a statement that a die to be used in a gambling game is fair, in which case we would expect about $\frac{1}{6}$ of the total number of throws of the die to fall into each of the 6 outcome categories numbered from 1 to 6. The expected frequencies might arise from a genetic model that specifies the proportions with which various phenotypes occur in the offspring of a particular type of mating. In such a case we would be interested in learning whether the observed phenotypic frequencies conform to the expected pattern or vary significantly from it. We might have a random sample of observations classified into a frequency table, and we might want to know whether the observed frequencies in the classes are in general agreement with the frequencies that would be expected if we were sampling from an underlying normal distribution having the same mean and variance as the data. We might be conducting a clinical trial to study the effect of several forms of treatment on a particular disease and classify each patient into one of several outcome categories. We would then wonder whether outcome was independent of treatment, or, in other words, whether the different treatments produced identical distributions of outcomes in the population from which the patients were selected. The observed frequencies of patients in the various treatment and outcome categories would then be compared with expected frequencies calculated under the hypothesis of independence.

The basic inferential problem to be considered in this chapter is to test the hypothesis H: The k observed frequencies $O_1, O_2, \ldots, O_k$ conform to a model that would produce expected frequencies $E_1, E_2, \ldots, E_k$. In other words, if H is true the observed frequencies differ from the expected frequencies only because of sampling variation and not because the model is false. Notice that, unlike the observed frequencies, the E_i's need not be whole numbers. They represent the long-run or average behavior of the class frequencies under repeated sampling from the underlying population. The test statistic for investigating this hypothesis will be of the following form:

$$\chi^2 = \sum_{i=1}^{k} \frac{(O_i - E_i)^2}{E_i}$$

Professor Karl Pearson showed in 1900 (41) that if the O_i are mutually independent random variables and if the E_i are all sufficiently large, then this statistic is approximately distributed according to the chi-square distribution when H is true. The number of degrees of freedom was a subject of lively controversy for some time until R. A. Fisher (17) showed that the correct value was $k - 1 - m$, where m is the number of parameters in the model, which must be estimated from the data. Thus the maximum number of degrees of freedom is $k - 1$, obtained when $m = 0$.

The critical region for a test at the α level of significance is readily determined. If H is false, we expect disagreement between the observed and expected frequencies. This will tend to make some of the terms $(O_i - E_i)$ large in absolute value; i.e., these differences will either be large positive or large negative numbers. The corresponding quantities $(O_i - E_i)^2$ will be large and positive, producing a relatively large positive value of χ^2. The critical region must then consist of large positive values of χ^2. Thus, and this is important, chi-square frequency tests *are inherently upper one-sided tests.*[1]

---

[1] The word "frequency" is important. In Chap. 8 we introduced two-sided tests with chi-square test statistics. Here we are referring only to those tests based upon the statistic

$$\chi^2 = \sum_{i=1}^{k} \frac{(O_i - E_i)^2}{E_i}$$

as chi-square frequency tests.

The critical value for an α-level test is $\chi^2_{1-\alpha}$, the $100(1 - \alpha)$ percentile of the chi-square distribution with $k - 1 - m$ degrees of freedom, and the critical region consists of values of χ^2 such that $\chi^2 \geq \chi^2_{1-\alpha}$.

We assume that the observations consist of a random sample of size n from some population, with each observation classified into one of k mutually exclusive and exhaustive classes. The observed frequencies are then counts of the numbers of observations falling into each of the classes, and random sampling guarantees that they are mutually independent. The independence assumption is crucial; although many users of chi-square frequency tests think of them as essentially assumption-free, misapplications can arise from violating the criterion of independence. This will be discussed in more detail in Sec. 9-5. Since the k classes are mutually exclusive and exhaustive, it follows that the sum of observed frequencies is n; i.e., $\sum_{i=1}^{k} O_i = n$. Furthermore, $\sum_{i=1}^{k} E_i = n$. To see this, notice, for example, that it is not sensible, in testing whether a die is true, on the basis of 60 tosses of the die, to expect 20 outcomes in each of the 6 categories. A true die should produce an expected frequency of 10 ($\frac{1}{6}$ of the total outcomes) in each outcome class, and the sum of expected frequencies should equal 60, the total number of tosses.

Since

$$\sum_{i=1}^{k} O_i = \sum_{i=1}^{k} E_i = n,$$

a computing formula for χ^2 is available.

$$\chi^2 = \sum_{i=1}^{k} \frac{(O_i - E_i)^2}{E_i} = \sum_{i=1}^{k} \frac{(O_i^2 - 2O_i E_i + E_i^2)}{E_i}$$

$$= \sum_{i=1}^{k} \left(\frac{O_i^2}{E_i} - 2O_i + E_i \right) = \sum_{i=1}^{k} \frac{O_i^2}{E_i} - 2 \sum_{i=1}^{k} O_i + \sum_{i=1}^{k} E_i$$

$$= \sum_{i=1}^{k} \frac{O_i^2}{E_i} - 2n + n = \sum_{i=1}^{k} \frac{O_i^2}{E_i} - n.$$

That is,

$$\chi^2 = \sum_{i=1}^{k} \frac{O_i^2}{E_i} - n.$$

On many desk calculators it is possible to calculate $\sum_{i=1}^{k} (O_i^2/E_i)$ in one continuous machine operation, copying nothing but the final value. We then subtract n from this value, obtaining χ^2.

We have remarked that expected frequencies must not be too small, but how small is too small? Sampling experiments by W. G. Cochran (7) suggest that some previous rules of thumb are unnecessarily and even dangerously severe in terms of increased type II error probability. Unfortunately, no single rule of thumb is completely adequate to cover all chi-square frequency tests, and we will state appropriate rules as each test is discussed. Notice that it is the size of expected frequencies rather than observed frequencies which determines the adequacy of the chi-square approximation. Although there may be an overall relationship between the size of the O_i and E_i, only the E_i directly affect the approximation.

9-2   The Case of All Expected Frequencies Specified Prior to Sampling

The die-tossing example discussed in the preceding section is an example of the kind of situation to be treated here. In that case, we wish to test the hypothesis that the die is true on the basis of a sample of n independent tosses. Since a true die has outcome probability $\frac{1}{6}$ for each of the outcomes, $1, 2, \ldots, 6$, we expect a frequency of $\frac{1}{6}$ times n or $n/6$ in each of the categories. Notice that as soon as n has been specified, the hypothesis permits calculation of all the expected frequencies prior to tossing the die. This feature characterizes the test of such a hypothesis. No parameters require estimation from the data; therefore, $m = 0$, and the test statistic has $k - 1$ degrees of freedom.

The term "degrees of freedom" can now be clarified. It refers to the number of expected frequencies that may be freely specified in the test statistic. Since $\sum_{i=1}^{k} E_i = n$, it follows that when $k - 1$ expected frequencies have been specified, the remaining expected frequency is then n minus the sum of the $k - 1$ specified expected frequencies. That is, we have $k - 1$ degrees of freedom in specifying the expected frequencies. Although we have considered the chi-square distribution in Chaps. 7 and 8 and have used this terminology there, the techniques being discussed in this chapter were discovered and used earlier and account for the expression, "degrees of freedom."

As a rule of thumb concerning minimum size of expected frequencies for this test, we suggest the following: *No expected frequency should be less than 1, and not over 20 per cent of expected frequencies should be less than 5.* If the rule is not satisfied, classes should be combined or otherwise redefined until the resulting E_i are of adequate size.

We shall consider a genetic example to illustrate the case of all expected frequencies specified before sampling.

EXAMPLE

A genetic model specifies that under a particular mating pattern, 4 phenotypes should occur in $9:3:3:1$ ratio (normal dihybrid segregation). Classification of 160 offspring produces observed frequencies of $100, 20, 35, 5$ in the respective classes. Are these frequencies consistent with the model?

The model states that the probabilities that an offspring will fall in each of the classes are $\frac{9}{16}$, $\frac{3}{16}$, $\frac{3}{16}$, and $\frac{1}{16}$, respectively. The expected frequencies will be $\frac{9}{16}(160)$, $\frac{3}{16}(160)$, $\frac{3}{16}(160)$, and $\frac{1}{16}(160)$, or $90, 30, 30$, and 10. Notice that these frequencies can be determined without knowledge of any observed frequencies, once the total number of offspring to be studied is fixed. Also, notice that the expected frequencies are of adequate size for the chi-square approximation. We will take $\alpha = 0.05$. The number of degrees of freedom is $k - 1 = 4 - 1 = 3$.

$$\chi^2 = \sum_{i=1}^{k} \frac{O_i^2}{E_i} - n = \frac{(100)^2}{90} + \frac{(20)^2}{30} + \frac{(35)^2}{30} + \frac{(5)^2}{10} - 160$$

$$= 167.77 - 160 = 7.77.$$

The critical value for 3 degrees of freedom is $\chi^2_{0.95} = 7.81$. The observed result, therefore, barely falls into the acceptance region. Since $\chi^2_{0.90} = 6.25$ and since $6.25 < 7.77 < 7.81$, we have $0.05 < P < 0.10$.

9-3   The Goodness of Fit Test for the Normal Distribution

In Chap. 8, specific tests for normality of an underlying distribution were discussed. These tests were sensitive to skewness and kurtosis. In this section we consider a general purpose procedure for testing any kind of departure from normality. Since the procedure is of an omnibus character, it cannot be expected to be highly sensitive to a particular type of non-normality, as were the procedures of Chap. 8. However, it can provide an overall assessment of the adequacy of fit of the normal distribution to a set of data grouped into a frequency table.

The n observations in the sample are classified into k classes, and a normal distribution is fitted to the data by using the procedures of Sec. 6-4-5. This procedure produces expected frequencies for each class in the table from a normal distribution having the same mean and variance as the sample. The hypothesis tested is H: The observations are a random sample from an underlying normal distribution. The distribution of the test statistic χ^2 when H is true is the chi-square distribution with $k - 1 - m$ degrees of freedom. Here, $m = 2$ since 2 parameters, the mean and variance (or standard deviation), must be estimated from the data. Thus there are $k - 1 - m = k - 3$ degrees of freedom. We shall carry through the test procedures for the 195 heights of Sec. 6-4-5.

In this procedure too conservative a rule of thumb concerning size of expected frequencies can be particularly dangerous. Departures from normality are frequently most pronounced in the tails of a distribution. This is also the region in which probabilities and therefore expected frequencies get small. In order to retain sensitivity to tail shape Cochran (7) suggests adoption of the rule: *No expected frequency should be less than 1.* Again, classes should be redefined, if necessary, adjacent classes being combined until the rule is satisfied, but with no more combining than is absolutely necessary to satisfy the rule. When one is combining adjacent classes the corresponding observed frequencies can, of course, merely be added to produce the new observed frequency, but this also applies to the expected frequencies. The resulting class intervals may be of different lengths, but this is of no disadvantage in carrying out the chi-square test.

EXAMPLE

Test the goodness of fit of the normal distribution to the set of 195 heights considered in Sec. 6-4-5, using the chi-square frequency test.

In Table 6-4 the expected frequencies are given as 3.32, 12.15, 31.73, 51.09, 50.74, 31.08, 11.74, and 3.16, corresponding to observed frequencies of 2, 16, 30, 48, 48, 39, 11, and 1, respectively.

The test statistic χ^2 will be compared with 11.07, the 95th percentile of the chi-square distribution with $k - 1 - 2 = 8 - 3 = 5$ degrees of freedom for a 5 per cent level test.

$$\chi^2 = \sum_{i=1}^{k} \frac{O_i^2}{E_i} - n = \frac{(2)^2}{3.32} + \frac{(16)^2}{12.15} + \cdots + \frac{(1)^2}{3.16} - 195$$

$$= 200.71 - 195 = 5.71.$$

Since $\chi^2 < \chi^2_{0.95}$, we accept the hypothesis that the data fit the normal distribution, and $0.30 < P < 0.40$. Recall that in Secs. 8-5-1 and 8-5-2 we tested the same data for normality, using the special purpose tests for skewness and kurtosis. Although there was no evidence of skewness, the observations did appear to be platykurtic. This effect is seen here as a depression of the observed frequencies relative to the corresponding expected frequencies for class intervals near the center of the distribution, as well as some tendency for tail observed frequencies to be too low. Since observed and expected frequencies must both total n, or 195 in this case, it follows that observed frequencies in some classes between the center and the tails must be slightly elevated, an effect that is also apparent. This example shows quite clearly that the chi-square frequency test, although possibly sensitive to severe nonnormality, will not necessarily detect special types of aberrations. In particular, if normality is being tested prior to inferential procedures on variances or standard deviations, a special purpose kurtosis test such as Geary's test (Sec. 8-5-2) is needed.

9-4 *Tests for Association in Contingency Tables*

Data in the form of frequencies are often cross-classified into tables with row and column classification systems. Such a cross-classification of a set of observed frequencies is called a contingency table. The example discussed earlier in this chapter, in which patients in a clinical trial are jointly classified by type of treatment and outcome, will produce this type of data. In an investigation in which a stratified random sample is selected and individuals in the sample are classified as participants or nonparticipants, we might wish to determine whether rate of nonparticipation is related to stratum. A cross-classification of sample subjects by stratum and response status produces a contingency table. A study of a possible relationship between fertility and blood groups might cross-classify women over age 45 by ABO group and number of pregnancies, yielding still another contingency table.

One of the most frequently tested hypotheses in statistical methods applied to biology is the hypothesis that states that in the population there is no association between the row and column classification variables. This hypothesis can be stated in several equivalent ways, as follows:

1. There is no association between the row and column classification variables.
2. The 2 classification variables are independent.
3. The probability that an observation falls into row i and column j equals the probability that it falls into row i times the probability that it falls into column j (see Sec. 4-10).
4. The population distribution of frequencies across the columns is the same within each row, i.e., does not depend upon the row.
5. Same as 4, but interchange the words row(s) and column(s).

In terms of the 3 examples, the hypothesis would state that outcome is unrelated to type of treatment, that population nonparticipation rate is the same in all strata, and that fertility as measured by number of pregnancies is unrelated to ABO blood group.

Notice that the contingency table hypothesis is truly a null hypothesis in that it is a statement of no association between rows and columns. It is frequently set up as a straw man to be knocked over by the data. For example, the clinical investigator will frequently suspect or even hope to find different outcome distributions for the various treatments under study. One treatment will frequently be a placebo or control while the others are presumably active. As we saw in Chap. 8, the decision to reject a tested hypothesis is stronger than the decision to accept, since the error probability associated with rejection is controlled at the value of α.

We shall need some notation in order to develop the contingency table test procedure. Suppose the table has r rows and c columns and the observed frequency in row i and column j—the (i, j) cell of the table—is denoted by O_{ij}. The general r by c contingency table with observed frequencies O_{ij} is represented in Table 9-1. The row marginal totals are denoted by $O_1., O_2., \ldots , O_r.$ and the column marginals by $O._1, O._2, \ldots , O._c$, the dot being used to indicate that the subscript it replaces has been summed out. That is, a row marginal total is obtained by summing across all the columns within a particular row; for example, $O_2. = \sum_{j=1}^{c} O_{2j}$. The grand total of observed frequencies is still n, the overall sample size, and we have

$$\sum_{i=1}^{r} \sum_{j=1}^{c} O_{ij} = \sum_{i=1}^{r} O_i. = \sum_{j=1}^{c} O._j = n.$$

If you are rusty on summation signs, double sums, and dot notation, take a look at Sec. 3-2 again.

Table 9-1

THE r BY c CONTINGENCY TABLE WITH OBSERVED FREQUENCIES O_{ij}
Column Classification

Row Classification	1	2	$\ldots$	j	$\ldots$	c	Row Totals
1	O_{11}	O_{12}	$\ldots$	O_{1j}	$\ldots$	O_{1c}	$O_1.$
2	O_{21}	O_{22}	$\ldots$	O_{2j}	$\ldots$	O_{2c}	$O_2.$
.	.	.		.		.	.
.	.	.		.		.	.
.	.	.		.		.	.
i	O_{i1}	O_{i2}	$\ldots$	O_{ij}	$\ldots$	O_{ic}	$O_i.$
.	.	.		.		.	.
.	.	.		.		.	.
.	.	.		.		.	.
r	O_{r1}	O_{r2}	$\ldots$	O_{rj}	$\ldots$	O_{rc}	$O_r.$
Column Totals	$O._1$	$O._2$	$\ldots$	$O._j$	$\ldots$	$O._c$	n

How about the expected frequencies? These are calculated under the presumption that the hypothesis is true. The hypothesis states that the row and column classifications are independent. In terms of probability, suppose we let

$$p_{ij} = P(\text{an observation falls into row } i \text{ and column } j),$$

$$p_{i\cdot} = P(\text{an observation falls into column } i),$$

and

$$p_{\cdot j} = P(\text{an observation falls into column } j).[2]$$

Recalling the definition of independence given in Sec. 4-10, which states that 2 events A and B are independent if $P(A \text{ and } B) = P(A) \cdot P(B)$, we can state the contingency table hypothesis in formal terms, using our notation, as follows:

$$H : p_{ij} = p_{i\cdot} p_{\cdot j} \quad \text{for } i = 1, 2, \ldots, r \quad \text{and} \quad j = 1, 2, \ldots, c.$$

Since the row and column classifications each consist of mutually exclusive and exhaustive classes, an observation must fall into some row and some column; i.e., $P(\text{an observation falls into some row}) = p_{1\cdot} + p_{2\cdot} + \cdots + p_{r\cdot} = 1$, and $P(\text{an observation falls into some column}) = p_{\cdot 1} + p_{\cdot 2} + \cdots + p_{\cdot c} = 1$. If H is true we must estimate the $p_{i\cdot}$ and $p_{\cdot j}$ from the observed frequencies. Clearly, we should estimate $p_{i\cdot}$, the probability of falling into row i, by the relative frequency with which our observations fall into that row, that is, by the quantity $O_{i\cdot}/n$. Similarly, $p_{\cdot j}$ is estimated by $O_{\cdot j}/n$. But the expected frequency in row i and column j is then the estimated value of $p_{i\cdot} \cdot p_{\cdot j}$ times the total frequency, or $(O_{i\cdot}/n) \cdot (O_{\cdot j}/n) \cdot n = (O_{i\cdot} \cdot O_{\cdot j})/n$. The multiplication by total frequency n converts the probability of the event to the frequency with which the event is expected to occur in n independent trials. In summary, we have

$$E_{ij} = \frac{O_{i\cdot} \cdot O_{\cdot j}}{n} = \frac{(i\text{th row total})(j\text{th column total})}{n}$$

for

$$i = 1, 2, \ldots, r \quad \text{and} \quad j = 1, 2, \ldots, c.$$

[2] The dot notation is justifiable here by the following argument: $P(\text{an observation falls into row } i) = P(\text{an observation falls into row } i \text{ and col. 1 or row } i \text{ and col. 2 or } \cdots \text{ or row } i \text{ and col. } c) = P(\text{an observation falls into row } i \text{ and col. 1}) + P(\text{row } i \text{ and col. 2}) + \cdots + P(\text{row } i \text{ and col. } c) = p_{i1} + p_{i2} + \cdots + p_{ic} = \sum_{j=1}^{c} p_{ij} = p_{i\cdot}$. A similar argument produces the $p_{\cdot j}$.

The test statistic is

$$\chi^2 = \sum_{i=1}^{r} \sum_{j=1}^{c} \frac{(O_{ij} - E_{ij})^2}{E_{ij}},$$

the single sum in the original definition being replaced by a double sum because of the cross-classification of observed frequencies. The computing formula becomes

$$\chi^2 = \sum_{i=1}^{r} \sum_{j=1}^{c} \frac{O_{ij}^2}{E_{ij}} - n.$$

Substituting the value of E_{ij} derived above, we obtain 2 alternative computing formulas:

$$\chi^2 = \sum_{i=1}^{r} \sum_{j=1}^{c} \frac{nO_{ij}^2}{O_{i.}O_{.j}} - n \quad \text{or} \quad \chi^2 = n\left(\sum_{i=1}^{r} \sum_{j=1}^{c} \frac{O_{ij}^2}{O_{i.}O_{.j}} - 1 \right).$$

The last 2 formulas, though advantageous in certain circumstances, do not directly involve the E_{ij}'s. It is helpful in interpreting data to be able to compare O_{ij} with E_{ij} directly. Furthermore, the criterion for adequacy of the chi-square approximation involves inspection of the expected frequencies.

For the $r \times c$ contingency table we suggest the following rule concerning minimum size of expected frequencies: *No expected frequency should be less than 1 and not over 20 per cent of expected frequencies should be less than 5.* Here again, classes for the row and column classifications are combined if necessary until the rule is satisfied. Notice that this rule is identical to that for the test of Sec. 9-2, in which all expected frequencies are specified prior to sampling. Also notice that if $r = c = 2$, the rule reduces to a statement that no expected frequency can be less than 5, since one expected frequency then constitutes 25 per cent of the E_{ij}'s.

To determine the number of degrees of freedom for the contingency table test, notice that we must estimate from the data the parameters $p_{i.}$ and $p_{.j}$, using $O_{i.}/n$ and $O_{.j}/n$ as the estimates. But since $\sum_{i=1}^{r} p_{i.} = 1$ and $\sum_{j=1}^{c} p_{.j} = 1$, it follows that if we estimate any $r - 1$ of the $p_{i.}$'s the remaining one is uniquely determined, and similarly estimation of any $c - 1$ of the $p_{.j}$'s determines all of them. Thus we are really only estimating a total of $r - 1 + c - 1 = r + c - 2$ parameters under H. Since the total number of classes in the contingency table is rc, it follows that the total number of degrees of freedom is

$$k - 1 - m = rc - 1 - (r + c - 2) = rc - r - c + 1 = (r - 1)(c - 1).$$

Another argument may be helpful in demonstrating that the correct number of degrees of freedom must be $(r - 1)(c - 1)$. We again say that the number of degrees of freedom is the number of expected frequencies that may be freely specified in the data, and we notice that estimation of $p_{i.}$ and $p_{.j}$ by $O_{i.}/n$ and $O_{.j}/n$ essentially fixes the marginal totals $O_{i.}$ and $O_{.j}$. The expected frequencies within the ith row sum to $O_{i.}$ and the expected frequencies within the jth column sum to $O_{.j}$, as shown by the following argument:

$$\sum_{j=1}^{c} E_{ij} = \sum_{j=1}^{c} \frac{O_{i.}O_{.j}}{n} = \frac{O_{i.}}{n} \sum_{j=1}^{c} O_{.j} = \frac{O_{i.}}{n} \cdot n = O_{i.}$$

and

$$\sum_{i=1}^{r} E_{ij} = \sum_{i=1}^{r} \frac{O_{i.}O_{.j}}{n} = \frac{O_{.j}}{n} \sum_{i=1}^{r} O_{i.} \cdot \frac{O_{.j}}{n} = \cdot n = O_{.j}.$$

Then we ask, for fixed row and column marginal totals, how many expected frequencies can be freely specified? Considering the first row, we see that the first $c - 1$ expected frequencies working across the table from left to right can be fixed, but then the last one is determined, since the total for row one is fixed. Similarly, the first $c - 1$ expected frequencies in the second row can be fixed. Continuing down the table row by row, we find that $c - 1$ expected frequencies can be fixed in each of the first $r - 1$ rows. In the last row, however, when we try to specify the expected frequency E_{r1} in the first column, we find that since the expected frequencies above it in that column have all been specified, it is determined. This is because the frequencies in the first column must add up to $O_{.1}$, the column marginal total. This argument applies to every expected frequency in row r. Thus, $c - 1$ expected frequencies may be freely specified in each of $r - 1$ rows, for a total of $(r - 1)(c - 1)$ in all, completing the argument.

EXAMPLE

As part of an epidemiologic study by Dr. Henry L. Taylor (49) of the University of Minnesota, concerning the effect of physical activity on cardiovascular disease, a sample consisting of 484 men over age 45 were invited to participate as study subjects. The men were classified jointly by their degree of physical activity on the job and degree of off-job activity, by using a series of responses to a standard questionnaire. On-job and off-job physical activity were each grouped into 4 classes ranging from class 1, least active, to class 4, most active. Test the hypothesis that there is no association between on-job and off-job activity classification. Table 9-2 gives the observed frequencies and Table 9-3 the expected frequencies for the data. Notice that to within

Table 9-2

OBSERVED FREQUENCIES FOR 484 MEN OVER AGE 45 CLASSIFIED JOINTLY
BY ON-JOB AND OFF-JOB LEVEL OF PHYSICAL ACTIVITY

Off-job Activity Class	On-job Activity Class				Row Total
	1	2	3	4	
1	21	9	9	2	41
2	46	32	26	3	107
3	53	32	38	14	137
4	87	57	42	13	199
Column Total	207	130	115	32	484

Table 9-3

EXPECTED FREQUENCIES CORRESPONDING TO OBSERVED
FREQUENCIES OF TABLE 9-2

Off-job Activity Class	On-job Activity Class				Row Total
	1	2	3	4	
1	17.5351	11.0124	9.7417	2.7107	40.9999
2	45.7624	28.7397	25.4236	7.0744	107.0001
3	58.5930	36.7975	32.5517	9.0579	137.0000
4	85.1095	53.4504	47.2831	13.1570	199.0001
Column Total	207.0000	130.0000	115.0001	32.0000	484.0001

errors due to rounding, the tables of observed and expected frequencies
have the same marginal totals.

Since only one of the 16 expected frequencies is less than 5, and no
E_{ij} is less than 1, the chi-square test can safely be applied.

$$\chi^2 = \frac{(21)^2}{17.5351} + \frac{(9)^2}{11.0124} + \cdots + \frac{(13)^2}{13.1570} - 484 = 493.66 - 484 = 9.66.$$

At the 5 per cent level of significance the critical value of χ^2 is $\chi^2_{0.95}$ for
$(r-1)(c-1) = (4-1)(4-1) = 9$ degrees of freedom, or 16.92.
The observed value of χ^2 is thus clearly within the acceptance region.

In fact, $0.30 < P < 0.40$. These data, therefore, provide no evidence of a relation between job-related and leisure time physical activity level.

In the case of a contingency table with 2 rows and 2 columns, it is common practice to "correct" the chi-square statistic in an attempt to adjust for the fact that the continuous chi-square distribution is used to approximate that of the discrete test statistic, the discreteness arising from the fact that the O_{ij} are discrete. There is general agreement that no such correction should be used for tables with larger numbers of rows or columns. The most commonly used correction in the 2 by 2 table reduces the absolute value (the sign is ignored) of the differences between observed and expressed frequencies by $\frac{1}{2}$ prior to the squaring operation. This correction is known as Yates's correction in honor of Dr. Frank Yates (62), who first proposed it. The Yates-corrected χ^2 statistic is

$$\chi^2_{\text{Yates-corrected}} = \sum_{i=1}^{2} \sum_{j=1}^{2} \frac{(|O_{ij} - E_{ij}| - \frac{1}{2})^2}{E_{ij}}.$$

Special computing formulas are available for 2 by 2 tables. These are

$$\chi^2_{\text{Yates-corrected}} = \frac{\left(|O_{11}O_{22} - O_{12}O_{21}| - \frac{n}{2}\right)^2 \cdot n}{O_{1.} \cdot O_{2.} \cdot O_{.1} \cdot O_{.2}};$$

$$\chi^2_{\text{uncorrected}} = \frac{(O_{11}O_{22} - O_{12}O_{21})^2 \cdot n}{O_{1.} \cdot O_{2.} \cdot O_{.1} \cdot O_{.2}}.$$

Although the use of Yates's correction is very common, there is increasing evidence that it produces an extremely conservative test; that is, that the corrected chi-square statistic fails to reject the hypothesis as often as it should. See, for example, Grizzle (24). While the issue has generated considerable controversy, we find Grizzle's arguments persuasive. We therefore recommend *against* the use of Yates's correction.

It can be shown that $\chi^2_{\text{uncorrected}}$ is the square of z', the test statistic discussed in Sec. 8-4-2 for testing the hypothesis that 2 binomial parameters p_1 and p_2 are equal. Recall that z' uses the overall proportion of successes in the 2 samples. Thus, in a sense, these 2 statistics are testing the same hypothesis. To see this, let column 1 correspond to population 1 and column 2 correspond to population 2. Let row 1 denote success and row 2, failure. Then the contingency table hypothesis of no association between rows and columns says that the distribution across rows (success and failure) is identical for the 2 columns (populations) or that the probability of success (falling into row 1) is the same in the 2 populations (columns 1 and 2) or that $p_1 = p_2$, providing the connection we sought. There is one difference between the procedures; z' can be positive or negative, according to whether $\hat{p}_1 - \hat{p}_2$ is positive or negative. However, $\chi^2_{\text{uncorrected}} = (z')^2$, and after squaring, the distinction between signs is lost. We know, of course, that χ^2 cannot be negative. Thus, the χ^2 test will take note of the size but not the direction of a difference.

EXAMPLE

In a clinical trial involving a potential hypotensive drug, patients are assigned at random either to receive the active drug or placebo. The trial is double blind. That is, neither the patient nor the examining physician knows which of the 2 treatments the patient is receiving. Patient response to treatment is categorized as favorable or unfavorable on the basis of degree and duration of blood pressure response. There are 50 patients assigned to each group. The data are presented in Table 9-4.

Table 9-4

RESULTS OF CLINICAL TRIAL OF
POTENTIAL HYPOTENSIVE DRUG

	Treatment		
Response	*Drug*	*Placebo*	*Row Totals*
Favorable	34	9	43
Unfavorable	16	41	57
Column totals	50	50	100

$$\chi^2_{\text{uncorrected}} = \frac{[(34)(41) - (9)(16)]^2 \cdot 100}{(50)(50)(43)(57)}$$

$$= \frac{(1{,}250)^2 \cdot 100}{6{,}172{,}500} = \frac{1{,}562{,}500}{61{,}275} = 25.50.$$

The critical value for a 5 per cent level test with one degree of freedom is 3.84. Thus the hypothesis of differential response to the 2 treatments is rejected, $P < 0.005$, and we conclude that the drug is having an effect.

Not all 2 by 2 tables can be tested for association by the conventional contingency table test. Suppose, for example, that the same patients are subjected to 2 different diagnostic procedures and are classified as positive or negative for a certain disease on the basis of each test result individually. We want to discover whether there is an association between positivity-negativity and diagnostic procedure, or, in other words, whether the positivity rates are different for the 2 procedures. The 2 × 2 table has the diagnostic procedures as its column classes and positivity-negativity as its row classes. *The fact that the procedures are carried out on the same patients invalidates the usual χ^2 test.* This is due to violation of the basic and essential assumption that the *n* observations in the full table must be independent. The responses of the same patient to the 2 different procedures cannot be assumed to be independent. In other words, the basic sampling unit is the *patient*, and the 2 by 2 table we have been discussing counts each patient twice, once for response to the first procedure, and once for response to the second.

McNemar (32), however, proposed a procedure for testing the hypothesis of no association for such data. This procedure is frequently called Mc-Nemar's test for correlated proportions. In order to carry out the test, the data must be recast to count each patient only once. The resulting 2 by 2 table has the patients' responses to the first procedure as row categories and their responses to the second procedure as column categories. In this table the observed frequencies are O_{ij}, indicating the number of patients in response category *i* for the first procedure and the number in category *j* for the second. The resulting test statistic has the chi-square distribution with one degree of freedom when the hypothesis of no difference in positivity rates to the 2 procedures is true. The statistic is

$$\chi^2 = \frac{(O_{12} - O_{21})^2}{O_{12} + O_{21}}.$$

As usual, the critical region is the upper tail of the chi-square distribution.

To see that this statistic really does test the difference between the sample proportions of patients positive in the 2 procedures, let the index value 1 denote positive, 2 negative; notice that the proportion of positives to the first procedure is then $(O_{11} + O_{12})/n$, and the proportion of positives to the second procedure is $(O_{11} + O_{21})/n$. These 2 sample proportions both involve the O_{11} patients who are positive to both procedures, and this is the fundamental reason for the correlation between them. The 2 proportions will be equal if O_{12} and O_{21} are equal, and this is the situation that should provide most support for the hypothesis that there is no difference between the procedures. Notice that this case corresponds to $\chi^2 = 0$, the value farthest from the critical region. On the other hand, if O_{12} and O_{21} are very different, then the sample proportions of positives are also very different, and in turn χ^2 is large and perhaps sufficiently so to fall into the critical region. The following numerical example should serve to clarify the situation.

EXAMPLE

A group of 75 patients are given 2 different diagnostic procedures designed to determine presence or absence of a particular disease entity. Test the hypothesis that the population proportions of positives to the 2 procedures are equal. The results are given in Table 9-5.

Table 9-5

RESULTS OF ADMINISTERING 2 DIAGNOSTIC PROCEDURES
TO A GROUP OF 75 PATIENTS

	Second Procedure		
First Procedure	*Positive*	*Negative*	*Total*
Positive	41	8	49
Negative	14	12	26
Total	55	20	75

$$\chi^2 = \frac{(8 - 14)^2}{8 + 14} = \frac{36}{22} = 1.64.$$

The critical value at the 5 per cent level is $\chi^2_{0.95} = 3.84$, and thus we accept H: the population proportions of positives for the 2 procedures are equal. $P \doteq 0.20$. Notice that the sample proportions of positives are $(41 + 8)/75 = 0.65$ for the first procedure and $(41 + 14)/75 = 0.73$ for the second. Furthermore, these proportions both include the 41 patients positive to both procedures.

9-5 Some Misapplications of Chi-square Frequency Tests

As we have seen, misapplications of chi-square frequency tests may result from expected frequencies that are too small or from violation of the assumption of independence. We have already discussed one method of dealing with a particular kind of violation of the independence assumption in a 2 by 2 contingency table—namely, McNemar's test for correlated proportions. In other cases the lack of independence may be more subtle.

For example, suppose we want to study the effect of instruction in tooth brushing technique in reducing the incidence of dental caries. Two schools are selected as being similar with respect to many variables, perhaps related to caries occurrence. They are also determined to have enrollments with very similar caries experience as indicated by distributions of counts of decayed, missing, and filled (DMF) teeth on the basis of dental examination of all the children. An instructional program in tooth brushing technique and oral hygiene is then instituted in one school, the other school being kept as an uninstructed control. At the end of 2 years another dental examination is conducted and a contingency table constructed with headings as indicated in Table 9-6.

Table 9-6

CONTINGENCY TABLE FOR EFFECT OF INSTRUCTION
ON DENTAL CARIES EXPERIENCE

	School A (instructed)	School B (uninstructed)
New caries activity (counts of new DMF teeth)		
No new caries activity (counts of caries-free teeth)		

This table will show counts of *teeth* as its O_{ij} values. The entries in row 1 are counts of DMF teeth on second examination which were non-DMF teeth on the first examination, and those in row 2 are counts of non-DMF teeth at first and second examination. Analyzing the data in such a table will be difficult, if not impossible, since the individual tooth is not the actual sampling unit. If a child enters the study then all his teeth enter the study, and it cannot be assumed that tooth status for, say, 2 adjacent teeth in the same mouth is as unrelated as status for 2 teeth, one each from 2 different mouths. In other words, teeth cluster together by the mouthful, and thus the independence assumption necessary to the χ^2 test and most other procedures is violated.

To avoid this difficulty we must be sure that the contingency table shows in its cells counts of children, the basic sampling units. This can be arranged in several different ways. The row classification might be changed to include the following classes: children with no new DMF teeth, children with exactly one new DMF tooth, children with exactly 2 new DMF teeth, children with more than 2 new DMF teeth. If necessary, in order to obtain expected frequencies of adequate size, row classes can be combined, but in any case the principle of counting children is satisfied. Alternatively, if a particular tooth, such as the lower right first molar, is defined as an indicator, the row classification might become: indicator tooth is a new DMF tooth, indicator tooth not a DMF tooth, again counting children in whom the indicator tooth is or is not affected.

Another example of misapplication of χ^2 concerns the selection of a stratified random sample of households in a community study. We might wish to see how nonparticipation varies from stratum to stratum in the sample. We then might use stratum as the column classification and the rows to show nonparticipants versus participants, giving in each cell of the table counts of the number of individuals falling into the particular combination of participation category and stratum. This is again incorrect, since the basic sampling unit is the household, not the individual. If a household is selected into the sample, then all members of the household are automatically selected. The decision to participate or not participate may well be made on a group basis. At least we would expect subjects within a household to be more similar in participation status than subjects from different households.

Again, the row classification should be redefined to insure that the table contains counts of households rather than individuals. For example, row classes might be households with no nonparticipants, households with at least one nonparticipant.

Although the chi-square frequency tests are attractive because of their breadth of applicability, we hope these examples show that the assumption of independence can be violated in rather subtle ways and that the data must be carefully set up to avoid such difficulties.

9-6   *The Variance Test for Goodness of Fit of the Poisson Distribution*

In Sec. 6-2 we saw that in a Poisson distribution the probability that the value x occurs is $e^{-\mu}\mu^x/x!$ for $x = 0, 1, 2, \ldots$. For a random sample of n such x values, the frequency O_i with which each value of x occurs could be determined and a corresponding set of expected frequencies calculated by multiplying n by the quantity $e^{-\bar{x}}\bar{x}^x/x!$; we use $\bar{x}$ to estimate μ and reason that, as in the case of fitting a normal distribution to a set of observations, if any Poisson distribution fits the data, the distribution having the same mean as the observations should fit best. It would be necessary to combine probabilities in a final class, and to let that class be, say, $x \geq x_L$ where x_L is some relatively large value of x depending on the data. We would then calculate

$$\chi^2 = \sum_{i=1}^{k} \frac{O_i^2}{E_i} - n$$

and compare it with $\chi^2_{1-\alpha}$, the $100(1 - \alpha)$ percentile of the χ^2 distribution with $k - 1 - 1 = k - 2$ degrees of freedom, since only the parameter μ is estimated from the data.

This is all well and good, but sampling investigations show that this is not a very powerful test. A more sensitive procedure makes use of the fact that the mean and variance of the Poisson distribution are equal, and then compares the sample mean and sample variance. Specifically if $x_1, x_2, \ldots, x_n$ are a random sample from a Poisson distribution, then

$$\chi^2 = \sum_{i=1}^{n} \frac{(x_i - \bar{x})^2}{\bar{x}}$$

is approximately chi-square distributed with $n - 1$ degrees of freedom. Chi-square can therefore be used as a test statistic for testing H: the observations are a random sample from a Poisson distribution. Notice that χ^2 uses every observation and has $n - 1$ degrees of freedom, whereas the conventional chi-square frequency test produces a statistic with $k - 2$ degrees of freedom, with k equal only to the number of groups used in the data.

Naturally, this "variance test" for goodness of fit of the Poisson distribution is of a rather different character from the frequency test discussed earlier. It is a special purpose test sensitive to discrepancies between the mean and variance of the distribution, whereas the frequency procedure is a general or omnibus test. Recall that a similar situation exists in testing for normality, where the special tests for skewness and kurtosis (Secs. 8-5-1 and 8-5-2) are compared to the general frequency test (Sec. 9-3). However, in dealing with discrete data, the comparison of mean and variance is a useful device for distinguishing the common distributions and seems in practice to detect the most important departures from the Poisson distribution.

There is an analogy between the variance test for the Poisson and some of our earlier work. Recall that for random samples from a normal distribution the sampling distribution of the quantity

$$\frac{(n - 1)s^2}{\sigma^2} = \frac{\sum_{i-1}^{n}(x_i - \bar{x})^2}{\sigma^2}$$

is the chi-square distribution with $n - 1$ degrees of freedom (Sec. 7-2-4). But since in the Poisson case the mean and variance are equal, it is reasonable to use $\bar{x}$ to estimate σ^2. Replacement of σ^2 by $\bar{x}$ produces the variance test statistic χ^2.

Note an important difference between the variance test and the frequency tests discussed earlier in this chapter. Since departures from Poisson distribution may show the variance to be either larger or smaller than the mean, *the variance test for goodness of fit of the Poisson can be a two-sided test.* The chi-square frequency tests, on the other hand, are invariably upper one-sided tests, since any disagreement between the O_i and E_i increases the size of the test statistic. In the variance test, however, upper one-sided, lower one-sided, or two-sided tests are all reasonable, depending upon the nature of the alternatives we wish to detect. Ordinarily, we are interested in departures of any sort from the Poisson and use the two-sided procedure.

The variance test was first proposed by R. A. Fisher (15) in his classical text *Statistical Methods for Research Workers*, but J. Berkson (4) did some more recent sampling studies, and W. G. Cochran (7) discusses the procedure as well as others similar to it.

EXAMPLE

In Table 6-1, 100 0.1 minute counts from a single radiologic source measured by Dr. John Nehemias are presented. The counts are grouped for ease of presentation and computation, and are presented again here. Test for fit of the Poisson distribution.

Count (x_j):   0   1   2   3   4   5

Frequency (f_j):   11   20   28   24   12   5

Now $n = 100$ and the x_i are the 100 individual observations, 11 of which are equal to 0, 20 equal to 1, etc. The expression $\sum_{i=1}^{n} (x_i - \bar{x})^2$ may be replaced by its grouped form $\sum_{j=1}^{c} f_j(x_j - \bar{x})^2$, where $c = 6$, the number of groups with nonzero frequency, and x_j is used to denote the distinct values of x. Notice that all the x_j are different, but the x_i involve many duplicate values; in other words, there are 6 x_j's and 100 x_i's. The computing formulas for grouped data discussed in Sec. 3-7-2 apply here, and we have

$$\sum_{j=1}^{c} f_j(x_j - \bar{x})^2 = \frac{n \sum_{j=1}^{c} f_j x_j^2 - \left(\sum_{j=1}^{c} f_j x_j \right)^2}{n};$$

$$\sum_{j=1}^{c} f_j x_j = (11)(0) + (20)(1) + \cdots + (5)(5) = 221;$$

$$\sum_{j=1}^{c} f_j x_j^2 = (11)(0) + (20)(1) + (28)(4) + \cdots + (5)(25) = 665;$$

$$\frac{n \sum_{j=1}^{c} f_j x_j^2 - \left(\sum_{j=1}^{c} f_j x_j \right)^2}{n} = \frac{(100)(665) - (221)^2}{100} = 176.59;$$

$$\bar{x} = \frac{\sum_{j=1}^{c} f_j x_j}{n} = \frac{221}{100} = 2.21;$$

$$\chi^2 = \frac{176.59}{2.21} = 79.90.$$

From Table A-6 we find that the critical values for a two-sided 5 per cent level test are, for $n - 1 = 99$ degrees of freedom, approximately $\chi^2_{0.025} = 74.2$ and $\chi^2_{0.975} = 130$, using the tabulated values for 100 degrees of freedom as an approximation. (The exact critical values in this case, obtained from more detailed tables, are 73.361 and 128.422.) We accept the hypothesis that the data fit the Poisson distribution. $0.10 < P < 0.20$. Notice that

$$s^2 = \frac{n \sum_{j=1}^{c} f_j x_j^2 - \left(\sum_{j=1}^{c} f_j x_j \right)^2}{n(n-1)} = \frac{176.59}{99} = 1.78,$$

a value less than $\bar{x} = 2.21$. Thus in the data themselves the mean and variance are approximately equal.

EXERCISES

9-1   Suppose a particular jaw trait is thought to be inherited in the ratio $1:2:1$ for homozygous dominant, heterozygous, and homozygous recessive. One hundred randomly selected children are distributed as follows:

Dominant	Heterozygous	Recessive
21	62	17

   a.   Does the trait follow the hypothetical pattern?
   b.   What assumptions about the data are needed?

9-2   For a certain genetic model four phenotypes should occur in $9:3:3:1$ ratio. Four hundred eighty offspring are classified with observed frequencies 290, 100, 80, 10. Are the data consistent with the genetic model?

9-3   In one of his experiments on plant variation, Mendel studied a number of plants to investigate seed morphology. He hypothesized that the round-angular ratio should be $3:1$. The observed results were

Round seed   336

Angular seed   101

Test to see whether there is any difference in the observed ratios with respect to the hypothesized ratio.

9-4 As part of a study 52 subjects were required to have physical examinations. All subjects had heart rates taken. It was hypothesized that these values were from a normal distribution. Test this hypothesis, using the following data.

Class Boundaries	Observed Frequencies
54–62	0
62–70	6
70–78	6
78–86	16
86–94	14
94–102	4
102–110	4
110–118	2
118–126	0

9-5 Test to see if the data of Exercise 3-10 come from a normal distribution.

9-6 A recent study found that mongolism in babies is associated with infectious hepatitis (I.H.) of the mother during pregnancy. Suppose a study of 2,000 randomly selected mothers-to-be yielded the following table after the births of their babies:

		Baby	
		Mongoloid	Nonmongoloid
Mother	I.H.	26	34
	No I.H.	4	1936

a. State the appropriate statistical hypothesis to test the study's finding.
b. Carry out the statistical analysis.

9-7 Suppose it is conjectured that eye color is related to occurrence of retinal detachment. One hundred randomly selected control individuals without the condition were compared with 100 individuals with detached retinas with the following results:

		Eye Color		
		Blue	Brown	Other
Detached retina	Yes	32	51	17
	No	37	40	23

State and test the appropriate statistical hypothesis.

9-8 One hundred persons are asked the following 2 questions:
1. Do you smoke heavily?
2. Do you cough in the morning?
The answers:

		Smoke Heavily	
		Yes	No
Cough	Yes	63	12
	No	7	18

State and test the appropriate statistical hypothesis.

9-9 It is felt that elevated blood pressure is strongly associated with occurrence of cerebral vascular accidents (CVA). The following data result for 200 men aged 45 to 60 followed for 5 years:

		CVA	
		Yes	No
Diastolic blood pressure	70–80	1	49
	80–90	4	46
	90–100	6	44
	100+	13	37

Carry out a statistical analysis of the hypothesis suggested.

9-10 Two communities, one with fluoridated water, the other without, are to be compared. Children aged 4 are given dental examinations with the following results:

		Fluoridated Water	
		Yes	No
Children—number of decayed or filled teeth	0	150	143
	1	10	13
	2	9	11
	3	6	14
	4+	4	6

Does fluoridated water seem to improve the status of children's teeth?

9-11　Suppose you have the following table of responses of 300 individuals. Test the data to see if mathematical ability and interest in statistics are associated.

		Ability in Mathematics		
		Low	Average	High
Interest in statistics	Low	50	25	15
	Average	25	60	5
	High	25	15	80

9-12　Suppose the following data represent counts of plankton in 100 small test aliquots of sea water.

Plankton Count per Aliquot	Frequency
0	36
1	40
2	18
3	5
4	1

Test the hypothesis that the data come from a Poisson distribution.

9-13　The data below are 0.1 minute radiologic counts from a single source. Test to see if they come from a Poisson distribution.

2	4	3	3	3	1	1	4	3	5
4	1	0	0	4	3	1	2	3	2
2	2	3	5	5	4	3	0	1	2
1	1	1	0	3	2	1	1	2	3
2	2	3	2	3	5	4	2	2	3
1	4	1	0	4	2	2	3	0	3
1	4	3	0	3	1	4	3	2	3
2	3	0	2	2	1	2	2	2	4
3	3	3	2	1	2	2	1	0	4
3	2	0	1	1	1	5	2	2	0

9-14　Show the following relationship holds in a 2×2 table:

$$(z')^2 = \text{``chi''}^2 \text{ (uncorrected).}$$

See Sec. 8-4-2 for the definition of z'.

10

Regression
and
Correlation

Regression analysis is a means of studying the variation of one quantity (dependent variable) at selected levels of another quantity (independent variable). The word regression was first used by Sir Francis Galton in describing certain genetic relationships.

10-1 Basic Ideas, Origin of the Term "Regression," and Regression toward the Mean

In his book, *Natural Inheritance*, Galton (20) stated the "law of universal regression," saying, "Each peculiarity in a man is shared by his kinsman, but *on the average* in a less degree." Under this law, for example, while tall fathers would tend to have tall sons, the sons would be on the average shorter than their fathers, and sons of short fathers, though having heights below the average for the entire population, would tend to be taller than their fathers. Pearson and Lee (44), in studying over 1,000 families, found that this effect did indeed hold.[1]

Today we call the characteristic of returning from extreme values toward the average of the full population "regression toward the mean." It can crop up in unusual places and be interpreted in strange ways. It probably is present in the well-known "sophomore jinx" phenomenon. The observation that professional baseball players, after being promoted to one of the

[1] In discussing one of their conclusions concerning the effects of selection upon changes in populations over time, the authors make the following interesting comment: "It emphasizes the all-important law that with judicious mating human stock is capable of rapid progress. A few generations suffice to modify a race of men, and the nations which breed freely only from their poorer stocks will not be dominant factors in civilization by the end of the century."

major league clubs, do less well on the average in their second or sophomore year than in their first year has been widely noted. Of course, only the players with the better performances are retained for a second year, and among these players are included some who had, for them, an unusually good first year. These players will, on the average, do less well in their sophomore year and serve to depress the second year average of the retained group. Notice this does not say that the better first year players are not really better players than the rest of the group, any more than the height example states that tall fathers do not have sons who are above average height. If a big league manager adopted the foolish practice of retaining his poorest first year players, he would find, no doubt, that they tended in the second year to do better on the average than they did in the first, though still probably more poorly than the average of the entire first year group.

This phenomenon may arise in longitudinal studies of human populations. If a group of men were examined and divided, say, into 5 equal-sized groups on the basis of systolic blood pressure at first examination, and reexamined a year later, it would not be surprising to find that the average level of the 20 per cent of men with highest initial readings decreased when these same men were examined a year later, whereas the lowest 20 per cent increased. Again, regression toward the mean. No doubt, this effect has been reported in other biological and medical studies as a significant new finding, when it is, in fact, inevitable and should be regarded as an essential feature of random variation. In statistics today, the term regression is used more broadly to describe relationships between variables.

10-2 The Straight Line

We do not mean to insult your intelligence with the heading to this section by suggesting that you do not know what a straight line is. It is just that we thought you might appreciate a little review of some of its algebraic characteristics. If we are wrong, skip this section. Although we are not in a position to arrange a refund of a portion of your purchase price as compensation for the unused section, perhaps you will receive sufficient ego satisfaction from reading material with which you are thoroughly familiar that the result of reading it anyway will be a net gain.

One thing you can say for a straight line—it is straight. So if you find 2 points on the line, you can draw the whole line. Mathematicians like to represent geometric objects by equations. If we have 2 variables x and y, a straight line relationship between them can be expressed by such an equation. It turns out that $y = c + dx$, where c and d are constants is such an equation.

For example, $y = 2 + x$, where $c = 2$, $d = 1$, represents a straight line. Any pair consisting of a value of x and a value of y satisfying the equation represents a point lying on the line. Such a pair of values is $x = 1$, $y = 3$, since $3 = 2 + (1)(1)$. A pair not satisfying the equation lies off the line, e.g., $x = 2$, $y = 3$, since $3 \neq 2 + (1)(2)$. When $y = c + dx$, we say that y is a linear function of x.

An equation such as $y = 2 + x$ has a corresponding graph. If we plot in an (x, y) plane all the pairs of values satisfying the equation, they lie in a line. To graph the line we need merely to locate some such pairs. Actually, as we have observed, 2 pairs is sufficient, but we shall locate some more just to check ourselves. You can verify that $x = 0$, $y = 2$; $x = -1$, $y = +1$; $x = 1$, $y = 3$; and $x = 2$, $y = 4$ all satisfy the equation. The graph of the line $y = 2 + x$ is presented in Fig. 10-1. These 4 pairs of values are plotted

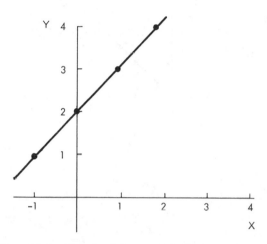

Figure 10-1 Graph of the straight line $y = 2 + x$.

as points, and the line is then drawn through them. Other examples of straight line or linear equations are $y = x$, in which $c = 0$, $d = 1$; $y = -3 + 2x$, in which $c = -3$, $d = 2$, and $y = 1 - 2x$, in which $c = 1$, $d = -2$. These lines are plotted in Fig. 10-2.

The constants c and d are called, respectively, the y intercept and slope of the line, and with good reason. The y intercept is the value of y at which the line crosses or intercepts the y axis, i.e., the value of y corresponding to $x = 0$. In the line $y = 2 + x$, when $x = 0$, $y = 2$ and in Fig. 10-1 we see that, sure enough, the line crosses the y axis at $y = 2$. This can be checked for the lines in Fig. 10-2 as well. In general, for the line $y = c + dx$, we see that when $x = 0$, $y = c$.

The slope is the amount of change in y when x increases by one unit. In the line $y = 2 + x$ we saw that when x increased from 0 to 1, y increased from 2 to 3; since $3 - 2 = 1$, we have a slope of 1, and of course $d = 1$ for this line. The slope (or y intercept, for that matter) of a line can be negative. In the line $y = 1 - 2x$, as x changes from 1 to 2, for example, y changes from -1 to -3 and slope $= -3 - (-1) = -2 = d$. The slope of a straight line

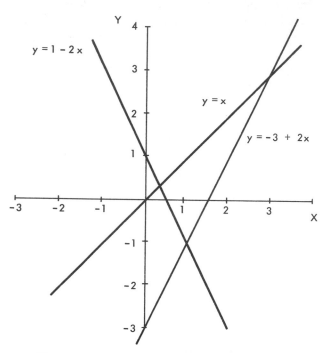

Figure 10-2 Other examples of straight lines.

is constant—in fact, that property can form both an intuitive and a mathematical basis for a formal definition of the concept of "straightness." It is easy enough to show that d must be the value of the slope for any line $y = c + dx$. As we move from any value x to $x + 1$, y changes from $c + dx$ to $c + d(x + 1)$ and $c + d(x + 1) - (c + dx) = c + dx + d - c - dx = d$.

A relation between x and y of the form $tx + uy = v$, where t, u, and v are constant with $u \neq 0$ also represents a straight line, since, solving for y, we have $y = (v/u) - (t/u)x$, and this is the equation of a line with y intercept v/u and slope $-t/u$. The equations $y = g$ and $x = h$, where g and h are constants, define lines parallel to the horizontal and vertical axes, respectively.

Now back to statistics.

10-3 Linear Regression: Assumptions and Examples

If x is the independent variable and y the dependent variable, a regression problem, as we shall consider it, is a problem in which, for a fixed value of x, y has some particular distribution of values. In other words, we are dealing with a series of populations, a different population of y values for each value of x. We say we are studying *the regression of y on x*.

Our analysis becomes simpler and our results more explicit if we make certain assumptions about the nature of the distribution of y (for fixed x). The 2 assumptions generally made are

1. For any x, the distribution of y is normal.
2. The variance of the distribution of y is the same for any chosen value of x. This is the assumption of homoscedasticity (see Sec. 7-2-5-2).

Note that we do not make any statement about the form of the mean y value for given x. We do not expect that this mean y will be the same for each value of x, or we would not choose to study different values of x, the distribution of y in that case being identical for all values of x.

If we assume that the mean value of y is a linear function of x (i.e., if we plotted the population mean of y for different values of x, these means would lie on a straight line) we have a case of *linear regression*. Figure 10-3 illustrates the situation.

If the mean y for fixed x, say $\mu_{y \cdot x}$, is a linear function of x, it must be of the form $\mu_{y \cdot x} = (y \text{ intercept}) + (\text{slope}) \cdot x$, where the y intercept and slope are parameters to be estimated statistically.

In performing the estimation we will choose a series of values of x and at each chosen x value select one or more individuals at random from the corresponding population of y's. The x's are assumed to be constants, not subject to sampling or measurement error. It will simplify all our calculations if we let $\bar{x}$ be the mean of all the x's selected, counting each x once for each value of y selected at that x value. For example, if we choose 20 y values at $x = 3$, the value 3 is used 20 times in computing $\bar{x}$.

If we rewrite $\mu_{y \cdot x} = (y \text{ intercept}) + (\text{slope}) \cdot x = (A - B\bar{x}) + Bx = A + B(x - \bar{x})$, all our work will be less complex. Thus, rather than estimating the slope and y intercept directly, we estimate A and B. Notice that in this form, the y intercept is $A - B\bar{x}$ and the slope is B. Alternatively, we could consider that we are talking about a line located relative to a pair of axes x' and y instead of x and y, for which $x' = x - \bar{x}$. Relative to those axes we see that $\mu_{y \cdot x'} = A + Bx'$, a line with y intercept A and slope B. To avoid confusion, we will continue to write $x - \bar{x}$ rather than x'.

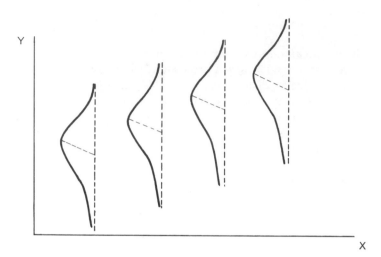

Figure 10-3   Linear regression of y on x.

It will be instructive now to pause and examine certain examples of regression problems. Although the assumptions of normality and equality of variances may not appear to be satisfied in these examples, it is frequently possible to transform y to a new variable y' more nearly satisfying these assumptions. With respect to the assumption of linearity of regression of y on x, we can possibly transform x or work with a nonlinear regression function.

EXAMPLES

1. The regression of measured hemoglobin concentration (y) on time (x) in preserved blood.
2. The regression of serum cholesterol value (y) on time following last meal (x).
3. The regression of dentists' incom~ (y) on number of years since graduation from dental school (x).
4. The regression of antibody titer (y) on dose of an immunizing agent (x).
5. The regression of ragweed pollen concentration (y) in a wind tunnel on operating wind speed (x) of the tunnel.
6. Regression of Lactobacillus acidophilus counts (y) on amount of starch and sugar (x) in a series of standard diets.

We will consider in detail an example in which ragweed pollen grains were sampled in a test chamber with a rotobar sampler. The grains are trapped by the leading edge of a rotating vane—the edge measures 1 mm by 1 inch—and counted under low power; a Whipple eyepiece disk is used, and the length of the sampling surface is traversed; one 1 mm by 1 mm microscopic field is counted at a time. It is of interest to determine whether there is some linear trend in counts per field along the surface. Thus, y is the number of ragweed grains in a field and x is the number of the field. The following data were obtained:

y	184	166	153	191	171	154	162	166	176	184	166	147
x	1	2	3	4	5	6	7	8	9	10	11	12

y	152	160	173	158	150	149	130	155	159	160	154
x	13	14	15	16	17	18	19	20	21	22	23

Data of this type are frequently represented in a so-called scatter diagram, which shows a dot for each pair of values (x, y). The scatter diagram for these data is shown as Fig. 10-4. Such a diagram is an important descriptive statistic and can provide much useful information.

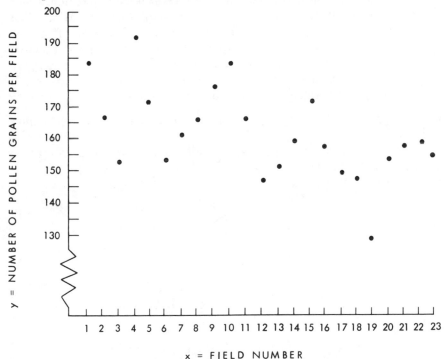

Figure 10-4 Scatter diagram of number of ragweed pollen grains per field on field number in a rotobar sampler.

10-4 Estimation of Slope and Intercept: Least Squares

It can be shown that if we have random samples of y values at each x value, if the sampling between different x values is independent, if our assumption of constant variance of y at each x is true, and if the population mean value of y is a linear function of x, the procedure known as *least squares* gives us very good estimates of A and B. Of course, we need to make the words "very good" more precise. We mean that by applying the principle of least squares we obtain estimates of A and B which are unbiased, which are linear functions of the observed y's, and which have the smallest variance among all other unbiased, linear estimates. Thus we have a form of minimum variance, unbiased estimation in accord with the criteria for good point estimates developed in Sec. 7-1. This result, a very important one in statistical theory, is known as the Gauss-Markov theorem.

Now just what is the principle of least squares? If from our sample of x and y values we wish to estimate the line on which the population means lie, we must first estimate A and B. The principle of least squares says that we should choose the line that minimizes the sum of squares of vertical deviations of the observed y values from the line. The line so chosen is called the least squares line of best fit. In other words, if $y_1, y_2, \ldots, y_n$ are our observed y values and if $\bar{y}_{x_i}$ is the estimated y value on the line of best fit corresponding to the value x_i of the independent variable, then we wish to minimize $\sum_{i=1}^{n} (y_i - \bar{y}_{x_i})^2$. Figure 10-5 presents our present situation graphically.

The next question is how do we actually estimate A and B so that we can draw the least squares line of best fit? If we let a and b denote the estimates, it can be shown that

$$a = \bar{y}$$

and

$$b = \frac{L_{xy}}{L_{xx}},$$

where

$$L_{xy} = n \sum_{i=1}^{n} x_i y_i - \left(\sum_{i=1}^{n} x_i \right) \left(\sum_{i=1}^{n} y_i \right)$$

and

$$L_{xx} = n \sum_{i=1}^{n} x_i^2 - \left(\sum_{i=1}^{n} x_i \right)^2$$

will give us the least squares line of best fit.

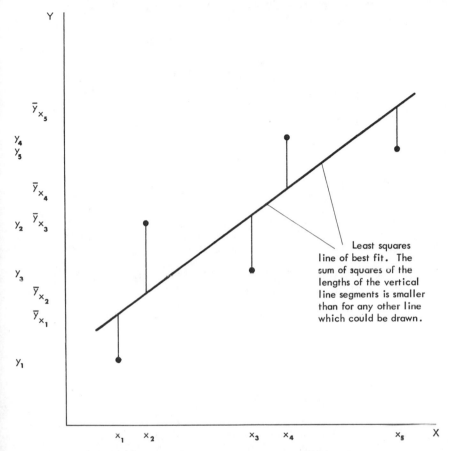

Figure 10-5   Least squares line of best fit.

Engineers, physicists, and others frequently use the least squares process as a method of smoothing observations, independent of the statistical questions of estimation and testing of hypotheses that are of interest to us. If we wish to obtain a line that will, in some sense, best approximate a scatter of points, the least squares line of best fit will frequently be satisfactory. In these cases the purely statistical assumptions (normality, random and independent sampling, equality of variances, and true linearity of the population regression line) become less relevant.

Returning to the statistical problems, we need in addition to the least squares line itself some estimate of how well the fitted or estimated line actually fits the data. For this purpose we use the sum of squares of deviations of the observed y values from their respective estimated values on the fitted line divided by a constant. That is, we use

$$s_{y \cdot x}^2 = \frac{1}{n-2} \sum_{i=1}^{n} (y_i - \bar{y}_{x_i})^2.$$

Of course, we now know that

$$\bar{y}_{x_i} = a + b(x_i - \bar{x}) = \bar{y} + b(x_i - \bar{x}).$$

As a computing formula we can use either

$$s_{y \cdot x}^2 = \frac{n-1}{n-2} (s_y^2 - b^2 s_x^2),$$

where s_y^2 is the sample variance of the y's if the x's are ignored and s_x^2 is the sample variance of the x's if the y's are ignored, or

$$s_{y \cdot x}^2 = \frac{L_{xx} L_{yy} - L_{xy}^2}{(n-2) n L_{xx}},$$

where

$$L_{yy} = n \sum_{i=1}^{n} y_i^2 - \left(\sum_{i=1}^{n} y_i \right)^2.$$

Recalling our assumption that the variance of the population of y values was the same at each value of x, suppose we let $\sigma_{y \cdot x}^2$ denote this common variance. Then it can be shown that $s_{y \cdot x}^2$ is an unbiased estimator of $\sigma_{y \cdot x}^2$. At first the divisor $n-2$ in the formula for $s_{y \cdot x}^2$ may seem strange. However, recall that we are estimating the 2 quantities A and B from the data and thus, roughly speaking, lose 2 "degrees of freedom" in our estimator, an idea in general agreement with the explanation of the degrees of freedom concept of Chap. 9.

The quantity $s_{y \cdot x}$ is frequently called the *standard error of estimate*, since it is a measure of the extent to which the y values estimated from the line deviate from those actually observed.

EXAMPLE

We will calculate a, b, $s_{y \cdot x}^2$, and $s_{y \cdot x}$ for the ragweed pollen sampling data of Sec. 10-3.

We find $n = 23$;

$$\sum_{i=1}^{n} x_i = 1 + 2 + \cdots + 23 = 276;$$

$$\sum_{i=1}^{n} y_i = 184 + 166 + \cdots + 154 = 3,720;$$

$$\sum_{i=1}^{n} x_i^2 = 1 + 4 + \cdots + 529 = 4,324;$$

$$\sum_{i=1}^{n} y_i^2 = (184)^2 + (166)^2 + \cdots + (154)^2 = 605,876;$$

$$\sum_{i=1}^{n} x_i y_i = (1)(184) + (2)(166) + \cdots + (23)(154) = 43,536;$$

$$L_{xx} = (23)(4,324) - (276)^2 = 23,276;$$

$$L_{yy} = (23)(605,876) - (3,720)^2 = 96,748;$$

$$L_{xy} = (23)(43,536) - (276)(3,720) = -25,392.$$

Notice that L_{xy} can be negative but L_{xx} and L_{yy} cannot. Although it is ordinarily possible and desirable on a desk calculator to calculate the L's in one continuous machine operation, we have shown the component sums and sums of squares and products.

$$a = \bar{y} = \frac{3,720}{23} = 161.74;$$

$$b = \frac{L_{xy}}{L_{xx}} = \frac{-25,392}{23,276} = -1.09091.$$

The equation of the fitted line is

$$y = a + b(x - \bar{x}) = 161.74 + (-1.09091)(x - \tfrac{276}{23})$$

$$= 161.74 + (1.09091)(12) - 1.09091x = 174.83 - 1.09091x.$$

We see that the fitted line will have negative slope, a reasonable result after inspecting the scatter diagram of Fig. 10-4, which shows a downward trend in counts as we proceed across the microscope slide.

$$s_{y \cdot x}^2 = \frac{L_{xx}L_{yy} - L_{xy}^2}{n(n-2)L_{xx}} = \frac{(23,276)(96,748) - (-25,392)^2}{(23)(21)(23,276)} = 142.96;$$

$$s_{y \cdot x} = 11.957.$$

The fitted line $y = 174.83 - 1.09091x$ is plotted over the scatter diagram in Fig. 10-6.

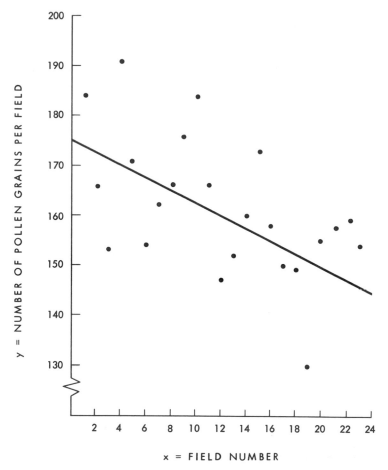

Figure 10-6 **Regression line fitted to scatter diagram of number of ragweed pollen grains on microscope field number.**

10-5 Confidence Intervals and Tests in Linear Regression

Statistical inferences concerning A, B, $\sigma^2_{y \cdot x}$, $\mu_{y \cdot x}$, and y_x, where the latter symbol denotes a predicted new value of y at the value x, can be made under the assumptions we have been using. For reference we restate those assumptions here.

1. The x's are free from error resulting either from sampling or from measurement; i.e., the x's are observable constants.
2. The distribution of y at each x is normal.
3. The population variance of y is the same at each value of x.
4. The population mean of y is a linear function of x.
5. At each x value we select a random sample (possibly a sample of size one) of y's, and y's observed at different values of x are independent.

When these assumptions are true, the following conclusions hold:

Conclusion 1. $t_A = \dfrac{(\bar{y} - A)\sqrt{n}}{s_{y \cdot x}}$ follows Student's t distribution with $n - 2$

degrees of freedom. Notice that $\bar{y} = a$.

Conclusion 2. $t_B = \dfrac{(b - B)s_x\sqrt{n - 1}}{s_{y \cdot x}}$ follows Student's t distribution with

$n - 2$ degrees of freedom.

Conclusion 3. $\chi^2 = \dfrac{(n - 2)s^2_{y \cdot x}}{\sigma^2_{y \cdot x}}$ follows the chi-square distribution with

$n - 2$ degrees of freedom.

Conclusion 4. $t_C = \dfrac{a + b(x_0 - \bar{x}) - \mu_{y \cdot x_0}}{s_{y \cdot x}\sqrt{\dfrac{1}{n} + \dfrac{(x_0 - \bar{x})^2}{(n - 1)s^2_x}}}$, where x_0 is a particular value

of x, follows Student's t distribution with $n - 2$ degrees of freedom.

These conclusions lead to the inferential procedures presented in succeeding sections.

10-5-1 INFERENCES CONCERNING A

Conclusion 1 provides the basis for confidence intervals and tests for A. The $(1 - \alpha)$-level confidence interval is derived from the probability statement

$$P(-t_{1-(\alpha/2)} < t_A < t_{1-(\alpha/2)}) = 1 - \alpha,$$

where $t_{1-(\alpha/2)}$ is the $100[1 - (\alpha/2)]$ percentile of the t distribution with $n - 2$ degrees of freedom. By a sequence of manipulations used repeatedly in Chap. 7, we find

$$P\left(\bar{y} - t_{1-(\alpha/2)} \frac{s_{y \cdot x}}{\sqrt{n}} < A < \bar{y} + t_{1-(\alpha/2)} \frac{s_{y \cdot x}}{\sqrt{n}}\right) = 1 - \alpha,$$

which provides the required interval.

To test $H:A = A_0$, use the test statistic

$$t_{A_0} = \frac{(\bar{y} - A_0)\sqrt{n}}{s_{y \cdot x}},$$

which has the t distribution with $n - 2$ degrees of freedom if H is true. The critical region for a two-sided α-level test is $t_{A_0} \leq t_{\alpha/2}$ or $t_{A_0} \geq t_{1-(\alpha/2)}$; for an upper one-sided α-level test, it is $t_{A_0} \geq t_{1-\alpha}$ and for a lower one-sided test it is $t_{A_0} \leq t_{\alpha}$.

10-5-2 INFERENCES CONCERNING B

Conclusion 2 leads to intervals and tests for B. A $(1 - \alpha)$-level confidence interval comes from the following probability statement:

$$P\left(b - t_{1-(\alpha/2)} \frac{s_{y \cdot x}}{s_x \sqrt{n-1}} < B < b + t_{1-(\alpha/2)} \frac{s_{y \cdot x}}{s_x \sqrt{n-1}}\right) = 1 - \alpha,$$

which is derived from the basic statement

$$P(-t_{1-(\alpha/2)} < t_B < t_{1-(\alpha/2)}) = 1 - \alpha.$$

To test $H:B = B_0$, use the test statistic

$$t_{B_0} = \frac{(b - B_0)s_x\sqrt{n - 1}}{s_{y\cdot x}},$$

which has a t distribution with $n - 2$ degrees of freedom under H. The critical region for a two-sided α-level test is $t_{B_0} \leq t_{\alpha/2}$ or $t_{B_0} \geq t_{1-(\alpha/2)}$; for an upper one-sided α-level test it is $t_{B_0} \geq t_{1-\alpha}$ and for a lower one-sided α-level test $t_{B_0} \leq t_\alpha$.

The hypothesis $B = 0$ is particularly interesting and important. It can be tested by setting $B_0 = 0$ in the expression for t_{B_0}, obtaining the test statistic

$$t_0 = \frac{bs_x\sqrt{n - 1}}{s_{y\cdot x}},$$

and selecting one of the critical regions in the preceding paragraph according to whether a two-sided or one-sided test is needed. A decision equivalent to that of the two-sided, α-level test can be reached by noting whether the $(1 - \alpha)$-level confidence interval for B includes or fails to include the value zero between its limits (see Sec. 8-2-2).

The importance of the hypothesis $B = 0$ can be seen by considering its implication when it is coupled with our assumptions of normality and constancy of variance of y across all values of x. If $B = 0$, then $\mu_{y\cdot x} = A$, the second term $B(x - \bar{x})$ vanishing, and we find that the population mean of y does not vary with x. Since a normal distribution is completely determined by its mean and variance, and since neither of these parameters varies with x, we can say that there is no regression. That is, since the population of y values is identical for all values of x, there is no need to include x in any problem involving study of y.

Notice, though, that this conclusion is based upon 2 assumptions, normality and equal variances, and the truth of the hypothesis $B = 0$. We have noted in Sec. 8-2-1 that when we accept a hypothesis we do not necessarily mean that it is mathematically true, but rather that the sample contains insufficient evidence to lead to rejection. The decision to reject a tested hypothesis is made with an error probability controlled at α, but the error probability associated with the decision to accept depends upon the true value of the parameter being tested, B in this case. Thus, we are well advised to go slowly in claiming that x is irrelevant when that claim is based upon the kind of inferential procedures available to us. However, an indication of the strength of the relationship between x and y is provided by these procedures.

10-5-3 INFERENCES CONCERNING $\sigma^2_{y\cdot x}$

Since $(n-2)s^2_{y\cdot x}/\sigma^2_{y\cdot x}$ follows the chi-square distribution with $n-2$ degrees of freedom (Conclusion 3), we can construct intervals and tests for $\sigma^2_{y\cdot x}$.

The $(1-\alpha)$-level confidence interval for $\sigma^2_{y\cdot x}$ comes from the probability statement

$$P\left[\frac{(n-2)s^2_{y\cdot x}}{\chi^2_{1-(\alpha/2)}} < \sigma^2_{y\cdot x} < \frac{(n-2)s^2_{y\cdot x}}{\chi^2_{\alpha/2}}\right] = 1-\alpha,$$

by reasoning practically identical with that of Sec. 7-2-4. An interval for $\sigma_{y\cdot x}$ comes from

$$P\left[\sqrt{\frac{(n-2)s^2_{y\cdot x}}{\chi^2_{1-(\alpha/2)}}} < \sigma_{y\cdot x} < \sqrt{\frac{(n-2)s^2_{y\cdot x}}{\chi^2_{\alpha/2}}}\right] = 1-\alpha.$$

To test $H:\sigma^2_{y\cdot x} = (\sigma^2_{y\cdot x})_0$, where $(\sigma^2_{y\cdot x})_0$ is a particular numerical value of $\sigma^2_{y\cdot x}$, we use the test statistic

$$\chi^2 = \frac{(n-2)s^2_{y\cdot x}}{(\sigma^2_{y\cdot x})_0}$$

with two-sided α-level critical region $\chi^2 \leq \chi^2_{\alpha/2}$ or $\chi^2 \geq \chi^2_{1-(\alpha/2)}$, upper one-sided α-level critical region $\chi^2 \geq \chi^2_{1-\alpha}$, and lower one-sided α-level critical region $\chi^2 \leq \chi^2_{\alpha}$.

10-5-4 INFERENCES CONCERNING $\mu_{y\cdot x}$

For a particular value of x, say x_0, Conclusion 4 says that

$$t_C = \frac{a + b(x_0 - \bar{x}) - \mu_{y\cdot x_0}}{s_{y\cdot x}\sqrt{\dfrac{1}{n} + \dfrac{(x_0 - \bar{x})^2}{(n-1)s^2_x}}}$$

follows the t distribution with $n-2$ degrees of freedom. Therefore,

$$P(-t_{1-(\alpha/2)} < t_C < t_{1-(\alpha/2)}) = 1-\alpha,$$

whence

$$P\left[a + b(x_0 - \bar{x}) - t_{1-(\alpha/2)}s_{y\cdot x}\sqrt{\frac{1}{n} + \frac{(x_0-\bar{x})^2}{(n-1)s^2_x}}\right.$$
$$\left. < \mu_{y\cdot x_0} < a + b(x_0-\bar{x}) + t_{1-(\alpha/2)}s_{y\cdot x}\sqrt{\frac{1}{n} + \frac{(x_0-\bar{x})^2}{(n-1)s^2_x}}\right].$$

This then provides a $(1 - \alpha)$-level confidence interval for the population mean y at the value $x = x_0$.

The length of the interval, obtained by subtracting the lower limit from the upper limit, is

$$2t_{1-(\alpha/2)}s_{y\cdot x}\sqrt{\frac{1}{n} + \frac{(x_0 - \bar{x})^2}{(n-1)s_x^2}}.$$

The role of x_0 in this quantity is interesting and important. Notice that when $x_0 = \bar{x}$, the term $(x_0 - \bar{x})^2/(n-1)s_x^2$ vanishes, and thus the interval is shorter in this case than for any other value of $\bar{x}$. The farther x_0 is located away from $\bar{x}$, the longer the interval becomes. This is quite reasonable when you consider that $\bar{x}$ is a central value of x, around which most of the sampling of y values occurs. Thus, we do a better job of estimating $\mu_{y\cdot x}$ near the center of the region in which we have sampled than we do out toward the periphery. In fact, the intervals for $\mu_{y\cdot x}$ can be represented as a pair of curves corresponding to various values of x_0, the curves bending away from the fitted regression line as x_0 gets farther from $\bar{x}$. This is illustrated in Figure 10-7.

10-5-5   Prediction of a New Value of y

In the preceding section, we were concerned with estimating the mean of the distribution of y at a fixed value of x. It is possible also to estimate a single new observation on the basis of a fitted regression line. Since single observations vary more than means, we would expect limits in this case to be wider than limits on a mean. We will construct an interval for y_{x_0}, a new value of y at the value $x = x_0$. Since this is not an interval for a parameter, we call it a prediction interval. A $(1 - \alpha)$-level prediction interval has the property that under repeated random sampling and drawing of an additional y value, the value will lie within the interval a proportion $1 - \alpha$ of the time. Notice that here both the interval and the new y vary for different samples. This repeated sampling involves selecting new values $y_1, y_2, \ldots, y_n$ at the values $x_1, x_2, \ldots, x_n$, fitting the regression line, calculating the prediction interval, and selecting a new y at $x = x_0$.

The limits of the $(1 - \alpha)$-level prediction interval are

$$a + b(x_0 - \bar{x}) \pm t_{1-(\alpha/2)}s_{y\cdot x}\sqrt{1 + \frac{1}{n} + \frac{(x_0 - \bar{x})^2}{(n-1)s_x^2}},$$

where in the double symbol ($\pm$), the plus sign gives the upper and the minus sign the lower limit. Notice that this interval is of length

$$2t_{1-(\alpha/2)}s_{y\cdot x}\sqrt{1 + \frac{1}{n} + \frac{(x_0 - \bar{x})^2}{(n-1)s_x^2}}$$

and is longer than the interval for $\mu_{y \cdot x_0}$ of the preceding section. This agrees with intuition, since prediction of a single value should be more difficult than prediction of a population mean. The greater difficulty is reflected in the longer interval. The effect of selecting a value of x_0 close to or far from $\bar{x}$ is similar, and again we can draw curves showing the prediction limits across a range of values of x_0, as illustrated in the following example.

EXAMPLE

We shall illustrate the application of the inferential techniques of the preceding 5 sections in the ragweed pollen-counting data of Sec. 10-3. Many of the necessary computations were completed in the example of Sec. 10-4.

95 per cent confidence limits for A:

$t_{0.975}$ for 21 degrees of freedom is 2.080.

$$\bar{y} \pm t_{0.975} \frac{s_{y \cdot x}}{\sqrt{n}} = 161.74 \pm 2.080 \cdot \frac{11.957}{4.7958} = 161.74 \pm 5.19.$$

lower limit: 156.55; upper limit: 166.93.

95 per cent confidence limits for B:

Now

$$s_x^2 = \frac{L_{xx}}{n(n-1)};$$

hence

$$s_x^2(n-1) = \frac{L_{xx}}{n} \quad \text{and} \quad s_x\sqrt{n-1} = \sqrt{\frac{L_{xx}}{n}}.$$

$$b \pm t_{0.975} \frac{s_{y \cdot x}}{s_x\sqrt{n-1}} = -1.09091 \pm 2.080 \cdot \frac{11.957}{\sqrt{1,012}}$$

$$= -1.09091 \pm 2.080 \cdot \frac{11.957}{31.812} = -1.09091 \pm 0.78180.$$

lower limit: -1.87271; upper limit: -0.30911.

Test of H: $B = 0$:

We shall do a two-sided 5 per cent level test.

By inspecting the 95 per cent confidence interval for B we see that it does not include zero, and this says that the hypothesis will be rejected.

However, we will proceed as if the interval had not been computed, in order to illustrate the general method. Now

$$t_0 = \frac{bs_x\sqrt{n-1}}{s_{y\cdot x}} = \frac{(-1.09091)(31.812)}{11.957} = -2.9024.$$

The critical values are ± 2.080; since t_0 does not lie between them, H is rejected. The downward trend in pollen counts as the slide is traversed is significant at the 5 per cent level.

95 per cent confidence limits for $\sigma_{y\cdot x}^2$:

With 21 degrees of freedom, $\chi_{0\cdot025}^2 = 10.283$ and $\chi_{0\cdot975}^2 = 35.479$.

$$(n-2)s_{y\cdot x}^2 = (21)(142.96) = 3002.16.$$

lower limit: $\dfrac{(n-2)s_{y\cdot x}^2}{\chi_{0\cdot975}^2} = \dfrac{3002.16}{35.479} = 84.62.$

upper limit: $\dfrac{(n-2)s_{y\cdot x}^2}{\chi_{0\cdot025}^2} = \dfrac{3002.16}{10.283} = 291.95.$

limits for $\sigma_{y\cdot x}$ are $\sqrt{84.62} = 9.199$ and $\sqrt{291.95} = 17.087.$

95 per cent confidence interval for $\mu_{y\cdot x}$ at $x = 10$:

$$a + b(x_0 - \bar{x}) \pm t_{0\cdot975}s_{y\cdot x}\sqrt{\frac{1}{n} + \frac{(x_0 - \bar{x})^2}{(n-1)s_x^2}}$$

$$= 161.74 + (-1.09091)(10 - 12) \pm (2.080)(11.957)\sqrt{\frac{1}{23} + \frac{(10-12)^2}{1012}}$$

$$= 163.92 \pm 24.871\sqrt{0.0434783 + 0.0039526} = 163.92 \pm 5.42.$$

lower limit: 158.50; upper limit: 169.34.

95 per cent prediction interval for y_x at $x = 10$:

$$a + b(x_0 - \bar{x}) \pm t_{0\cdot975}s_{y\cdot x}\sqrt{1 + \frac{1}{n} + \frac{(x_0 - \bar{x})^2}{(n-1)s_x^2}}$$

$$= 161.74 + (-1.09091)(10 - 12)$$

$$\pm (2.080)(11.957)\sqrt{1 + 0.0434783 + 0.0039526}$$

$$= 163.92 \pm 25.45.$$

lower limit: 138.47; upper limit: 189.37.

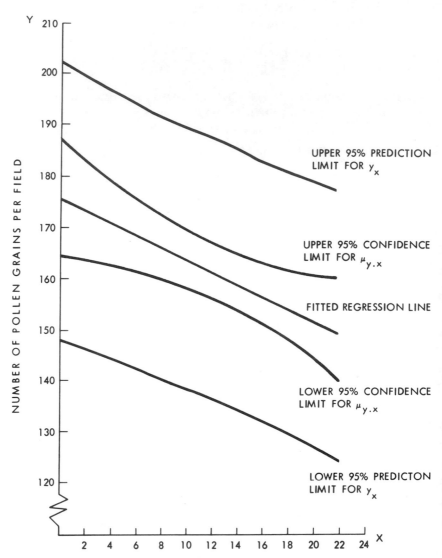

Figure 10-7 **Regression line, confidence bands for** $\mu_{y,x}$**, and prediction bands for** y_x **for number of ragweed pollen grains on microscope field number.**

Notice that these limits are wider than those for $\mu_{y \cdot x}$ at $x = 10$. Figure 10-7 shows 95 per cent confidence bands for $\mu_{y \cdot x}$ and 95 per cent prediction bands for y_x obtained by calculating confidence and prediction limits at several values of x. Notice the narrowing effect as x gets close to $\bar{x}$.

10-6   The Linear Correlation Coefficient in Linear Regression

We have seen that the variance $s_{y \cdot x}^2$ gives us an indication of the extent to which the fitted regression line fails to fit the observations. It would seem reasonable to ask for some measure of the extent to which the fitted line fits the data. Recall that

$$s_{y \cdot x}^2 = \frac{n-1}{n-2} (s_y^2 - b^2 s_x^2).$$

Factoring out s_y^2, we have

$$s_{y \cdot x}^2 = \frac{n-1}{n-2} \cdot s_y^2 \left(1 - b^2 \frac{s_x^2}{s_y^2}\right).$$

Letting

$$b^2 \frac{s_x^2}{s_y^2} = r^2,$$

we have

$$s_{y \cdot x}^2 = \frac{n-1}{n-2} s_y^2 (1 - r^2),$$

or, solving for r^2,

$$r^2 = \frac{(n-1)s_y^2 - (n-2)s_{y \cdot x}^2}{(n-1)s_y^2}.$$

But

$$s_{y \cdot x}^2 = \sum_{i=1}^{n} \frac{(y_i - \bar{y}_{xi})^2}{n-2} = \sum_{i=1}^{n} \frac{[y_i - a - b(x_i - \bar{x})]^2}{n-2}$$

and

$$s_y^2 = \frac{\sum_{i=1}^{n} (y_i - \bar{y})^2}{n-1},$$

whence

$$r^2 = \frac{\sum_{i=1}^{n} (y_i - \bar{y})^2 - \sum_{i=1}^{n} [y_i - a - b(x_i - \bar{x})]^2}{\sum_{i=1}^{n} (y_i - \bar{y})^2}.$$

Since $r^2 = b^2(s_x^2/s_y^2)$, which is nonnegative, it follows that

$$\sum_{i=1}^{n} (y_i - \bar{y})^2 - \sum_{i=1}^{n} [y_i - a - b(x_i - \bar{x})]^2 \geq 0.$$

Let us examine these relations in detail. $SS_y = \sum_{i=1}^{n} (y_i - \bar{y})^2$ measures the total variability in y, ignoring x. $SS_R = \sum_{i=1}^{n} [y_i - a - b(x_i - \bar{x})]^2$ measures the variability of y about the fitted regression line. Now if we take x into account we can possibly "explain" part of the variability in y, and thus it seems entirely reasonable that $SS_y \geq SS_R$. But we have seen that $SS_y - SS_R \geq 0$, which says the same thing. In other words, we certainly cannot increase the variability of y by accounting for a possible relationship with x.

Another way of explaining this is that $SS_y = \sum_{i=1}^{n} (y_i - \bar{y})^2$ measures the variation of the y's about the "flat" line $y = \bar{y}$. Certainly if we do not require this line to be flat but let its slope vary as the data indicate, we can do no worse in explaining variability in y. SS_R measures the variation of y about a line whose slope best fits the data in the least squares sense.

Thus $SS_y - SS_R$ measures the reduction in variation of y by fitting the regression line and $(SS_y - SS_R)/SS_y$ measures the proportionate reduction. But $(SS_y - SS_R)/SS_y = r^2$. Since r^2 is a proportion, it must lie between 0 and 1, whence it follows that $-1 \leq r \leq 1$.

The quantity r is called the *coefficient of linear* correlation between x and y, and we can see that either r^2 or r measures the strength of the linear relationship between x and y. Note that r is a so-called "pure" number, in the sense that it is unit-free.[2]

The computation of r uses the following relationships:

$$\frac{s_x^2}{s_y^2} = \frac{L_{xx}/n(n-1)}{L_{yy}/n(n-1)} = \frac{L_{xx}}{L_{yy}};$$

$$b^2 = \frac{L_{xy}^2}{L_{xx}^2}.$$

Therefore,

$$r^2 = b^2 \frac{s_x^2}{s_y^2} = \frac{L_{xy}^2}{L_{xx}^2} \cdot \frac{L_{xx}}{L_{yy}} = \frac{L_{xy}^2}{L_{xx}L_{yy}}$$

and hence

$$r = \frac{L_{xy}}{\sqrt{L_{xx}L_{yy}}}.$$

Alternatively,

$$r = \frac{\sum_{i=1}^{n} (x_i - \bar{x})(y_i - \bar{y})}{\sqrt{\sum_{i=1}^{n} (x_i - \bar{x})^2 \cdot \sum_{i=1}^{n} (y_i - \bar{y})^2}}.$$

[2] r is sometimes called the Pearson, product-moment correlation coefficient.

People frequently make the error of interpreting a moderately large r as indicating a strong relationship between x and y. The real strength of the relationship is best indicated by r^2, the proportion of variance of y "explained" by linear regression on x. Table 10-1 indicates that r^2 is never numerically larger than r, and is ordinarily much smaller.

Table 10-1

VALUES OF r AND r^2

r	r^2
± 0.00	0.00
± 0.10	0.01
± 0.20	0.04
± 0.30	0.09
± 0.40	0.16
± 0.50	0.25
± 0.60	0.36
± 0.70	0.49
± 0.80	0.64
± 0.90	0.81
± 1.00	1.00

If $r = \pm 1$, then all the points in the scatter diagram lie exactly on a straight line; i.e., there is a perfect linear relationship between x and y. If $r = 0$, there is no linear relationship.

Note further that r measures the strength only of *linear* relationships between x and y—nonlinear relationships would more appropriately be measured by other quantities.

10-7 Correlation and the Bivariate Normal Distribution

In our discussion of linear regression we considered x fixed and y variable. In that discussion all questions of probability were associated with y; x itself was not thought of as having any probability distribution.

We now turn to a class of problems in which both x and y have probability distributions—in fact, they have a joint distribution. This joint distribution gives the probabilities that x and y will simultaneously take on sets of values.

10-7-1 THE BIVARIATE NORMAL DISTRIBUTION

This distribution is a direct extension of the normal distribution. The distribution is depicted in Dixon and Massey [(9), Fig. 11.4, p. 203], and is shaped somewhat like a fireman's hat. It has a single mode.

The actual mathematical form of the distribution is somewhat complicated and is presented here only for reference purposes:

$$f(x, y) = \frac{1}{2\pi\sigma_x\sigma_y\sqrt{1 - \rho^2}} \, e^{-\frac{1}{2(1-\rho^2)}\left[\left(\frac{x-\mu_x}{\sigma_x}\right)^2 - 2\rho\left(\frac{x-\mu_x}{\sigma_x}\right)\left(\frac{y-\mu_y}{\sigma_y}\right) + \left(\frac{y-\mu_y}{\sigma_y}\right)^2\right]}$$

We notice several things in this formula. As might be expected, μ_x and μ_y denote the means of x and y, respectively, and σ_x and σ_y denote their standard deviations. The quantity ρ is called the correlation coefficient of x and y. It is closely related to the quantity r discussed in Sec. 10-6.

The *marginal distributions* of x and y give probabilities for x and y individually, ignoring the remaining variable. As might be expected, the marginal distributions are normal. For example, the marginal distribution of x is normal with mean μ_x, variance σ_x^2.

The conditional distribution of y given x gives probabilities that y will take certain values for a preassigned value of x. It can be seen that this provides a connection between the subject of bivariate distribution and the regression problems considered earlier.

The conditional distribution of y given x in the bivariate normal distribution is a normal distribution with mean $\mu_y + (\rho\sigma_y/\sigma_x)(x - \mu_x)$ and variance $\sigma_y^2(1 - \rho^2)$.

Now we can notice many important connections between linear regression and the bivariate normal conditional distributions. In the bivariate normal distribution, the conditional distribution of y for fixed x is normal, it has a variance $\sigma_y^2(1 - \rho^2)$, which does not depend on the value of x (homoscedasticity), and its mean is a linear function of x (linearity of true regression). Thus the *assumptions* we made about the population in the case of simple linear regression are mathematically true for the bivariate normal conditional distribution.

Now let us examine the mean and variance of this conditional distribution in detail. In the regression setting we called this variance $\sigma_{y\cdot x}^2$. Thus we see that in this case $\sigma_{y\cdot x}^2 = \sigma_y^2(1 - \rho^2)$. Recalling that in Sec. 10-6 we found

$$s_{y\cdot x}^2 = \frac{n-1}{n-2} s_y^2(1 - r^2),$$

and using the symbol $\doteq$ to denote "is approximately equal to," we find that

$s_{y \cdot x}^2 \doteq s_y^2(1 - r^2)$. Furthermore,

$$\rho^2 = \frac{\sigma_y^2 - \sigma_{y \cdot x}^2}{\sigma_y^2},$$

and is thus the proportionate reduction in the variance of y due to regression on x, again a strong analogy to the regression situation.

Turning to the mean of this conditional distribution, we note that it is $\mu_y + (\rho\sigma_y/\sigma_x)(x - \mu_x)$. If we let $A = \mu_y$ and $B = \rho\sigma_y/\sigma_x$, we can reduce this mean to the form $A + B(x - \mu_x)$, which is similar to the regression function considered earlier, except that $\bar{x}$ has been replaced by μ_x. This is reasonable, since we are here concerned with the entire population of x and y values.

We further note that $B = \rho\sigma_y/\sigma_x$, which is analogous to the relation $b = r(s_y/s_x)$ arising from the definition $r^2 = b^2(s_x^2/s_y^2)$.

We can see from this discussion that our linear regression results, including tests, estimation, and confidence intervals, carry over to the bivariate normal conditional distribution.

10-7-2   ESTIMATION AND TESTS FOR THE CORRELATION COEFFICIENT

A random sample in the case of the bivariate normal distribution will be a set of n pairs of observations, $(x_1, y_1), (x_2, y_2), \ldots, (x_n, y_n)$.

We shall use $r = L_{xy}/\sqrt{L_{xx}L_{yy}}$ as an estimator for ρ. Since we know that for the bivariate normal distribution the relation between x and y, as indicated by the mean y value for a given x, is linear, ρ measures the strength of this relationship between x and y. ρ lies between -1 and $+1$, the sign indicating whether the relationship is direct (when ρ is positive, y tends to increase with increasing x) or inverse (when ρ is negative, y tends to decrease with increasing x). When ρ is equal to 0, there is no relation between x and y, and it can be shown that in the bivariate normal distribution, this implies that the random variables are independent.

Let us now consider the problem of constructing a confidence interval for ρ. Three pieces of information are needed to construct the interval—the confidence level, n, and r.

Table A-12 permits direct graphic determination of 95 per cent and 99 per cent confidence limits. The use of the table is illustrated by the following example. Suppose that we have computed $r = 0.40$ on the basis of a random sample of 20 pairs of observations from a bivariate normal distribution and want to construct 99 per cent confidence limits for ρ. We read the limits as -0.05 and $+0.71$.

Experimenters frequently want to test the hypothesis $\rho = 0$ to determine whether there is any linear relationship between x and y. The test is completely equivalent to the test of $H : B = 0$ discussed in Sec. 10-5-2, and we will not discuss it further. Alternatively, we can determine confidence limits for ρ from Table A-12, noting whether zero is included within the interval.

The sampling distribution of r was discovered by R. A. Fisher in 1915 (16). This distribution depends only on ρ and n. The distribution of r is symmetric for $\rho = 0$, but skewed otherwise—negatively skewed for $\rho > 0$ and positively skewed for $\rho < 0$. Thus if ρ is positive, r can frequently underestimate it quite badly, and if ρ is negative, r can overestimate it quite badly.

Partly because of this skewness, Fisher introduced the transformation

$$z = \frac{1}{2} \log_e \frac{1 + r}{1 - r}$$

and showed that z was approximately normally distributed with mean $\frac{1}{2} \log_e [(1 + \rho)/(1 - \rho)]$ and variance $1/(n - 3)$. This transformation is very effective for moderate-sized samples. Since the adequacy of the approximation depends both on ρ and n, it is difficult to say how large n should be, but if $n \geq 20$, the transformation will give good results. Table 14 of Pearson and Hartley (43) is useful for making the transformation from r to z. We see from the table, for example, that an r of 0.332 corresponds to a z of 0.3451.

Additional tests and examples for the bivariate normal correlation coefficient are to be found in Pearson and Hartley, pp. 28–32.

E X E R C I S E S

10-1 The following data represent diastolic blood pressures taken during rest. The x values denote the length of time in minutes since rest began, and the y values denote diastolic blood pressures.

x:	0	5	10	15	20
y:	72	66	70	64	66

a. Construct a scatter diagram.
b. Find a, b, $s_{y \cdot x}^2$, $s_{y \cdot x}$.
c. Write the equation of your fitted line and draw it on the scatter diagram.

10-2 Using the results of Exercise 10-1, find a

a. Ninety-five per cent confidence interval for A.
b. Ninety per cent confidence interval for B.

c. Ninety-five per cent confidence interval for $\mu_{y \cdot x}$ for $x = 10$.
d. Ninety per cent prediction interval for Y for $x = 10$.

10-3 Using the results of Exercise 10-1, test the following hypotheses:

a. $H:A = 68.0$.
b. $H:B = 0$.
c. $H:\sigma^2_{y \cdot x} = 9.0$.

10-4 The following data represent systolic blood pressure readings on 30 females preselected by age (covering ages 40–85):

Subject Number	Age (x)	Systolic (y)
1	42	130
2	46	115
3	42	148
4	71	100
5	80	156
6	74	162
7	70	151
8	80	156
9	85	162
10	72	158
11	64	155
12	81	160
13	41	125
14	61	150
15	75	165
16	53	135
17	77	153
18	60	146
19	82	156
20	55	150
21	71	158
22	76	158
23	44	130
24	55	144
25	80	162
26	63	150
27	82	160
28	53	140
29	65	140
30	48	130

a. Draw a scatter diagram of y on x.
b. Find a, b, $s^2_{y \cdot x}$, $s_{y \cdot x}$.
c. Compute your fitted line and draw it on the scatter diagram.
d. Find a 95 per cent confidence interval for A.

    e.   Find a 90 per cent prediction interval for y for $x = 65$.

    f.   Test the hypothesis $H:B = 0$.

10-5   Using the data of Exercise 10-4,

    a.   Compute r, the coefficient of correlation.

    b.   Find a 95 per cent confidence interval for B.

    c.   Test the hypothesis $H:\rho = 0$.

10-6   Thirteen pingue-fed sheep [Aanes (1)] died as a result of this poisoning. Below are recorded weight x in pounds and time of death y in hours:

Weight (x)	Time of Death (y)
46	44
55	27
61	24
75	24
64	36
75	36
71	44
59	44
64	120
67	29
60	36
63	36
66	36

    a.   Examine the regression of y on x by finding: scatter diagram, a, b, and fitted line.

    b.   Do you feel that there is a significant relationship between the weight of the sheep and the time of death? Validate your argument by appropriate statistical tests.

    c.   Compute a 90 per cent confidence interval for $\mu_{y \cdot x}$ for $x = 60$ lb.

    d.   Compute a 90 per cent prediction interval for y for $x = 60$ lb.

    e.   Find a 95 per cent confidence interval for B.

    f.   Discuss ways of dealing with the uncooperative sheep who refused to die before 120 hours. Does he affect any of the regression assumptions? which ones?

10-7   Extreme values can have the effect of changing r to a large extent.

    a.   Recompute r for the data of Exercise 10-4, removing subjects numbered 3 and 6.

    b.   Compare its magnitude with the r computed in Exercise 10-5.

10-8   From the formula

$$s_{y \cdot x}^2 = \frac{n-1}{n-2}(s_y^2 - b^2 s_x^2),$$

derive the formula

$$s_{y \cdot x}^2 = \frac{L_{xx}L_{yy} - L_{xy}^2}{(n-2)nL_{xx}}.$$

10-9 (This exercise requires calculus.) Minimize the expression

$$\sum_{i=1}^{n} (y_i - y_{x_i})^2 = \sum_{i=1}^{n} \{y_i - [a + b(x_i - \bar{x})]\}^2$$

as a function of a and b showing that the values of a and b at the minimum are $\bar{y}$ and L_{xy}/L_{xx}.

The Analysis
of
Variance

11-1 The General One-factor Analysis of Variance

11-1-1 INTRODUCTION

In many experimental situations samples are selected from several different populations. A frequent problem in such situations is to determine whether there are any differences among the population means. More precisely, we frequently want to test the null hypothesis that there is no difference in the means of the populations from which the separate samples have been drawn. Examples of situations of this general type include:

1. Study of several different lots of vaccine in terms of effect on anti-body titer in monkeys.
2. Serum cholesterol determinations on aliquots from the same pool of serum by several different examiners to see if there is a significant difference among the examiners' means.
3. Analysis of water samples drawn at several different sampling points in a city supply to see if there is significant variation in mean water quality among sampling points.

The technique used most frequently to test the above hypothesis is the analysis of variance. It is based on the following question. Is there significantly more variation among the group means than there is within the groups? The collected or pooled variation within groups is used as a standard of comparison, because it measures the inherent observational variability in the data. A difference in means should be large relative to this inherent variability if it is to be meaningful. The term "analysis of variance" is used because, as we shall show, the total variability in the set of data can be broken up into the sum of the variability among the sample means and the variability within samples. Notice that although the technique is called the analysis of variance, it is basically a method for studying variation among means, this variation being measured, as usual, by a variance.

11-1-2   ANALYTIC DEVELOPMENT

We must develop a little notation to treat this problem:

k = number of populations sampled.
i = index of sampled populations.
j = index of the observation within sample.
n_i = the number of observations in sample i.
n = the total number of observations.
x_{ij} = the jth observation in the ith sample.
T = the sum of all observations.
T_i = the sum of the observations in sample i.
$\bar{x}$ = the mean of all the observations.
$\bar{x}_{i.}$ = the mean of the observations in sample i.
μ_i = the mean of population i.

Except for division by a certain constant, which we will discuss later, a good measure of the total variability in the complete set of data is

$$\sum_{i=1}^{k} \sum_{j=1}^{n_i} (x_{ij} - \bar{x})^2.$$

This says that we take each observation, subtract the mean, square the difference, and sum across all the observations. We recognize this as being the same sequence of operations we go through in finding the variance except for dividing by a constant, though we ordinarily use a more efficient computing formula.

A good measure of the variability among the sample means is

$$\sum_{i=1}^{k} \sum_{j=1}^{n_i} (\bar{x}_{i.} - \bar{x})^2.$$

This in effect measures the amount of variability left in the data if each observation is replaced by the mean of its sample. That is, we ask how much variability is left if we remove all the within-sample variability?

The measured variability within a particular sample, say the first sample, would be $\sum_{j=1}^{n_1} (x_{1j} - \bar{x}_1.)^2$.

In the ith sample this variability would be $\sum_{j=1}^{n_i} (x_{ij} - \bar{x}_i.)^2$.

Finally, the total variability within the samples would be obtained by summing this expression across samples to obtain $\sum_{i=1}^{k} \sum_{j=1}^{n_i} (x_{ij} - \bar{x}_i.)^2$.

It seems quite reasonable to suppose that the total variability in the full set of data should be the sum of the variability within samples plus the variability among sample means. That is,

$$\sum_{i=1}^{k} \sum_{j=1}^{n_i} (x_{ij} - \bar{x})^2 = \sum_{i=1}^{k} \sum_{j=1}^{n_i} (x_{ij} - \bar{x}_i.)^2 + \sum_{i=1}^{k} \sum_{j=1}^{n_i} (\bar{x}_i. - \bar{x})^2$$

$$= \sum_{i=1}^{k} \sum_{j=1}^{n_i} (x_{ij} - \bar{x}_i.)^2 + \sum_{i=1}^{k} n_i (\bar{x}_i. - \bar{x})^2,$$

the second equals sign arising from the fact that the terms in the second sum do not change with j.

In line with our objective, comparison of the size of the 2 terms on the right should help us to determine the significance of the variability among means. The question is: what standard of comparison is appropriate? Remember that we want to handle various numbers of samples and various numbers of observations within samples.

If we can make certain assumptions about the data, a standard is available. These assumptions are

1. Each sample is a random sample from the corresponding population, and observations from different populations are independent.
2. The measurement variable x is normally distributed in each of the k populations.
3. The populations all have the same variance (homoscedasticity).

To get a standard of comparison for the among- and within-sample variation, we divide by a constant related to the number of observations in each sum of squares. In fact, this constant is the number of degrees of freedom and has an interpretation analogous to that given in Sec. 9-2. A degree of freedom is lost for each parameter estimated in the data. For the within-sample variance we are estimating the population mean μ_i by the sample mean $\bar{x}_i.$. Since there are k such means, we lose a total of k degrees of freedom from the original n observations, leaving $n - k$ degrees of freedom. For the variation among samples we begin with k sample means and lose only a single degree of freedom for the overall mean estimated by $\bar{x}$, leaving

$k - 1$ degrees of freedom in all. An alternative explanation examines the number of terms of the form $(x_{ij} - \bar{x}_{i.})$ and $n_i(\bar{x}_{i.} - \bar{x})$ that may be freely specified. Since for each i, $\sum\limits_{i=1}^{n_i} (x_{ij} - \bar{x}_{i.}) = 0$, only $n_i - 1$ such terms can be specified, giving $\sum\limits_{i=1}^{k} (n_i - 1) = n - k$ in all. Similarly, since

$$\sum_{i=1}^{k} n_i(\bar{x}_{i.} - \bar{x}) = 0,$$

only $k - 1$ terms of the form $n_i(\bar{x}_{i.} - \bar{x})$ can be specified.

When a sum of squares is divided by its number of degrees of freedom, the result is called a *mean square*. The mean squares for within samples and among samples are, respectively,

$$\frac{\sum\limits_{i=1}^{k} \sum\limits_{j=1}^{n_i} (x_{ij} - \bar{x}_{i.})^2}{n - k} \quad \text{and} \quad \frac{\sum\limits_{i=1}^{k} n_i(\bar{x}_{i.} - \bar{x})^2}{k - 1}.$$

Now let us consider how to assemble the mean squares within and among samples into a test statistic. If the hypothesis of no difference between the populations' means is true, then we expect the sample means $\bar{x}_{i.}$ from these populations to be relatively similar. Thus, the difference between each sample mean and the pooled sample mean $\bar{x}$ will be small, as will be $(\bar{x}_{i.} - \bar{x})^2$. Finally, the among-samples mean square will be relatively small. However, if the hypothesis is false, we would expect $\bar{x}_{i.} - \bar{x}$ to be large at least for some samples, resulting in a relatively large value for the among-samples mean square.

Notice that $\bar{x}_{i.}$ is a sample mean of n_i observations and will, therefore, be subject to a variance only $1/n_i$ times as large as that of a single observation. The multiplication of $(\bar{x}_{i.} - \bar{x})^2$ by n_i in the among-sample sum of squares adjusts for this variance reduction.

Now we must define "relatively small or large," and therefore we need a standard or norm with which to compare the among-samples mean square. The within-samples mean square, measuring the pooled variability of the observations within samples, yields a measure of the "naturally" occurring experimental variation. That is, the within-samples mean square is the pooled or residual variance in the system. Although this variability may be subject to control prior to collection of the data, for example, by refinements in the technique of measurement, it represents the variation still remaining at the time of analysis. As a test, it is plausible to compare the among-samples mean square to that value measuring variability inherent in

Table 11-1

ANALYSIS OF VARIANCE TABLE

Source	Sum of Squares	Degrees of Freedom	Mean Square	Computing Formula	F
Among samples	$\sum_{i=1}^{k} n_i(\bar{x}_{i\cdot} - \bar{x})^2$	$k-1$	$\dfrac{\sum_{i=1}^{k} n_i(\bar{x}_{i\cdot} - \bar{x})^2}{k-1}$	$\dfrac{\sum_{i=1}^{k}\dfrac{T_i^2}{n_i} - \dfrac{T^2}{n}}{k-1}$	$\dfrac{\left(\sum\limits_{i=1}^{k}\dfrac{T_i^2}{n_i} - \dfrac{T^2}{n}\right)\Big/(k-1)}{\left(\sum\limits_{i=1}^{k}\sum\limits_{j=1}^{n_i} x_{ij}^2 - \sum\limits_{i=1}^{k}\dfrac{T_i^2}{n_i}\right)\Big/(n-k)}$
Within samples	$\sum_{i=1}^{k}\sum_{j=1}^{n_i}(x_{ij} - \bar{x}_{i\cdot})^2$	$n-k$	$\dfrac{\sum_{i=1}^{k}\sum_{j=1}^{n_i}(x_{ij} - \bar{x}_{i\cdot})^2}{n-k}$	$\dfrac{\sum_{i=1}^{k}\sum_{j=1}^{n_i} x_{ij}^2 - \sum_{i=1}^{k}\dfrac{T_i^2}{n_i}}{n-k}$	
Total	$\sum_{i=1}^{k}\sum_{j=1}^{n_i}(x_{ij} - \bar{x})^2$	$n-1$			

the investigation, the within-samples mean square. If the ratio of these is larger than could reasonably be attributed to chance, we reject the hypothesis. If this ratio is not too large, we accept the hypothesis.

Thus the ratio of the among-samples mean square to the within-samples mean square is our test statistic. If the hypothesis of no difference in the mean of the k populations is true, this statistic follows the F distribution with $k - 1$ numerator degrees of freedom and $n - k$ denominator degrees of freedom. That is,

$$F = \frac{\sum\limits_{i=1}^{k} n_i(\bar{x}_{i.} - \bar{x})^2 \Big/ (k - 1)}{\sum\limits_{i=1}^{k} \sum\limits_{j=1}^{n_i} (x_{ij} - \bar{x}_{i.})^2 \Big/ (n - k)}.$$

We see that F is large if the group means are quite different, and thus the test must be an upper one-sided test. The critical region for an α-level test is $F \geq F_{1-\alpha}$.

The F value we have derived is intuitively reasonable, but it is not much good for computing purposes. A better computing formula is

$$F = \frac{\left(\sum\limits_{i=1}^{k} \dfrac{T_i^2}{n_i} - \dfrac{T^2}{n} \right) \Big/ (k - 1)}{\left(\sum\limits_{i=1}^{k} \sum\limits_{j=1}^{n_i} x_{ij}^2 - \sum\limits_{i=1}^{k} \dfrac{T_i^2}{n_i} \right) \Big/ (n - k)}.$$

The results of the analysis of variance are frequently summarized in tabular form, as shown in Table 11-1.

EXAMPLE

A new analgesic drug has been proposed by a pharmaceutical house. It is desired to compare the effect of the drug with aspirin and placebo for use in treating simple headache. The variable measured is the number of hours a patient is free from pain following administration of the drug. In a small pilot study, two patients are given placebo, 4 are given the new drug, and 3 are given aspirin. The data are as follows:

Placebo: 0.0   1.0

Drug: 2.3   3.5   2.8   2.5

Aspirin: 3.1   2.7   3.8

Are the group means significantly different at the 5 per cent level?

$$k = 3;$$

$$n_1 = 2; \qquad n_2 = 4; \qquad n_3 = 3; \qquad n = 9;$$

$$T_1 = 1.0; \qquad T_2 = 11.1; \qquad T_3 = 9.6; \qquad T = 21.7;$$

$$\sum_{i=1}^{k} \sum_{j=1}^{n_i} x_{ij}^2 = (0.0)^2 + (1.0)^2 + \cdots + (3.8)^2 = 63.97;$$

$$\sum_{i=1}^{k} \frac{T_i^2}{n_i} = \frac{(1.0)^2}{2} + \frac{(11.1)^2}{4} + \frac{(9.6)^2}{3} = 62.02;$$

$$\frac{T^2}{n} = \frac{(21.7)^2}{9} = 52.32;$$

$$\text{Among-samples mean square} = \frac{62.02 - 52.32}{2} = 4.85;$$

$$\text{Within-groups mean square} = \frac{63.97 - 62.02}{9 - 3} = 0.325;$$

$$F = 14.92;$$

$$F_{0.95} = 5.14.$$

Since $F > F_{0.95}$, the difference in sample means is significant at the 5 per cent level. In fact, $0.001 < P < 0.005$. Notice that this result tells us only that there is some difference between the means, not where that difference lies. Inspection of the data tells us that the bulk of the significance is undoubtedly due to the difference between the placebo and the 2 drugs. There are procedures for identifying specific differences between means after the analysis of variance has established that a difference exists somewhere, but these procedures will not be presented here.

11-2   The General Two-factor Analysis of Variance

"Most experimental work today is based on the rule: 'Keep all variables constant but one,' an ancient and erroneous dictum which guarantees a high degree of inefficiency. One well-designed experiment, taking account of all relevant factors, is worth dozens or even hundreds of experiments which study one factor at a time keeping the others constant." [Mood (37), p. 358.]

11-2-1 INTRODUCTION, EXAMPLES, HYPOTHESES, AND ASSUMPTIONS

The analysis of variance is a commonly used technique for analyzing experiments that involve several factors, and is applicable in many different experimental situations of varying degrees of complexity.

The term "factor," when applied to the analysis of variance, refers to a quantity used to classify or otherwise to distinguish experimental units, the effect of which we wish to assess. A factor occurs in the experiment at several "levels." These levels are specific fixed values or states of the factor. For example, in the preceding section we discussed a trial of a new analgesic drug. Here the factor was type of treatment, and it occurred at 3 levels: placebo, new drug, and aspirin.

In a 2-factor experiment a pair consisting of a fixed level of the row factor and a fixed level of the column factor is called a *cell*. Suppose the first or row factor occurs ar r levels and the second or column factor at c levels. Then the experiment contains rc cells.

EXAMPLE

1. An experiment to test the effect in terms of antibody response of 4 different vaccine preparations and of 6 different amounts of a certain vaccine additive. Here the 2 factors are *type of vaccine* and *amount of additive*. The first factor occurs at 4 levels; the second at 6 levels ($r = 4$, $c = 6$).
2. An experiment to study simultaneously genetic and dietary factors in scurvy production in guinea pigs. Suppose we choose 3 varieties of guinea pigs and 5 different diets for study ($r = 3$, $c = 5$).
3. A study of the effect in terms of prolongation of life on 2 types of leukemia under 5 different methods of treatment ($r = 2$, $c = 5$).

Notice that the main objective of the analysis of variance is to assess the influence of each factor individually and in combination upon some response variable. In the 3 examples the response variables are, respectively, antibody level; some measure of scurvy activity, perhaps a measure of extent and severity of hemorrhagic activity; length of survival after initiation of treatment.

Specifically, we will test 3 hypotheses:

H_R: The row factor has no effect. That is, changing the level of the row factor does not change the population mean of the response variable. In the first example, this would state that changes in vaccine preparation have no influence on antibody level.

H_C: The column factor has no effect. Changing the level of the column factor does not change the population mean of the response variable. In the example, changing the amount of additive does not affect antibody level.

H_I: The row and column factors do not interact. In other words, the effect of changing the level of the row factor is the same at each level of the column factor. In the vaccine example, if the third type of vaccine produces in the population an overall average antibody response one unit greater than the first type, then H_I states that it produces this same amount of increase for each amount of additive. The influence of rows and columns in the statement of H_I is symmetric, and thus it is equivalent to regard the hypothesis as stating that any column effect is constant across the rows. In the absence of interaction, the response at level i of the row factor and level j of the column factor is the result of the individual action of these levels, with no joint action additional to that explainable by knowledge of the amount of the individual actions. When H_I is true, we say that the row and column factors are *additive*. Notice that no experiment keeping all factors constant but one could provide information about interaction.

To test H_R, H_C, and H_I we make certain assumptions. These are

1. The response variable is normally distributed.
2. The "residuals," i.e., the values obtained by subtracting from each observation the overall population mean, the row effect, the column effect, and any interaction effect, are independent and have the same variance.

Fortunately, the analysis of variance is quite robust to violations of the assumptions of normality and equal variances. Furthermore, it is often possible to transform or rescale the response variable to decrease the degree of departure from these assumptions, although such transformations will influence the size and extent of interaction effects.

11-2-2 RANDOM AND FIXED FACTORS

In testing H_R, H_C, and H_I the nature of the individual factors is important. The issue is whether the factor levels actually present in the experiment can be regarded as a sample from some large population of levels. That is, can the r levels of the row factor used in the study be regarded as a sample of levels, or are they the only relevant levels? To be more formal, we state the following definitions:

Definition

If, for a factor, the levels included in an experiment can be regarded as a random sample from some very large or infinite population of factors, the factor is said to be a *random factor*.

Definition

If the included levels of a factor are the only relevant levels, then the factor is said to be a *fixed factor*.

Before one proceeds to test the significance of a factor in an analysis of variance involving more than one factor, that factor should be classified as either fixed or random. Although this decision will partly reflect the basic objectives of the research, we will give some examples of factors that are relatively easy to classify.

If we are investigating differences among subjects, it is rare that the subjects in the experiment are the only ones to whom inferences are desired. Ordinarily we wish to consider subjects to be a sample from some large population. The factor "subjects" will then be considered a random factor. In an investigation of interobserver variability, usually we wish to extend inferences to a much larger set of observers than the specific individuals participating in the study; "observers" would, therefore, ordinarily be a random factor. Similarly, if we are investigating variation over time of some response variable and collect observations during several weeks, "weeks" should probably be considered a random factor, since we would usually hope to make inferences to many weeks other than those used in the study itself.

Notice that in a strict sense the levels of a random factor are not often selected at random from some population of levels. In such cases we have the problem, as we have had throughout this book, of asking, "From what sort of population might these units (levels) be reasonably considered to be a random sample?" The important thing here is the intent to make inferences to some large population of levels, balanced by the need for care in not extending inferences too far. Here again, statistics and the field of application must be combined to produce an inference that is neither too broad nor too narrow. Naturally, in those relatively rare situations when levels can actually be selected at random, we are all for the idea. But to label a factor "random" only when this situation applies is to be ridiculously pedantic.

Fixed factors, as we have seen, have all their pertinent levels included in the experiment. If we are comparing the 2 sexes with respect to some response variable, the factor "sex" will be fully represented in the study by inclusion of its 2 levels, male and female, and will thus be a fixed factor. Similarly, if we classify the study population into individual age groups and compare subjects from all the groups, the factor "age" will be fixed. In a clinical trial in which an active drug and a placebo are being compared, the treatment factor is fixed, since its 2 relevant levels are both present in the experiment. Similarly, if several dosages of a drug are being compared, the factor dosage would ordinarily be considered fixed. While inferences to other dosages might be desired, these inferences would usually be made by assuming some particular mathematical form for a dose-response curve, with the included dosages carefully selected to provide reasonable coverage of the curve. Inferences to other dosages would then not be based on considering the selected dosages to be a random sample.

In a 2-factor experiment the decision of whether both factors are fixed, both are random, or one is fixed and the other random is an important issue and must be determined before the analysis of variance can be performed. A valid analysis depends heavily upon the nature of the factors in the experiment. In the jargon of experimental statistics, the experiment in which one factor is fixed and the other random is often called a mixed-model experiment.

11-2-3 DEFINITIONS AND NOTATION

We shall discuss the most general 2-factor analysis of variance; i.e., we shall suppose that at any fixed combination of levels of the 2 factors we have several observations. That is, any particular "treatment" combination has several observations associated with it. A dot appearing in place of a subscript indicates that the missing subscript has been summed out. As usual, a bar over a symbol means that an appropriate mean has been calculated.

i = index of row (level of first factor).

j = index of column (level of second factor).

k = index of individual observation within a cell.[1]

r = number of rows (number of levels of first factor).

c = number of columns (number of levels of second factor).

m = number of observations per cell (note that in this discussion this is the same for every cell).

x_{ijk} = the kth observation in the ith row and the jth column.

$\bar{x}_{ij.}$ = the mean of the observations in cell (i, j), i.e., in row i and column j.

$\bar{x}_{i..}$ = the mean of the observations in row i.

$\bar{x}_{.j.}$ = the mean of the observations in column j.

$\bar{x}$ = the mean of all the observations.

[1] Notice that this is a different use of the letter k than in the section on the one-factor analysis of variance. It is one of the sadder facts of life for mathematicians, statisticians, and students that our alphabet contains only 26 letters, and that some of them have to be used to represent more than one entity. You are really old hands at this—after all, x has been used to represent many things by now.

T_{ij} = the sum (total) of all the observations in cell (i, j), i.e., in row i, column j.

R_i = the sum of all the observations in row i.

C_j = the sum of all the observations in column j.

T = the sum of all the observations.

$n = rcm$ = the total number of observations.

Table 11-2 gives a schematic representation of data from such an experiment.

11-2-4 Assessing Main Effects and Interaction

The effect of a factor acting alone is called the main effect of that factor. Further, the effect of a factor is the total effect of all its levels. You will notice, then, that we are using the word "effect" in a particular way, to indicate the differential effect of one level of a factor as compared to the overall response level. Thus in the data the average response of the ith level of the row factor is $\bar{x}_{i..}$, the overall average response is $\bar{x}$, and the "effect" of the ith row is $\bar{x}_{i..} - \bar{x}$. Similarly, the effect of column j is $\bar{x}_{.j.} - \bar{x}$.

The main effect of the row factor might be measured by adding together all the squares of the effects of the individual rows,[2] i.e., by summing $(\bar{x}_{i..} - \bar{x})^2$. However, we shall proceed in a fashion similar to that used in the one-factor experiment. The total variation present in this experiment is obtained by adding $(x_{ijk} - \bar{x})^2$ over all observations, producing

$$\sum_{i=1}^{r} \sum_{j=1}^{c} \sum_{k=1}^{m} (x_{ijk} - \bar{x})^2.$$

If we ask how much variation remains when each observation is replaced by the mean of its row, we obtain the desired measure of the row main effect or of row-to-row variation, the row sum of squares:

$$\sum_{i=1}^{r} \sum_{j=1}^{c} \sum_{k=1}^{m} (\bar{x}_{i..} - \bar{x})^2 = cm \sum_{i=1}^{r} (\bar{x}_{i..} - \bar{x})^2.$$

The summations over j and k are replaced by coefficients c and m, respectively, since the quantity being summed does not involve j or k. Similar reasoning produces the column sum of squares:

$$rm \sum_{j=1}^{c} (\bar{x}_{.j.} - \bar{x})^2.$$

---

[2] As usual, we square before summing to avoid cancellation of positive and negative effects.

Table 11-2

SCHEMATIC REPRESENTATION OF DATA FROM A TWO-FACTOR EXPERIMENT

Levels of Column Factor

$i =$ \ $j =$	1	2	3	...	c	Sum	Mean
1	x_{111} x_{112} . . . x_{11m}	x_{121} x_{122} . . . x_{12m}	x_{131} x_{132} . . . x_{13m}	...	x_{1c1} x_{1c2} . . . x_{1cm}	R_1	$\bar{x}_{1\cdot\cdot}$
2	x_{211} x_{212} . . . x_{21m}	x_{221} x_{222} . . . x_{22m}	x_{231} x_{232} . . . x_{23m}	...	x_{2c1} x_{2c2} . . . x_{2cm}	R_2	$\bar{x}_{2\cdot\cdot}$
3	x_{311} x_{312} . . . x_{31m}	x_{321} x_{322} . . . x_{32m}	x_{331} x_{332} . . . x_{33m}	...	x_{3c1} x_{3c2} . . . x_{3cm}	R_3	$\bar{x}_{3\cdot\cdot}$
. . .	. . .	. . .	. . .	. . .	. . .	. . .	. . .
r	x_{r11} x_{r12} . . . x_{r1m}	x_{r21} x_{r22} . . . x_{r2m}	x_{r31} x_{r32} . . . x_{r3m}	...	x_{rc1} x_{rc2} . . . x_{rcm}	R_r	$\bar{x}_{r\cdot\cdot}$
Sum	C_1	C_2	C_3	...	C_c	T	
Mean	$\bar{x}_{\cdot1\cdot}$	$\bar{x}_{\cdot2\cdot}$	$\bar{x}_{\cdot3\cdot}$	...	$\bar{x}_{\cdot c\cdot}$		$\bar{x}$

Levels of Row Factor

Measurement of the interaction effect follows similar lines. The interaction effect of the ith row with the jth column is the extent to which the effect of row i acting within column j differs from its overall effect. The effect of row i within column j is the difference between the average response level of row i within column j, $\bar{x}_{ij.}$, and the overall average response level for observations in column j, $\bar{x}_{.j.}$, and is thus $\bar{x}_{ij.} - \bar{x}_{.j.}$. As we have seen, the overall effect of row i is $\bar{x}_{i..} - \bar{x}$. The interaction effect is, therefore, the effect of row i within column j minus the overall effect of row i or $(\bar{x}_{ij.} - \bar{x}_{.j.}) - (\bar{x}_{i..} - \bar{x}) = x_{ij.} - \bar{x}_{.j.} - \bar{x}_{i..} + \bar{x}$, or, if we rearrange the two middle terms, $\bar{x}_{ij.} - \bar{x}_{i..} - \bar{x}_{.j.} + \bar{x}$. Again, these terms are squared to avoid cancellation and added over all rows and columns. Multiplying by m, we obtain the interaction sum of squares: $m \sum\limits_{i=1}^{r} \sum\limits_{j=1}^{c} (\bar{x}_{ij.} - \bar{x}_{i..} - \bar{x}_{.j.} + \bar{x})^2$.

The constants cm, rm, and m, which form the respective multipliers for the row, column, and interaction sums of squares, have the effect of keeping the measures of variation of commensurate size. We would expect the quantity $\bar{x}_{i..}$ to be less variable than the quantity x_{ijk}—in fact, if each x_{ijk} is subject to variance σ^2, then $\bar{x}_{i..}$ as the mean of cm such quantities has variance σ^2/cm. In this light the multiplication of the quantities $(\bar{x}_{i..} - \bar{x})^2$ by cm in the row sum of squares makes more sense. Similar arguments account for the constants rm and m in the column and interaction sums of squares.

The number of degrees of freedom for rows is $r - 1$, for columns is $c - 1$, and for interaction is $(r - 1)(c - 1)$. The values $r - 1$ and $c - 1$ for rows and columns are reasonable extensions of the conclusions in the one-factor analysis of variance. The value $(r - 1)(c - 1)$ for number of interaction degrees of freedom is reminiscent of the value obtained in Sec. 9-4 for the r by c contingency table and arises from the fact that the quantity $\bar{x}_{ij.} - \bar{x}_{i..} - \bar{x}_{.j.} + \bar{x}$ sums to zero within any individual row or column, leaving $(r - 1)(c - 1)$ "independent" terms in all.

An additional sum of squares is available by summing the squares of terms of the form $x_{ijk} - \bar{x}_{ij.}$. The resulting quantity

$$\sum_{i=1}^{r} \sum_{j=1}^{c} \sum_{k=1}^{m} (x_{ijk} - \bar{x}_{ij.})^2$$

is called the within-cell sum of squares, or merely the within sum of squares. It measures the amount of variation among observations whose row and column levels are identical and thus determines the amount of "error" or "residual" variation not due to the action of the 2 factors under study. Notice that this is a special use of the term "error" and simply connotes variation due to all sources uncontrolled by the experiment. The number of degrees of freedom corresponding to the within-cell sum of squares is

$rc(m - 1)$. This is reasonable when you consider that there should be $m - 1$ degrees of freedom for each cell, since the cell mean in the population is estimated by $\bar{x}_{ij}$. with a resultant loss of one degree of freedom. There are rc cells and thus $rc(m - 1)$ within-cell degrees of freedom.

Again, mean squares are formed by dividing each sum of squares by the corresponding number of degrees of freedom. To test a particular hypothesis, the mean square corresponding to the hypothesis is divided by another mean square to form a statistic following the F distribution under the hypothesis. The divisor mean square depends upon whether the factors are both fixed, both random, or one of each. This will shortly be made explicit.

As in the case of the one-factor experiment, computing formulas are available to simplify the calculation of sums of squares. Here, however, an additional simplification is possible if each sum of squares is algebraically multiplied by $n = rcm$. Somewhat surprisingly, this has the effect of removing all divisions in such sums of squares—the divisions implicit in such quantities as $\bar{x}_{ij}$, $\bar{x}_{i..}$, $\bar{x}_{.j.}$, and $\bar{x}$, each of which is a sample mean of some sort. The quantity resulting from the multiplication of a sum of squares by n is called a large sum of squares. Division by degrees of freedom produces a large mean square and the ratio of 2 large mean squares is still the F test statistic, since the multiplications by n now cancel algebraically in numerator and denominator.

For example, if the row sum of squares $cm \sum\limits_{i=1}^{r} (\bar{x}_{i..} - \bar{x})^2$ is multiplied by n and expanded, we obtain the large sum of squares for rows $r \sum\limits_{i=1}^{r} R_i^2 - T^2$, a quantity denoted by L_R. Similarly, the large sum of squares for columns is $c \sum\limits_{j=1}^{c} C_j^2 - T^2$, and the large sum of squares for interaction is

$$L_I = rc \sum_{i=1}^{r} \sum_{j=1}^{c} T_{ij}^2 - r \sum_{i=1}^{r} R_i^2 - c \sum_{j=1}^{c} C_j^2 + T^2.$$

The within-cell large sum of squares is

$$L_W = n \sum_{i=1}^{r} \sum_{j=1}^{c} \sum_{k=1}^{m} x_{ijk}^2 - rc \sum_{i=1}^{r} \sum_{j=1}^{c} T_{ij}^2.$$

The large mean squares are denoted by M with the appropriate subscript. Explicitly, for rows $M_R = L_R/(r - 1)$, for columns $M_C = L_C/(c - 1)$, for interaction $M_I = L_I/(r - 1)(c - 1)$ and for within $M_W = L_W/[rc(m - 1)]$.

The F tests depend, as we have noted, upon the kind of model involved: fixed model (both factors fixed), random model (both factors random), or mixed model (one factor fixed, the other random). In all cases, to test H_R (no row main effect), M_R is the numerator of the F ratio; to test H_C (no column main effect), M_C is the numerator; to test H_I (no interaction), M_I is the numerator.

The denominators of the F ratios depend upon the model. In the fixed model, M_W is the denominator for all 3 test statistics, while in the random model M_I is the denominator of the F ratios for testing H_R and H_C, and M_W is the denominator for H_I. In the mixed model, if it is assumed for simplicity that the data are so arranged that the row factor is fixed and the column factor random, M_I is the denominator for H_R, M_W the denominator for H_C, and M_W the denominator for H_I. Data from mixed model experiments should be organized so that the fixed factor is the row factor.

The results of this section are summarized in the analysis of variance table, Table 11-3. All the analysis of variance F tests are one-sided tests, having as critical region $F > F_{1-\alpha}$.

Notice that if $m = 1$ the results are affected. In particular, since there is only one observation per cell, it is impossible to estimate within-cell variability, and we see that in this case $rc(m - 1)$, the number of within-cell degrees of freedom, is zero. Any test statistic with M_I as its denominator is unaffected, and the corresponding test continues to be valid. Ordinarily, for those test statistics for H_R or H_C with M_W as denominator, we use M_I instead to produce an approximate test. These approximations become exact if there is, in fact, no interaction between rows and columns in the underlying populations being sampled; if interaction is present, however, these approximate tests tend to understate the significance of the data. That is, the presence of population interaction when $m = 1$ tends to make it more difficult to find significant main effects for tests in which M_W must be replaced by M_I. Thus the true level of significance for such a test is smaller than the nominal level. It is impossible to test the hypothesis H_I of no interaction when $m = 1$.

11-2-5 A NUMERICAL EXAMPLE

One method of determining effectiveness of influenza vaccines is to inject a quantity of vaccine into groups of mice, wait a suitable period until antibody has had an opportunity to form, then sacrifice the mice, pool their serum, form successively increasing dilutions of serum, mix each dilution with live virus, and reinject the virus-serum mixture into other living organisms. The highest dilution of serum capable of neutralizing the pathologic action of the virus in these organisms is taken as the observed value—the higher the dilution, the more effective the vaccine. The response variable is often taken to be the highest numbered tube in the dilution series at which neutralization occurs.

The data in Table 11-4 represent such responses to 4 different vaccines, to each of which has been added one of 6 different amounts of a mineral oil, adjuvant preparation designed to increase antibody production. Each

Table 11-3

TWO-FACTOR ANALYSIS OF VARIANCE WITH m OBSERVATIONS PER CELL

Source	Sum of Squares	Large Sum of Squares	Degrees of Freedom	Large Mean Square*	F (fixed model)	F (random model)	F (rows fixed columns random)
Rows	$cm\sum_{i=1}^{r}(\bar{x}_{i\cdot\cdot} - \bar{x})^2$	$L_R = r\sum_{i=1}^{r} R_i^2 - T^2$	$r - 1$	M_R	$\dfrac{M_R}{M_W}$	$\dfrac{M_R}{M_I}$	$\dfrac{M_R}{M_I}$
Columns	$rm\sum_{j=1}^{c}(\bar{x}_{\cdot j\cdot} - \bar{x})^2$	$L_C = c\sum_{j=1}^{c} C_j^2 - T^2$	$c - 1$	M_C	$\dfrac{M_C}{M_W}$	$\dfrac{M_C}{M_I}$	$\dfrac{M_C}{M_W}$
Interaction	$m\sum_{i=1}^{r}\sum_{j=1}^{c}(\bar{x}_{ij\cdot} - \bar{x}_{i\cdot\cdot} - \bar{x}_{\cdot j\cdot} + \bar{x})^2$	$L_I = rc\sum_{i=1}^{r}\sum_{j=1}^{c} T_{ij}^2 \\ - r\sum_{i=1}^{r} R_i^2 - c\sum_{j=1}^{c} C_j^2 + T^2$	$(r-1)(c-1)$	M_I	$\dfrac{M_I}{M_W}$	$\dfrac{M_I}{M_W}$	$\dfrac{M_I}{M_W}$
Within cells	$\sum_{i=1}^{r}\sum_{j=1}^{c}\sum_{k=1}^{m}(x_{ijk} - \bar{x}_{ij\cdot})^2$	$L_W = n\sum_{i=1}^{r}\sum_{j=1}^{c}\sum_{k=1}^{m} x_{ijk}^2 \\ - rc\sum_{i=1}^{r}\sum_{j=1}^{c} T_{ij}^2$	$rc(m-1)$	M_W			
Total	$\sum_{i=1}^{r}\sum_{j=1}^{c}\sum_{k=1}^{m}(x_{ijk} - \bar{x})^2$	$L_T = n\sum_{i=1}^{r}\sum_{j=1}^{c}\sum_{k=1}^{m} x_{ijk}^2 - T^2$	$n - 1$				

* A large mean square is a large sum of squares divided by the corresponding number of degrees of freedom.

Table 11-4

INFLUENZA ANTIBODY RESPONSES TO 4 VACCINES
WITH 6 DIFFERENT AMOUNTS OF ADDITIVE

Vaccine	Amount of Additive						Row Sum (R_i)	Row Mean $(\bar{x}_{i..})$
	I	II	III	IV	V	VI		
A	5	2	3	7	3	7		
	6	4	3	4	8	8	87	4.83
	5	4	6	3	6	3		
B	3	3	5	2	6	4		
	2	6	7	7	3	7	82	4.56
	4	3	6	4	4	6		
C	5	5	6	5	9	3		
	2	3	7	6	7	6	95	5.28
	2	6	4	7	4	8		
D	2	4	2	7	5	5		
	4	2	2	2	6	2	59	3.28
	2	3	2	3	2	4		
Column Sum (C_j)	42	45	53	57	63	63	323 (T)	
Column Mean $(\bar{x}_{.j.})$	3.50	3.75	4.42	4.75	5.25	5.25		4.49 $(\bar{x})$

CELL TOTALS (T_{ij})

Vaccine	Amount					
	I	II	III	IV	V	VI
A	16	10	12	14	17	18
B	9	12	18	13	13	17
C	9	14	17	18	20	17
D	8	9	6	12	13	11

vaccine—additive amount combination is studied in 3 different mouse serum pools. Table 11-4 also presents the various sums needed in the analysis of variance. Perhaps the most important part of the table, however, is the presentation of row and column means. Such a presentation should accompany any analysis of variance. The tests of significance are sensitive to differences in such means, and insight into the meaning of significant results is provided by the means themselves.

Table 11-5 gives the results of the analysis of variance. Notice that the large sums of squares for row and column main effects have the advantage that on most modern electric desk calculators they can be computed in a single continuous machine operation. They are, however, not necessarily the best form in which to program the analysis of variance for computation on an electronic computer. Before the F ratios can be computed, a decision concerning the nature of the factors must be made. The vaccines are presumably the only relevant ones for inclusion in the experiment, and the row

Table 11-5

ANALYSIS OF VARIANCE OF INFLUENZA ANTIBODY RESPONSES

Source	Large Sum of Squares	D.F.	Large Mean Square	F	P
Vaccines	2,867	3	955.67	4.210	$0.01 < P < 0.025$
Amounts	2,381	5	476.20	2.098	$0.05 < P < 0.10$
Vaccines—amount interaction	2,719	15	181.27	0.799	$P > 0.50$
Within cells	10,896	48	227.00		
Total	18,863	71			

(vaccines) factor is, therefore, fixed. The amounts of additive were chosen carefully to cover the range of interest and could not be considered a random sample from any population. Therefore, both factors are fixed, and in accordance with Table 11-3 the hypotheses H_R, H_C, and H_I are tested by F ratios for which M_W is the denominator.

The significance of the vaccine factor indicates that the vaccines apparently differ in their ability to elicit antibody production. However, the differential effect of the various amounts of additive is not significant. The small F ratio for interaction indicates that the differential response to the various vaccines is essentially constant for the various amounts of additive. For example, we see from Table 11-4 that the mean response to vaccine D is somewhat lower than that to the other vaccines. The small interaction effect indicates that vaccine D is lower in response by roughly the same amount no matter how much additive is used.

11-3 The Latin Square

The analysis of variance can be extended to examine the effects of more than 2 factors. The inclusion of additional factors, however, rapidly increases the size of the experiment and the number of observations required. For example, if a third factor with b levels is investigated, then the number of observations required for a complete experiment is increased over the 2-factor experiment from rcm to $brcm$, i.e., a b-fold increase.

To avoid this difficulty, particular experimental arrangements called incomplete designs are available for application in some instances. These designs have the property that not all combinations of levels of the factors are present in the experiment.

The only example of an incomplete design we shall discuss is the Latin square. This is a means of assessing the main effects of 3 factors in a 2-way array. Table 11-6 gives an example of a Latin square with 5 rows and 5

Table 11-6

FIVE BY FIVE LATIN SQUARE

A	C	E	B	D
B	A	C	D	E
E	D	B	C	A
C	E	D	A	B
D	B	A	E	C

columns. Notice that the square consists of an arrangement of 5 Latin letters in a 5 by 5 array. The 3 factors under study are a row factor, a column factor, and a Latin letter factor. Each factor has the same number of levels, 5 in this case—the levels of the Latin letter factor being denoted by the letters A, B, C, D, E. Notice that the presence of a particular letter in a particular row and column indicates that that combination of factor levels is present in the experiment and will produce one observation for analysis. For example, the A in the upper left corner of the square indicates that row level 1, column level 1, and Latin letter level 1 occur in combination in the experiment. However, the third levels of the factors do not occur together, since in row 3, column 3 we find the letter B instead of C. In fact, in this square there are only 25 positions, indicating that only 25 combinations of factor levels are present, whereas there are $5^3 = 125$ combinations of levels potentially available (we dust off the good old multiplication principle of Sec. 4-4-1).

The essential defining property of the Latin square is that each Latin letter occurs exactly once in each row and each column of the square. Take a long look at Table 11-6 and verify for yourself that this is true.

The saving in number of observations is made at some cost. In particular, interactions cannot be tested in the Latin square. In fact, the situation is even more serious. The presence of interactions in the population can distort the Latin square tests for main effects. Ordinarily this distortion is in the direction of making true main effects more difficult to detect. For a very thorough but technical discussion of this problem, see Wilk and Kempthorne (61). These authors also discuss the influence of having various combinations of fixed and random factors in the Latin square.

Before exhibiting the analysis of variance for an r by r Latin square, we will need some notation. Again let R_i be the sum of observations in row i and C_j be the sum for column j. Let L_k be the sum for the kth Latin letter; L_k is determined by picking out all the observations in the positions in the square at which the kth letter is found and adding them up. Again T denotes the grand total of all observations. We test 3 hypotheses: H_R, no row main effect in the population; H_C, no column main effect in the population, and H_L, no Latin letter main effect in the population. As in the 2-factor analysis of variance, we calculate large sums of squares denoted by the letter L with appropriate subscript,[3] and large mean squares, obtained from the large sums of squares by dividing by the corresponding numbers of degrees of freedom. The standard of comparison for a large mean square, i.e., the denominator of the F ratio for testing the corresponding hypothesis, is M_D, the large mean square for deviations (or for denominator, if you prefer). The large sum of squares for deviations is denoted by L_D and is found by subtracting L_R, L_C, and L_L from L_T, the large sum of squares for total. $L_T = \sum x_{ijk}^2 - T^2$, where x_{ijk} denotes the observation for row i, column j, and the particular Latin letter found in that row and column, and where the sum denoted by $\sum$ with no limits is taken over all of the r^2 observations.

The assumptions made in the F tests for the Latin square analysis of variance are similar to those made in the 2-factor situation. That is, the response variable is assumed normal, and residuals, after the overall mean and row, column, and letter effects have been accounted for, are assumed independent, with constant variances. Further, we assume no interactions between rows, columns, and letters. In practice, the assumptions of normality and equal variances do not seem particularly restrictive, and moderate violations are probably not serious.

---

[3] No notational confusion should arise in practice, since the L_k, i.e., $L_1, L_2, \ldots, L_r$, are subscripted with numbers and denote the various Latin letter totals, whereas an L subscripted with an upper case letter denotes a large sum of squares.

Table 11-7 presents the analysis of variance for the Latin square consisting of r rows, r columns, and r Latin letters. Each of the 3 F tests has $r - 1$ degrees of freedom in the numerator and $(r - 1)(r - 2)$ in the denominator, and again is an upper, one-sided test with critical region $F \geq F_{1-\alpha}$.

Table 11-7

ANALYSIS OF VARIANCE FOR THE r BY r LATIN SQUARE

Source	Large Sum of Squares	Degrees of Freedom	Large Mean Square	F
Rows	$L_R = r \sum\limits_{i=1}^{r} R_i^2 - T^2$	$r - 1$	M_R	$\dfrac{M_R}{M_D}$
Columns	$L_C = r \sum\limits_{j=1}^{r} C_j^2 - T^2$	$r - 1$	M_C	$\dfrac{M_C}{M_D}$
Latin letters	$L_L = r \sum\limits_{k=1}^{r} L_k^2 - T^2$	$r - 1$	M_L	$\dfrac{M_L}{M_D}$
Deviations	$L_D = L_T - L_R - L_C - L_L$	$(r - 1)(r - 2)$	M_D	
Total	$L_T = n \sum x_{ijk}^2 - T^2$	$r^2 - 1$		

As a numerical example of a Latin square experiment, suppose that we are investigating the interobserver variation in taking systolic blood pressure readings. Suppose that 5 observers and 5 subjects agree to participate in the study. Each observer takes 3 systolic blood pressure readings in quick succession on a given subject, and the recorded observation is the average of these 3 readings. Since the order in which the blood pressure is taken on a given subject may conceivably be important, e.g., the first set of 3 readings on a subject may be slightly higher than the last 3, order of observation is another factor the investigator wishes to control.

Then the 3 factors, subjects, order of observation, and observers, can conveniently be the row, column, and Latin letter factors, respectively, in a Latin square. The Latin square property that each letter (observer) occurs exactly once with each row (subject) and column (order) provides a useful economy of time and effort and guarantees that in any single observation period each observer is collecting data from some subject and not a subject he has previously observed. For instance, if the Latin square of Table 11-6 is used for the experiment, during the third observation period (third column), subject 1's blood pressure is taken by observer E, subject 2's pressure by observer C, etc.

Table 11-8 represents data from such an experiment, using the design configuration of Table 11-6.

Table 11-9 gives the analysis of variance for the data in Table 11-8. We notice that the among-subjects effect is highly significant, as might be expected ($P < 0.0005$). This merely says that different people may tend to have different blood pressures. The effect of order of observation does not

Table 11-8

DATA FROM A LATIN SQUARE EXPERIMENT TO STUDY INTEROBSERVER
VARIABILITY IN SYSTOLIC BLOOD PRESSURE DETERMINATION*

Subject	Order				
	1	2	3	4	5
1	126	121	118	113	108
2	137	130	135	137	129
3	126	127	117	129	119
4	113	118	113	112	104
5	117	115	116	110	118
Subject totals	586	668	618	560	576
Subject means	117.2	133.6	123.6	112.0	115.2
Order totals	619	611	599	601	578
Order means	123.8	122.2	119.8	120.2	115.6
Observer totals	603	586	616	602	601
Observer means	120.6	117.2	123.2	120.4	120.2
	$T = 3008$; $\bar{x} = 120.3$				

* Each observation is the average of 3 systolic blood pressure measurements taken in quick succession by the same observer on the same subject.

reach significance. The interobserver effect, the main point of this investigation, is slight if we judge by the small F value, 1.212, associated with this effect. Inspection of the subject, order, and observer means in Table 11-8 reveals that the observed degree of variability among means seems consistent with the results of the F tests. As we noted earlier, it is generally useful when one is performing an analysis of variance to calculate and inspect the means for each of the levels of the various factors.

The Latin square analysis of variance produces F tests with $(r-1)(r-2)$ denominator degrees of freedom. It is a good rule of thumb that F tests with 6 or fewer degrees of freedom in the denominator are so insensitive to true

Table 11-9

ANALYSIS OF VARIANCE FOR LATIN SQUARE EXPERIMENT ON INTEROBSERVER
VARIABILITY IN SYSTOLIC BLOOD PRESSURE DETERMINATIONS

Source	Large Sum of Squares	Degrees of Freedom	Large Mean Square	F	P
Subjects	36,536	4	9,134.00	19.545	$P < 0.0005$
Order	4,776	4	1,194.00	2.555	$0.05 < P < 0.10$
Observers	2,266	4	566.50	1.212	$0.25 < P < 0.50$
Deviations	5,608	12	467.33		
Total	49,186	24			

effects as to be of limited practical value. Thus, we see that if $r = 4$, $(r - 1)(r - 2) = 6$, and our rule is just violated. As a result, we would recommend that the Latin square design for smaller than a 5 by 5 square should not in practice be used alone. Smaller squares can, however, be the basis for more complicated designs using several such squares.

Finally, a Latin square to be used in practice should be selected at random from the large number of squares of that size available. The introduction to the collection of tables by Fisher and Yates (18) gives a detailed description of the selection process for squares of different sizes.

11-4 Concluding Remarks

The analysis of variance comprises a collection of techniques that are among the most useful devices statistics can offer the experimental scientist. There is insufficient opportunity here for us to do more than provide an introduction to this extensive and fascinating area.

There are techniques available for extending the analysis of variance F tests. Notice, for example, that the F value for variation among subjects in the Latin square of Sec. 11-3 was highly significant. This tells us that not all the subjects have similar systolic blood pressures but does not tell us where the differences lie. Does one subject have a considerably higher or lower pressure than all the others, who are essentially similar, or do the subjects divide into 2 groups of 2 and 3 subjects, respectively, with similar blood pressures? Examination of the subject means in Table 11-8 is helpful but still does not tell us which pairs of individual subjects differ significantly. Work by Scheffé (47), Tukey (51), Duncan (12), Dunnett (13), and others has provided us a variety of techniques for answering such questions. The field is summarized by Miller (36), who gives an extensive list of references.

It is possible to transform the response variable in some cases in order to make the data more nearly normal and homoscedastic. That is, instead of analyzing x itself we might work with $\log x$ or $\sqrt{x}$. Such transformations really involve nothing more than changing the scale of measurement of the response variable, and are analogous to changing the template scale on a measuring guage. A suitable choice of transformation can improve the degree of satisfaction of assumptions such as those of normality and equal variances. Bartlett (2) discusses methods for choosing an appropriate transformation. It is important to realize that interactions can be affected by transformation and can, in fact, be either substantially increased or reduced in size.

EXERCISES

11-1 As a general surgeon on a hospital staff, your observations have led you to believe that certain kinds of pain may be alleviated by the mere offering of a drug, potent or not. To test your hypothesis 2 pain-killing drugs and a placebo are randomly administered to patients immediately following tonsillectomies. You record the time in hours until the patient complains of pain. The results are

Placebo	Drug A	Drug B
2.2	2.8	1.1
0.3	1.4	4.2
1.1	1.7	3.8
2.0	4.3	2.6
3.4	4.3	0.5

Carry out the appropriate statistical analysis.

11-2 A group of 8 mice with mammary adenocarcinoma are treated by irradiation of the tumor by a regime of 667 r 3 times a week. A biopsy is taken from each tumor 48 hours after cessation of radiation and a series of 3-minute mitotic counts made. The results are

			Mouse Number				
1	2	3	4	5	6	7	8
19	73	50	11	1	26	12	47
26	70	59	10	11	15	12	47
			12	12	11	9	
					11		

Is there a significant difference between the mean mitotic counts of the various mice?

11-3 In a particular hospital laboratory 3 technicians are making serum cholesterol determinations in milligrams per centimeter. To test the similarity of these technicians, sera from 5 normal subjects are split into $\frac{1}{6}$ aliquots to be tested twice by each technician. The data are as follows:

		Observer		
		1	2	3
	1	190, 193	187, 186	192, 190
	2	172, 170	164, 166	168, 169
Subject	3	180, 178	176, 177	178, 181
	4	206, 204	200, 201	203, 205
	5	175, 173	172, 173	176, 177

a. State and discuss whether each factor is fixed or random.
b. Carry out the appropriate 2-factor analysis of variance.

11-4 Consider the following study comparing 3 vitamins. Seven sets of triplets at age 1 year were gathered for study. Each child of a given family was randomly assigned to one of the 3 vitamin regimes for a 2 year period. One indicator of overall effect of the vitamins was felt to be growth. Suppose that the following data are weights gained in pounds:

		Vitamin		
		A	*B*	*C*
	1	11.2	9.3	10.4
	2	9.7	12.0	11.5
	3	8.2	9.4	8.9
Family	4	9.1	10.1	7.9
	5	11.0	10.3	10.8
	6	7.3	9.1	8.4
	7	8.2	8.3	10.1

Test to see if the vitamins are producing the same mean weight gain.

11-5 As an investigation of the effect of smoking on physical activity, consider the following experiment. Twenty-seven individuals were classified into one of 3 groups by smoking history and randomly assigned to one of the stress tests: bicycle ergometer, treadmill, or step tests. The time until maximum oxygen uptake was recorded in minutes. The results are

		Bicycle	Treadmill	Step
			Test	
Smoking History	Nonsmokers	12.8	16.2	22.6
		13.5	18.1	19.3
		11.2	17.8	18.9
	Moderate smokers	10.9	15.5	20.1
		11.1	13.8	21.0
		9.8	16.2	15.9
	Heavy smokers	8.7	14.7	16.2
		9.2	13.2	16.1
		7.5	8.1	17.8

a. State whether each factor is fixed or random.
b. Test the various hypotheses by the appropriate 2-factor analysis of variance.

11-6 Preparatory to a study involving blood pressure readings, an investigation of interobserver variability was conducted. An 8 by 8 Latin square design was used to assign the subjects to the observers. Thus the 3 factors were subjects (rows), observers (columns), and order (letter) of assigning subjects to observers. The following are the design configuration and systolic blood pressures read:

		1	2	3	4	5	6	7	8
					Observers				
Subjects	1	A	D	C	B	E	F	G	H
	2	B	C	D	A	F	E	H	G
	3	D	A	B	C	G	H	E	F
	4	C	B	A	D	H	G	F	E
	5	E	F	G	H	A	D	C	B
	6	F	E	H	G	B	C	D	A
	7	G	H	E	F	D	A	B	C
	8	H	G	F	E	C	B	A	D

Observers

		1	2	3	4	5	6	7	8
	1	128	108	110	106	100	102	112	110
	2	122	100	120	128	108	130	120	110
	3	110	98	110	120	102	108	108	104
Subjects	4	96	96	90	106	96	90	98	98
	5	120	128	130	128	110	132	128	134
	6	140	128	130	130	126	108	142	140
	7	110	108	110	106	114	110	114	118
	8	102	118	108	110	114	110	122	110

a. Carry out the tests of the various hypotheses.

b. What assumptions are made in completing the analysis?

11-7 Show that the conventional t test for equality of 2 normal means, assuming equal variances (Sec. 8-3-2), is equivalent to the analysis of variance test; i.e., show that $t^2 = F$, where F is the 1 factor analysis of variance statistic and

$$ t = \frac{\bar{x}_{1\cdot} - \bar{x}_{2\cdot}}{s_p \sqrt{\dfrac{1}{n_1} + \dfrac{1}{n_2}}} . $$

11-8 Verify that the same interaction sum of squares would result if we were to look at the effect of column j within row i as in the argument of Sec. 11-2-4.

11-9 Show that $\bar{x}_{ij\cdot} - \bar{x}_{i\cdot\cdot} - \bar{x}_{\cdot j\cdot} + \bar{x}$ sums to zero within any row or column.

12

Distribution-free and Nonparametric Methods

The basic methods of statistics make a variety of assumptions about the data to be analyzed. These assumptions are of 2 general types—assumptions about the state of nature and assumptions about experimental practice.

Normality of distribution, equality of population variances, and linearity of regression are examples of assumptions concerning the state of nature. Such assumptions are to a considerable extent outside the control of the statistician and the investigator. To be sure, there are devices for altering either the data or the type of analysis to improve the degree to which such assumptions are satisfied. For example, we might transform a variable under study and deal with its logarithm instead of its original value. We might account for nonlinearity of regression by fitting a polynomial or some other nonlinear regression curve (techniques not discussed in this book). However, in the end we are making assumptions about underlying distributions—either about their form or about their parameters.

The second class of assumptions concerns the form of the study and the way data are collected. Under this heading we include such things as the type of sampling, whether subjects are allocated to treatment and control groups at random, and whether observations are independent.

The techniques we shall discuss in this chapter are designed to bypass or avoid assumptions about the state of nature. Although there are some methods for avoiding assumptions about the nature of the study and data collection, these are not as extensively developed and generally require more mathematical background than we are assuming. Fortunately, assumptions in the latter class—experimental practice—are often under the direct control of the investigator, and with the collaboration of a statistician or at least with a general understanding of the modern principles of experimentation, he can so arrange his data collection as to satisfy or nearly satisfy them.

We have seen that assumptions about the state of nature often either specify a certain underlying form of distribution, e.g., the normal distribution, or make statements about parameters, e.g., equality of variances. Partly for these reasons, the techniques to be presented in this chapter are called either distribution-free or nonparametric methods. Furthermore, recall that the techniques that we have discussed for estimation and hypothesis testing in Chaps. 7, 8, 10, and 11 are designed to provide inferences about parameters: for example, intervals and tests for normal means, inferences about regression coefficients, and tests of the hypothesis that several normal distributions have the same mean. The chi-square frequency techniques of Chap. 9 are somewhat different and are, in fact, distribution-free. They are not presented as a part of this chapter, because their importance as a group justifies separate treatment.

We frequently want to make inferences that are inherently nonparametric. These include investigations of hypotheses such as: these 2 samples are drawn from populations having identical distributions, or these observations have been drawn at random with respect to order of selection (e.g., no tendency for the larger values to occur first). We may wish to estimate the median of a distribution by a confidence interval, specifying only that the underlying distribution is continuous, but with no other distributional assumption.

As a class, distribution-free or nonparametric methods have advantages and disadvantages. One of their chief attractions is that they are generally quicker and easier to apply than their competing parametric alternatives. They require less calculation or in some instances essentially no calculation. Their ease of application has made them attractive to individuals who are easily whelmed or even overwhelmed by formulas and symbols. And, of course, their chief advantage is their inherently greater range of applicability because of their milder assumptions.

Distribution-free or nonparametric methods also have a number of important disadvantages which in practice often outweigh their advantages. They sometimes pay for freedom from assumptions by a marked lack of sensitivity. Important effects can be entirely missed by a nonparametric technique because of its lack of power or because it produces confidence intervals that are too wide. To put it another way, some of these techniques are very blunt instruments. In many cases it is unwise to substitute one of these techniques, either because of computational ease or because the data do not quite satisfy some distributional assumption. The competing distribution-tied procedure may be quite robust to departures from that assumption.

An additional related disadvantage is that distribution-free techniques often disregard the actual scale of measurement and substitute either ranks or even a more gross comparison of relative magnitude. This is not invariably bad, but is often the basis for the loss of power or sensitivity referred to in the preceding paragraph.

The final disadvantage is one that is gradually being corrected by research in statistical methodology. At present, we have available a much larger body of techniques in distribution-free hypothesis testing than in distribution-free estimation. To some extent, this is an inevitable accompanying feature of techniques that disregard or weaken the underlying scale of measurement, since estimation is often tied to that scale, whereas hypothesis testing is concerned with selecting one of 2 competing decisions or actions.

There are 4 situations in which the use of a distribution-free or nonparametric technique is indicated:

1. When quick or preliminary data analysis is needed. Maybe you just want to see roughly how things are going with the data. Maybe somebody is about to leave for Atlantic City to present a paper tomorrow, and his chief thought it would be a nice idea to include some statistics in the paper, and you are feeling unusually tenderhearted or browbeaten today. Maybe your computer is suffering an acute exacerbation of some rare electronic disease. Maybe your computer programmer is manifesting that chronic, epidemic, occupational disability: temporal disorientation with gross distortion of task-time relationships. Maybe you are doing research in a "primitive" setting without access to modern computational aids.

2. When the assumptions of a competing distribution-tied or parametric procedure are not satisfied and the consequences of this are either unknown or known to be serious. Notice that violation of assumptions per se is not a sufficient reason to reject one of the classical procedures. As we have emphasized throughout this book, these procedures are often insensitive to such violations. For example, the *t* tests and intervals for normal means are relatively insensitive to nonnormality. We think that one of the most common errors made today is the underapplication of classical statistical methods because of the overzealous adherence to the letter of the law with respect to assumptions. Remember that assumptions will rarely (never?) be exactly satisfied. It is, therefore, as important to have a grasp of the consequences of violations of assumptions as to know what the assumptions are. A great deal of theoretical research in statistics today is directed toward investigating the robustness of statistical procedures.

3. When data are only roughly scaled; for example, when only comparative rather than absolute magnitudes are available. In dealing with clinical data, perhaps patients can only be classified as better, unchanged, or worse. Perhaps only ranks, i.e., largest, second largest, . . . , smallest, are available.

4. When the basic question of interest is distribution-free or nonparametric in nature. For example, do we have a random sample, or are these 2 samples drawn from populations with identical distributions?

Having delivered, we hope, sufficient caveats to make you wary of overapplication of these simple and attractive techniques, we can now proceed to discuss individual methods in detail.

12-1 The Wilcoxon Rank Sum Test

This procedure was first proposed by Wilcoxon in 1945 (60) and is designed to test the hypothesis that 2 random samples have been drawn from populations having identical distributions, if it is assumed that the samples are independently drawn, i.e., that observations in one sample are independent of those in the other. Notice that this hypothesis is similar to that of the t test of Sec. 8-3-2, if the assumptions of that test hold, that is, if we have 2 independent, random samples from normally distributed populations having the same variance. When we test the hypothesis that the means are equal, we are in effect testing the hypothesis that the underlying distributions are identical, normal distributions, since they are completely determined by their means and variances. The rank sum test eliminates the assumptions of normality and equal variances.

The test is performed by combining the samples and arranging the entire, combined set of observations in order of magnitude from the smallest to the largest observation. Rank numbers are then assigned, the smallest observation being given a rank of 1, the next smallest a rank of 2, etc., the largest being given a rank of $n_1 + n_2$, where n_1 and n_2 are the sample sizes.

The test statistic is T_1, the sum of the ranks of the first sample. This number is compared with the critical values in Table A-13. If it is between the values, the hypothesis is accepted; otherwise it is rejected. That is, if we let T_l and T_r be the lower and upper tabulated values, respectively, the critical region is $T_1 \leq T_l$ or $T_1 \geq T_r$; the acceptance region is $T_l < T_1 < T_r$.

The rationale for the test is that if the first sample tends to consist mainly of observations that are smaller than those of the second sample, then its rank values in the combined sample will be small, producing a small value of T_1. This provides evidence that values from the first distribution tend to be located to the left of the second—a denial of the tested hypothesis. On the other hand, if the first sample contains mainly observations larger than those in the second, a large value of T_1 will result, indicating a difference in distributions in the opposite direction.

Notice that the rank sum test does not use the actual values of the observations, only their relative magnitudes. In spite of this, theoretical investigations have established that even when all the assumptions of the t test hold, the rank sum procedure has power only slightly less than that of the t test. That is, even on the t test's home ground, this procedure performs very well.

One complication that often arises concerns the treatment of tied observations—that is, observations having identical values. They should intuitively be assigned identical rank numbers, but what common value should we choose? The best procedure and the one most commonly used is to assign to each of the tied observations the average of the rank numbers that would have been assigned if the observations had not been tied. Suppose, for example, that 2 observations are tied at a value of 120 and that in the combined sample these observations, if untied, would have been assigned ranks of 10 and 11. Then each observation is given a rank of $10\frac{1}{2}$. Remember, though, that the next larger observation in the sample must now be assigned a rank of 12.

It is customary to assume in the rank sum test that the observations come from continuously distributed populations. In this event, the probability of a tie will be zero. However, since even continuous variables are recorded in discrete form, an occasional tie might occur because of rounding. The rank sum test should probably not be applied to data with a great many ties, since the derivation of its distributional properties makes no explicit allowance for them. Putter (46) has examined the treatment of ties in this and several other nonparametric procedures.

To illustrate the application of the rank sum test, we will use the data of Sec. 7-2-5-3 on total histidine excretions in milligrams per 24-hour urine sample for 5 men and 10 women on unrestricted diets. The observed values were

Men: 229, 236, 435, 172, 432

Women: 197, 224, 115, 74, 138, 135, 107, 204, 200, 138

The combined, ordered sample (observations from male sample are underscored), with assigned rank numbers, is given below:

74, 107, 115, 135, 138, 138, 172, 197, 200, 204, 224, 229, 236, 432, 435

1, 2, 3, 4, 5, 6, 7, 8, 9, 10, 11, 12, 13, 14, 15

$$T_1 = 7 + 12 + 13 + 14 + 15 = 61^1$$

Inspecting Table A-13, we see that for $n_1 = 5$, $n_2 = 10$, the critical values of T_1 for a two-sided test with $\alpha = 0.01$ are $T_l = 19$ and $T_r = 61$. Thus, we can conclude that the sex difference is significant with $P \leq 0.01$. We might be tempted to conclude that $P = 0.01$, but this would not be correct. T_1 is discrete, and thus as it moves between adjacent values, such as 60 and 61, the

---

[1] A useful check on the calculation of T_1 is available. If T_1 and T_2 are, respectively, the sums of ranks for the 2 samples, then $T_1 + T_2 = [(n_1 + n_2)(n_1 + n_2 + 1)]/2$. Now, for these data $T_1 = 61$, as we have seen, and we find $T_2 = 1 + 2 + \cdots + 11 = 59$. But $[(n_1 + n_2)(n_1 + n_2 + 1)]/2 = [(15)(16)]/2 = 120$, and as a check $T_1 + T_2 = 61 + 59 = 120$.

corresponding two-sided significance level actually can be shown to move from 0.012 to 0.008. Thus a value of T_1 equal to one of the tabulated critical values is significant at or below the corresponding level of significance.

When we constructed a confidence interval for the difference between means for these same data in Sec. 7-2-5-3, we assumed normality but used the Welch approximation because of the rather large discrepancy between the sample variances. We found that the 95 per cent confidence limits were -0.2 and 295.4, which just include zero, causing us to say that the difference in means was not significant with $P = 0.05$. If we had applied the corresponding significance test of Sec. 8-3-3 to the data, we would have found that the test statistic

$$t = \frac{\bar{x}_1 - \bar{x}_2}{\sqrt{\dfrac{s_1^2}{n_1} + \dfrac{s_2^2}{n_2}}} = \frac{300.8 - 153.2}{57.5048} = 2.566.$$

The number of degrees of freedom from the Welch approximation would be 5, as calculated in Sec. 7-2-5-3, and from Table A-5 we find $P \doteq 0.051$. This is a larger value of P than that attained by the rank sum test for the same data. The procedures are different, make different assumptions, and may produce different results. In our experience, however, it is unusual that the results differ greatly.

An alternative form of the Wilcoxon rank sum test was proposed by Mann and Whitney in 1947 (31). Although these authors realized and stated the relation between the test statistic T_1 and their proposed statistic U,[2] other authors have apparently overlooked the fact that the tests are therefore equivalent. It seems to us that the test should properly be called the Wilcoxon rank sum test. In any case, it is important to be aware that when the Mann-Whitney U test is mentioned in the literature, the rank sum test is completely equivalent to it, and that the procedures give identical results, although they are based upon different test statistics.

The rank sum test as we have presented it is a two-sided test. However, it can be readily modified to provide one-sided critical regions. In fact, the critical region $T_1 \leq T_l$ is a one-sided critical region at significance level $\alpha/2$, where α is the tabulated two-sided level of significance. For example, if T_l comes from that portion of Table A-13 for two-sided, 5 per cent level tests, then it is the critical value for a one-sided 2.5 per cent level test. Similarly, the critical region $T_1 \geq T_r$ is suitable for upper one-sided tests at the $\alpha/2$ level. Notice that Table A-13 extends to values of n_2 larger than the maximum tabulated for n_1. Since the labeling of the samples is arbitrary, redefining the labels may be necessary for the larger values of n_1.

[2] $U = n_1 n_2 + \dfrac{n_1(n_1 + 1)}{2} - T_1.$

12-2　The Sign Test

In this section we discuss perhaps the simplest of all statistical tests—the sign test. It is suitable for paired data, and we would recommend that you take another look at the discussion of pairing in the first part of Sec. 7-2-5-4. The only assumption we make in the sign test is that different pairs of observations are independent. Notice that we do not assume that the 2 observations making up a pair are independent. In fact, a frequent application of the sign test and other procedures for paired data is to before-after measurements on the same subjects, and such readings can never be assumed to be independent.

We suppose that there are n independent pairs, and we look only at the direction of change in the 2 readings of each pair. That is, if $x_{1j}, j = 1, 2, \ldots, n$ is the first observation from the jth pair and x_{2j}, the second observation from the same pair, we record only the sign of the difference $x_{1j} - x_{2j}$. The determination of which is the first and which the second reading is completely arbitrary, so long as the definition is used consistently throughout the full set of data.

The hypothesis tested by the sign test is that the first members and second members of the pairs come from populations having the same median, or that the differences come from a population having median zero. If this is true, then positive and negative differences are equally likely, and we would expect about half the differences to be positive.

More formally, if we let n_+ be the number of plus signs, i.e., the number of positive differences out of the n pairs, then under the hypothesis n_+ will be binomially distributed with $p = \frac{1}{2}$. In the language of Sec. 6-1, a success is a plus sign, the probability of a success on a single trial is $\frac{1}{2}$, and we have n independent trials, a pair being a trial. Thus, the theoretical background for the sign test is already familiar to us, based on an understanding of Sec. 6-1. The critical region would consist of values of n_+ so far from $n/2$ that in aggregate they occur with probability α. Again, since n_+ is discrete, we will usually be unable to find critical values giving level of significance exactly equal to α and will take values corresponding as closely as possible to the α level without exceeding it.

Table A-14 gives two-sided critical values for the sign test. For example, notice that for $n = 25$ we read for α (two-sided) = 0.05 the critical values 7, 18. This means that if n_+ is between these values, i.e., from 8 to 17 inclusive, we accept the hypothesis of no difference in medians at the 5 per cent level. If, however, $n_+ \leq 7$ or $n_+ \geq 18$, we reject H at that level. One-sided tests are directly available from Table A-14. For example, if $n = 25$ the lower-one-sided critical region for $\alpha = 2.5$ per cent is $n_+ \leq 7$. The upper one-sided critical region is $n_+ \geq 18$.

If the members of a pair are tied, i.e., take the same value, then the corresponding difference is zero, which has no sign. Such differences should be eliminated from the sample, and n should be correspondingly reduced. In the presence of ties, n therefore becomes the number of nontied pairs. Again, it is often assumed that the observations are continuously distributed, making the occurrence of tied values unlikely.

We will apply the sign test to data of O'Rourke, et al. (40) on left coronary artery blood flow in dogs in control and pericardial tamponade states. In each dog, left coronary flow in milliliters per minute per 100 grams of left ventricle was measured during a control period and during pericardial tamponade induced by injecting saline solution into the pericardium. The data are presented in Table 12-1.

Table 12-1

LEFT CORONARY FLOW* IN 14 DOGS IN STATES OF CONTROL
AND PERICARDIAL TAMPONADE†

							Dog Number							
	1	2	3	4	5	6	7	8	9	10	11	12	13	14
Control	125	52	91	97	82	53	71	128	71	63	66	107	144	106
Tamponade	69	82	37	58	30	33	66	82	84	27	47	38	95	73
Sign of difference	+	−	+	+	+	+	+	+	−	+	+	+	+	+

* In milliliters per minute per 100 grams of left ventricle.
† Data reproduced with permission from R. A. O'Rourke, D. P. Fischer, E. E. Escabar, V. L. Bishop, and E. Rapaport (40).

There are no ties in the data, so $n = 14$ and $n_+ = 12$. This is larger than the value of 7 that would be expected if the control and tamponade readings came from populations having the same median, and we check Table A-14 to determine significance. Using a 5 per cent two-sided test, we find critical values of 2 and 12. Since $n_+ = 12$, we conclude that the difference is significant at the 5 per cent level. However, the same critical values apply to the 2 per cent level, and we can, therefore, say that $0.01 < P \leq 0.02$, noting that $n_+ = 12$ fails to attain significance at $\alpha = 0.01$. The fact that the critical values for $\alpha = 0.02$ and $\alpha = 0.05$ are identical in this case is a consequence of the discreteness of n_+. In fact, the exact significance level corresponding to critical values 2 and 12 is 0.012, the level for critical values 3 and 11 is 0.058, and we see that the values $\alpha = 0.02$ and $\alpha = 0.05$ are between these values. In other words, as n_+ moves from 12 to 11, we jump across both the 2 per cent and 5 per cent levels for a two-sided test. Finally, we conclude that under the conditions of this experiment pericardial tamponade is influencing left coronary blood flow.

A final general word about the sign test: notice that the actual values of the observations are used only to determine the direction of the differences. Neither the values of the differences themselves nor their relative magnitudes are used in testing significance. This points up the often overlooked fact that even very small differences, if they are consistent in direction and based on independent pairs of observations, can yield statistical significance. The sign test is one of the most useful devices available for obtaining a quick, rough assessment of a set of data. It involves no computation (unless you include counting in your definition of computation), but is, of course, a relatively insensitive procedure. In its place, we like it very much.

12-3   *The Wilcoxon Signed Rank Test*

Wilcoxon (59) proposed a test for paired data which makes the same assumptions as the sign test, namely independence of the n pairs of observations, and which is related to the rank sum test. Again the tested hypothesis is that the first and second members of the pairs come from populations having the same median, or that the pair differences come from a population having median zero. Ties, i.e., differences equal to zero, are treated as in the sign test—eliminated from the data before analysis, with consequent reduction in sample size. Thus, n again becomes the number of nontied pairs.

To perform the test, the n differences are arranged in order of size, their signs being ignored completely. Rank numbers are then assigned to these absolute differences, rank 1 being given to the smallest difference, rank 2 to the next smallest, etc., rank n to the largest. The signs of the original differences are then restored to the rank numbers, and T_+, the sum of the positive rank numbers, is the test statistic. In the event that ties occur among the differences, the same procedure as in the rank sum test (Sec. 12-1) is used. The tied differences are each given the average of the rank numbers that would have been assigned had the differences not been tied.

Table A-15 gives critical values of T_+. For example, if $n = 20$ and we want a two-sided test at the 5 per cent level, the critical values are 52 and 158. That is, if $T_+ \leq 52$ or $T_+ \geq 158$, the hypothesis is rejected. One-sided critical regions are available directly from the table. For instance, in this case the critical region $T_+ \leq 52$ provides a lower one-sided test with $\alpha = 0.025$, and the region $T_+ \geq 158$ provides an upper one-sided test, also at the 2.5 per cent level.

Notice that the signed rank test uses somewhat more sample information than the sign test. It is based on the relative sizes of the differences in addition to their signs.

As an illustration we shall apply the signed rank test to the data on left coronary arterial blood flow in dogs before and during pericardial tamponade, presented in Table 12-1. Differences with signs:

$$+56, -30, +54, +39, +52, +20, +5, +46,$$

$$-13, +36, +19, +69, +49, +33.$$

Differences without signs in order of size:

$$5, 13, 19, 20, 30, 33, 36, 39, 46, 49, 52, 54, 56, 69.$$

Rank numbers with signs of original differences attached:

$$+1, -2, +3, +4, -5, +6, +7, +8, +9, +10, +11, +12, +13, +14;$$

$$T_+ = 1 + 3 + 4 + \cdots + 14 = 98.$$

As a check, if we let T_- be the sum of negative rank numbers, expressed positively then $T_+ + T_-$ should equal $[n(n + 1)]/2$. $T_- = 2 + 5 = 7$; $[n(n + 1)]/2 = [(14)(15)]/2 = 105$; $T_+ + T_- = 98 + 7 = 105$. The critical values of T_+ for a 5 per cent level two-sided test are, from Table A-15, 21 and 84. Therefore, the hypothesis that control and tamponade readings come from populations having the same median is rejected. In fact, $P < 0.01$. Recall that in Sec. 12-2 when we applied the sign test to these data we found $0.01 < P \leq 0.02$. The fact that the signed rank test finds significance at the 1 per cent level may be a reflection of its somewhat greater sensitivity.

12-4   Distribution-free Confidence Interval for the Median of a Continuous Distribution

The determination of a distribution-free confidence interval for the median of a continuously distributed random variable is a particularly simple operation, involving only ranking the observations in order of size. In fact, once the observations have been so ranked, Table A-16 gives the *serial numbers* of the observations that constitute the 95 per cent and 99 per cent limits. Notice that the values from the table are not the confidence limits themselves, but only the position of the observations in the ordered sample which themselves are the limits.

For example, the 14 values of left coronary arterial flow during pericardial tamponade presented in Table 12-1 are, when arranged in order of size,

27, 30, 33, 37, 38, 47, 58, 66, 69, 73, 82, 82, 84, 95.

Table A-16 says that for $n = 14$, 95 per cent confidence limits for the median are given by the 3rd and 12th observations in this list. That is, the limits are 33 and 82. The corresponding 99 per cent limits are the 2nd and 13th observations, i.e., 30 and 84, and are, as we expect, somewhat wider because of the increased confidence level.

Because of its lack of assumptions about the nature of the underlying distribution, the distribution-free confidence interval for the median will ordinarily be somewhat longer than, say, the corresponding confidence interval for the mean of a normal distribution at the same level of confidence. Of course, the mean and median of a normal distribution are identical because of symmetry, so we are estimating the same parameter by both methods. In practice, unless we are in one of the 4 situations indicating the use of distribution-free methods, which were presented early in this chapter, we would stick to the interval for the normal mean. Just to indicate the comparison of intervals and their lengths, however, we shall find the distribution-free limits for the median based on the data for percentage of calcium content of sound teeth, presented in Sec. 7-2-3. The distribution-free limits are

95 per cent confidence level: 35.15 and 36.39;
99 per cent confidence level: 34.20 and 36.63.

The limits assuming normality are, respectively, 35.15 and 36.17 and 34.93 and 36.39. Thus, we see that the distribution-free intervals are somewhat longer.

12-5 A Test for Randomness: Total Number of Runs Above and Below the Median

A great many tests have been proposed to determine whether the observations in a sample have arisen in a random order. Strictly speaking, it is not entirely logical to call such procedures tests of randomness, since they cannot in fact determine whether a particular series of observations was or was not selected by a random process. Each test is sensitive to one or more of the kinds of regularities that we often find associated with a nonrandom process.

The concept of a run is important to the test that we shall discuss here. If we have a sequence of symbols of 2 kinds, then a run is defined to be a maximum-length unbroken subsequence of symbols of one kind. That is, consider the following sequence of plus and minus signs: $+ + + - - - +$ $- - - - + +$. This sequence begins with a run of 3 plus signs, followed by runs of 2 minus signs, 1 plus sign, 4 minus signs, and 2 plus signs. We shall be concerned with the total number of runs, which in this case is 5.

If we have sample observations recorded *in the order in which they were selected, not necessarily in order of size*, we find the sample median (See Sec. 3-3-2) and for each observation record a plus sign if it is above the median, a minus sign if it is below. If the sample size is odd, one observation will equal the median. Since this observation will produce neither a plus nor minus sign, it is eliminated, thereby reducing the sample size by one. The test statistic is u, the total number of runs.

Table A-17 gives critical values for u for testing the randomness hypothesis—that is, for testing the hypothesis that the plus signs are equally likely to be found in any positions in the sequence. Clearly a random sample should show no tendency for large values or small values to occur together, which would produce a small value for u. For instance, if the observations steadily decreased in size, with all the large values occurring first, then there would be only 2 runs, a run of plus signs followed by a run of minuses. On the other hand, if a large value was inevitably followed by a small one which was in turn followed by a large one, etc., then u would be too large. This situation or one approaching it could arise either because of some compensatory self-adjusting mechanism in a measuring instrument or because of a purposive attempt to "mix up" the observations, for example.

The critical values in Table A-17 are such that if u is between the tabled values, the randomness hypothesis is accepted, and if it falls outside the values or equals one of them, the hypothesis is rejected at the corresponding significance level for a two-sided test. One-sided tests are available as usual: if u is less than or equal to the lower critical value, the result is significant at the corresponding one-sided level; if u is greater than or equal to the upper value, we have significance for an upper one-sided test.

To illustrate the run test we shall use the 50 kidney weights of Table 3-4. Recording these in the order in which they occurred, we have:

374, 309, 323, 288, 301, 345, 358, 340, 329, 309, 363, 358,

355, 361, 265, 311, 388, 240, 260, 288, 252, 332, 403, 277,

208, 322, 307, 379, 319, 369, 305, 387, 349, 303, 293, 356,

350, 470, 362, 288, 323, 329, 327, 311, 256, 310, 342, 247,

329, 358.

The median of the ungrouped observations is 325 gm, and the corresponding sequence of plus (above the median) and minus (below the median) signs is

$+, -, -, -, -, +, +, +, +, -, +, +, +, +,$

$-, -, +, -, -, -, -, +, +, -, -,$

$-, -, +, -, +, -, +, +, -, -, +,$

$+, +, +, -, -, +, +, -, -, -, +, -, +, +.$

For this sample, $u = 23$ and from Table A-17 we find the critical values for a two-sided, 5 per cent level test to be 18 and 34, since n', the number of observations above (or below) the median, is 25. Thus, the hypothesis of randomness is accepted, and we see from the table that $P > 0.10$.

Anomalies upsetting the test based on total number of runs may sometimes arise. For example, the sequence of observations 7, 6, 7, 3, 7, 8 has 3 observations equal to the median value of 7. If the 7's are eliminated, then there will be an unequal number of plus and minus signs. In this case an extension of Table A-17 to account for unequal numbers is needed. Such extensions can be found in the *Documenta Geigy* [(10), p. 129)], or in Swed and Eisenhart (48). These references also discuss several other tests for randomness.

12-6 *Distribution-free Tolerance Limits*

As in Sec. 7-3, we may be interested in estimating the location of a certain proportion of a population with a certain level of confidence. We would use Table A-19 in preference to Table A-18 if we could not assume that the underlying distribution is normal.

To use Table A-19 we must decide what proportion Π of the population we wish our limits to cover and our level of confidence $1 - \alpha$. The 2 numbers (a, b) in the body of the Table A-19 are the ath smallest and bth largest ordered values in a random sample of size n such that at least a proportion Π of the population will be between a and b with confidence $1 - \alpha$.

We shall illustrate the use of this table by several examples:

1. Consider the 50 kidney weights of Table 3-1. Arranging these observations in order of size, we have

208, 240, 247, 252, 256, 260, 265, 277, 288, 288,
288, 293, 301, 303, 305, 307, 309, 309, 310, 311,
311, 319, 322, 323, 323, 327, 329, 329, 329, 332,
340, 342, 345, 349, 350, 355, 356, 358, 358, 358,
361, 362, 363, 369, 374, 379, 387, 388, 403, 470.

If we want those 2 values which bracket at least 90 per cent of the population with confidence 95 per cent, we see from Table A-19 that we need (1, 1), i.e., the first and last observations (the first smallest and the first largest). These are 208 and 470. Thus we are 95 per cent confident that at least 90 per cent of all kidney weights lie between 208 and 470 mg.

2. Suppose we wish to find two values based on a random sample of size $n = 80$ such that at least 90 per cent of the population will be covered with confidence 95 per cent.

Using that portion of Table A-19 for $1 - \alpha = 0.95$, $\Pi = 0.90$, and $n = 80$, we see that between second smallest and second largest values of our sample we have 95 per cent confidence that 90 per cent of the population will be found.

3. Suppose we want to know how small a sample will be sufficient to insure 95 per cent confidence that at least 95 per cent of the population will lie between the 2 most extreme values of our sample.

Using the values $1 - \alpha = 0.95$, $\Pi = 0.95$, and looking in the body of the table for 1, 1, we note that n must be at least 95; that is, we would need at least 95 observations to insure that all our requirements are satisfied.

EXERCISES

12-1 In order to test reaction time of certain facial muscles to a tap on the massetur muscle, suppose the following data are collected. The values represent the reaction time measured in microseconds to a tap on the opposite side of the face. Six individuals were tapped on the left side with time to contraction of the right massetur being noted. Six other persons were tapped with the sides reversed:

Left: 10.2,   9.8,   10.1,   14.3,   7.2,   13.8
Right: 10.4,   8.7,   11.1,   9.3,   13.2,   10.1

By means of the rank sum test, see if these samples come from populations with the same distributions.

12-2 In a study to determine the importance of weight gained by a mother during pregnancy, suppose the following data were obtained. For 7 babies of mothers with no dietary restrictions the birth weights (in pounds) were

$$8.1, 9.2, 6.5, 7.3, 8.1, 6.8, 5.9.$$

For 6 babies of mothers whose total weight gain was restricted to 15 pounds, the corresponding figures were

$$6.9, 6.8, 8.4, 10.2, 9.1, 6.1.$$

Use the rank sum test to see if these samples come from populations with the same distributions.

12-3   Suppose the following data represent age at vaccination in months for infants seen during a certain month by 2 pediatricians practicing in the same area of a large city.

Age at vaccination (months)

Pediatrician 1:   13.2, 11.1, 17.2, 19.3, 8.1, 13.9, 9.2, 12.7
Pediatrician 2:   9.4, 10.2, 8.7, 10.3, 14.1, 12.6, 16.2

Use the rank sum test to verify whether or not these samples come from populations with identical distributions.

12-4   Suppose the following are data obtained on 10 men aged 45 to 55. The values are cholesterol readings taken after 12 hours of fasting and repeated one hour after eating.

Subject	Fasting	After Eating
1	180	185
2	210	225
3	195	200
4	220	225
5	210	200
6	190	180
7	225	235
8	260	265
9	200	195
10	210	220

Test to see if the differences may come from a population having median zero by the sign test. Discuss the result.

12-5   Suppose the same 10 subjects discussed in Exercise 12-4 had supine systolic blood pressures taken at the onset and conclusion of a $\frac{1}{2}$-hour rest period. Use the sign test to see if the following first and second members of the pairs come from populations having the same median.

Subject	Before	After
1	140	132
2	136	134
3	160	162
4	172	170
5	154	146
6	150	154
7	138	142
8	146	138
9	152	144
10	128	124

12-6 Test the hypothesis of Exercise 12-4, using the signed rank test.

12-7 Test the hypothesis of Exercise 12-5, using the signed rank test.

12-8 Find a 95 per cent confidence interval for the median of the population of patients given the diuretic agent, using the data of Exercise 7-8.

12-9 Using the serum cholinesterase indices of Exercise 3-10 as presented (read down each column and continue at the top of the next column), test to see if the data are arranged in random order, using the runs test.

12-10 Using the data of Exercise 10-4, check to see if the data were in random order by applying the runs test to the systolic blood pressures.

12-11 Suppose $n = 4$ in the signed rank procedure; i.e., 4 pairs are involved.

    a. Generate all 16 possible totals for T_+, the sum of the positive ranks.

    b. Show that under the tested hypothesis the probabilities for the T_+ are as follows:

T_+:	0	1	2	3	4	5	6	7	8	9	10
$P(T_+)$:	$\frac{1}{16}$	$\frac{1}{16}$	$\frac{1}{16}$	$\frac{1}{8}$	$\frac{1}{8}$	$\frac{1}{8}$	$\frac{1}{8}$	$\frac{1}{8}$	$\frac{1}{16}$	$\frac{1}{16}$	$\frac{1}{16}$

13-1 Introduction

When one is considering quantitative aspects of human populations, it is common to think of tabulations of figures such as those compiled from a decennial census. This is one kind of population study. However, in recent times this point of view has been widened to include techniques now generally termed demographic methods.

These techniques are of value to investigators with various biological and medical interests. In this chapter a few of the common techniques will be presented, including rates and ratios, standardization, and life tables.

In Sec. 3-3 attention was called to the fact that descriptive statistics apply to groups rather than individuals. Similarly, measures of populations of people are to be considered as statements about the whole rather than a particular individual. Demography involves investigation of techniques to measure groups of people and thus is defined as the quantitative study of human populations with respect to events such as birth, death, marriage, morbidity, and migration. Demographic methods encompass 2 aspects of this quantification—"static" techniques concerned with characteristics of a population at a fixed time, e.g., census methods, and "dynamic" techniques concerned with the changing nature of a population, e.g., vital events: birth, death, marriage, divorce, and migration. We are concerned with both static and dynamic characteristics, since they are equally necessary in determining rates, the basic measures of vital statistics.

13-2 Historical Remarks

Collecting statistical information on human populations is not a 19th or 20th century invention. We know of at least one biblical illustration. But for a census, Christ would not have been born in Bethlehem. There are suggestions that enumerations of populations were made in China, Babylon, and Egypt prior to 2500 B.C. Ancient Greek and Roman records indicate that censuses were frequently made for special purposes. However, these head counts seem to have been ad hoc, lacking in continuity, and without a definite periodicity. The first preplanned periodic census was initiated in the United States in 1790, with Great Britain following in 1801. These countries have subsequently had a regular decennial census, with the exception of Great Britain in 1941, a war year.

Principal motives for ancient enumerations were the needs for assessing military strength or regulating taxation. With the modern era of census taking governmental needs have still dominated; however, more personal data are frequently collected, including sex, age, marital status, and often occupation.

As noted, demographic information cuts across 2 axes. The census primarily refers to static properties of the population. However, the dynamic aspect is probably of greater interest to the medical or biological investigator. The history of data collection for vital events is as varied as that for censuses. There appears to have been no regular recording of births, deaths, or marriages on a large scale prior to the Middle Ages. In England, starting in the middle of the 16th century, clerics were assigned to obtain vital statistics for their parishes. After fits and starts, regular complete national collection was instituted in 1875. The record for the remainder of Europe is even less inspiring. For the United States as a whole such information was irregularly kept until 1900, when certain city and state systems were of a sufficient caliber for the federal government to begin annual death data collection. These cities and states constituted the first United States National Death Registration Area. Birth statistics collected in these same areas to be used as model information for the United States were initiated in 1915. By 1933 all states were involved in routine birth and death certification. At present the United States National Marriage Registration Area consists of 35 states and is expanding.

13-3 Vital Rates

13-3-1 RATES AND RATIOS

By this time you are so at home with various arithmetic manipulations that a lengthy introduction to the topic of a rate is unwarranted. A rate is a proportion. Specifically, a rate is a fraction such that the numerator c is included within the denominator $c + d$ so that a rate has the form $c/(c + d)$. The numbers c and d are never negative, and thus a rate is a number between

0 and 1. Notice that this is a specialized use of the word rate. This definition, appropriate to vital statistics and demography, does not include such concepts as velocities, growth rates, and the like. We would have been happier if a different, less familiar term had been devised originally, but at this late date attempting a change in terminology would be a bit like trying to fill the Grand Canyon with a teaspoon.

We are interested in rates so that comparisons between groups of unequal size can be made. If we look only at the absolute number of events of interest in several groups, we have no meaningful way of relating these groups. However, considering events of interest per population at risk removes this difficulty. We shall illustrate this below, but first we shall define several of the basic vital rates.

These rates fall into one of 2 categories—crude and specific rates. The word "crude" implies that the total number of events are used in the computation, whereas "specific" means that only the events in a particular category of age, sex, race, particular disease, or other classification variable are used. Although we have noted that rates are proportions always lying between 0 and 1, in vital statistics these rates are usually expressed per some convenient base, such as 1,000 or 1,000,000, a large enough multiple of 10 being chosen as base so that the resulting rate is greater than 1. This may not always be possible when several rates are to be compared, for certainly they should all be expressed to the same base. To illustrate the concept of base, consider percentage, in which a rate is expressed per 100. For example, $\frac{37}{925} = 0.04$, when expressed as a percentage, is 4 per cent or $\frac{37}{925} \times 100$. A base is used for 2 reasons: first, persons are loath to work with the extremely small numbers that frequently arise in rate computation. Second, an answer is often more easily understood if stated as parts per some round number, such as 100 or 1,000, rather than per 1.

Certain vital rates are used so often that we feel it is important to list and define them to insure that there will be no confusion in their use.

Death Rates

$$\text{Crude death rate} = \frac{\text{total deaths}}{\text{mid-period total population}} \times 1,000.$$

$$\text{Crude death rate for cause (e.g., cancer)} = \frac{\text{total deaths for cause (e.g., cancer)}}{\text{mid-period population}} \times 100,000.$$

$$\text{Specific (e.g., age) death rate} = \frac{\text{deaths in the specific (e.g., age) group}}{\text{population in the specific (e.g., age) group}} \times 1,000.$$

Race and sex specific rates are defined analogously. Classification variables used in defining specific rates may be combined. For example, we might consider a race-sex specific mortality rate for, say, white females.

Certain death rates relating to mortality of the very young use total live births as denominator because laws in the United States require each birth to be recorded on a birth certificate (it is estimated that 99.7 per cent actually are), whereas census enumerations of infants can be expected to be incomplete. Babies are sometimes forgotten in the stating of household members.

$$\text{Infant mortality rate} = \frac{\text{deaths of children} < 1 \text{ year of age}}{\text{total live births}} \times 1,000.$$

$$\text{Neonatal mortality rate} = \frac{\text{deaths of children} < 28 \text{ days of age}}{\text{total live births}} \times 1,000.$$

These rates may be made specific by limiting deaths and live births to one sex, race, or cause.

Birth Rates

$$\text{Crude birth rate} = \frac{\text{total live births}}{\text{total population}} \times 1,000.$$

$$\text{Specific (e.g., by race) birth rate} = \frac{\text{specific (e.g., by race) live births}}{\text{specific (e.g., by race) population}} \times 1,000.$$

Finally, demographers wishing to state whether the population is growing or declining may use the vital index: (total live births/total deaths) × 100.

Let us now consider a detailed example of the use of rates. Consider an epidemiologist who is interested in the effect of climatic conditions on mortality. He decides to study deaths in Alaska and Arizona, feeling that this will allow comparison of a cold, damp climate with a hot, dry one. Using 1960 as his reference year, he notes that there were 1,316 deaths in Alaska and 10,121 in Arizona. Obviously this does not properly indicate that there is a lower force of mortality in Alaska than Arizona. The populations at risk must be considered. In 1960 there were 226,167 residents of Alaska and 1,302,161 of Arizona. Using deaths per population as an indication of mortality, we obtain crude death rates of (1,316/226,167) × 1,000 = 5.8 per 1,000 for Alaska and (10,121/1,302,161) × 1,000 = 7.8 per 1,000 for Arizona.

Our epidemiologist might now be surprised to find Alaska having the smaller death rate. However, a little knowledge of the populations of these states perhaps will cause him to adjust his interpretation. Alaska, a newer state, has tended to attract a younger population. The dry warm climate of Arizona has, on the other hand, attracted many older persons. Table 13-1 gives deaths, population, and mortality rates by the age structure of these states, while Table 13-2 exhibits the differential age pattern by presenting age distributions. These values are abstracted from *Vital Statistics of the United States* (56). Note that there are proportionately more persons under 25 in Alaska than in Arizona (53.6 per cent versus 49.2 per cent). Also, Arizona has a greater percentage of persons 65 and older (Alaska, 2.4 per cent, versus Arizona, 6.9 per cent). Thus we would expect relatively more deaths in Arizona than in Alaska, suggesting that we should include age specific rates in making the overall assessment of climatic conditions on mortality.

Table 13-1

DEATHS, POPULATIONS AND SPECIFIC DEATH RATES BY AGE
FOR 1960—ALASKA AND ARIZONA

		Alaska			*Arizona*	
Age	*Deaths*	*Population*	*Death Rate per 1,000 Population*	*Deaths*	*Population*	*Death Rate per 1,000 Population*
<1	306	7,101	43.1	1,174	34,599	33.9
1–4	57	27,092	2.1	236	132,367	1.8
5–14	40	46,110	0.9	138	285,830	0.5
15–24	59	40,722	1.4	286	186,789	1.5
25–34	72	39,672	1.8	325	169,878	1.9
35–44	126	31,981	3.9	568	173,029	3.3
45–54	173	18,957	9.1	1,049	136,573	7.7
55–64	150	9,146	16.4	1,621	92,871	17.5
65–74	149	3,745	39.8	2,287	63,634	35.9
75–84	143	1,354	105.6	1,762	22,499	78.3
85+	41	287	142.9	675	4,092	165.0
Total	1,316	226,167	5.8	10,121	1,302,161	7.8

13-3-2   ADJUSTED RATES

Carrying on with our example, we shall indicate how we can adjust rates for age. The techniques are equally appropriate as adjustments for sex, race, etc. There are 2 types of adjustment procedures, direct and indirect.

In the direct method we need the following basic information:

1.   The specific (e.g., by age) rates for each population.
2.   A standard population.

The procedure is to apply the specific rates for each population to the standard population. This technique shows the frequency of the characteristic (e.g., death) to be expected if the population being studied had the same

Table 13-2

PERCENTAGE DISTRIBUTION OF POPULATION BY AGE
ALASKA AND ARIZONA, 1960

Age	Alaska	Arizona
<1	3.1	2.7
1–4	12.0	10.2
5–14	20.4	22.0
15–24	18.1	14.3
25–34	17.6	13.0
35–44	14.1	13.3
45–54	8.4	10.5
55–64	4.0	7.1
65–74	1.7	4.9
75–84	0.6	1.7
85+	0.1	0.3

distribution (e.g., age) as the standard population, while retaining its own observed specific rates. The feature of this technique is that it allows the rates to be compared so that the idiosyncracies or differences in the distributions with respect to the specific characteristic (e.g., age) are removed.

Returning to our example, we shall adjust the crude rates for age differences. For computational ease we generally use as standard population some meaningful extant population set to total 1,000,000. In our example we use the 1960 United States population (Table 13-3). The process of setting

Table 13-3

TOTAL POPULATION FOR THE UNITED STATES, 1960, AND
STANDARD MILLION POPULATION BASED ON THE
UNITED STATES 1960 POPULATION BY AGE

Age	United States 1960 Population	Standard Million
<1	4,126,403	22,883
1–4	16,195,413	89,812
5–14	35,474,882	196,727
15–24	24,089,957	133,591
25–34	22,821,888	126,559
35–44	24,076,192	133,515
45–54	20,625,775	114,381
55–64	16,707,225	92,650
65–74	10,848,086	60,158
75–84	4,496,032	24,933
85+	863,922	4,791
Total	180,325,775	1,000,000

it to total 1,000,000 amounts to finding the proportionate part of the total population in each age group and multiplying that proportion by 1,000,000. Thus each age segment is given its appropriate proportion of the standard population of 1,000,000 persons. For example, as presented in Table 13-3, there were 24,089,957 persons between the ages of 15 and 25 (limits 15–24) out of a total population of 180,325,775. Thus (24,089,957/180,325,775) × 1,000,000 = 133,591 persons in the standard million are in this age group. Table 13-3 presents the full standard million population.

Certainly the choice of standard population can affect the values of the adjusted rates. We should choose this population so that it agrees with the individual populations as closely as is reasonable. We would not, after all, want the adjustment process to produce rates so radically different in size from the original rates that the adjusted values would not even appear to relate to the phenomenon under study, e.g., mortality. Frequently we use an average of the specific populations as the standard. We have chosen the entire United States population, since it forms a stable base and contains the 2 specific populations. Common sense and the purpose of the analysis play roles in the selection of the standard population.

To obtain the age adjusted death rates, we simply multiply the age specific death rates by the standard age specific population. This product represents the number of age specific deaths we would expect if the observed death rates were in effect in the standard population. For example, if we apply the infant mortality rate for Alaska, 43.1 per 1,000, to the appropriate age group in the standard population, 22,883 persons, the resulting expected number of deaths is 22,883 × (43.1/1,000) = 986. Table 13-4 lists all these values for Alaska and Arizona.

To find the overall (crude) adjusted rate we now sum the expected number of deaths across all ages. This sum is the total number of deaths for a population of 1,000,000 adjusted for age; i.e., this sum is an adjusted death rate per 1,000,000. Generally we have expressed such rates per 1,000. If we divide the rate per 1,000,000 by 1,000, we will make the final rate per 1,000; e.g., for Alaska we have 10,560 expected deaths per 1,000,000. Expressing this per 1,000, we obtain a rate of 10,560/1,000 = 10.6 per 1,000. The result for Arizona is 9.3 per 1,000. We notice that a very interesting change has occurred. The rate for Alaska has risen from 5.8 to 10.6 and, even though the adjusted Arizona rate has also increased from 7.8 to 9.3, it is smaller than the adjusted value for Alaska. This implies that when the differences in age makeup of the populations in Alaska and Arizona are removed, the force of mortality in Alaska is actually greater than that of Arizona. Both rates were higher after adjustment since the populations in both Alaska and Arizona are different from that of the United States with respect to age. Arizona's rate changes less (7.8 to 9.3) since its population is closer in age distribution to the United States than is Alaska's distribution.

Table 13-4

EXPECTED DEATHS IN THE UNITED STATES, 1960, STANDARD
MILLION POPULATION USING AGE SPECIFIC RATES FOR
ARIZONA AND ALASKA, 1960

Age	Alaska	Arizona
1	986	776
1–4	189	162
5–14	177	98
15–24	187	200
25–34	228	241
35–44	521	441
45–54	1,041	881
55–64	1,519	1,621
65–74	2,394	2,160
75–84	2,633	1,952
85+	685	791
Total	10,560	9,322
Age adjusted rate per 1,000	10.6	9.3

Finally, through these shifts in rates, we can safely conclude that
adjusting for age is worthwhile in assessing the force of mortality in these
states. We might wish to adjust the rates further for sex, race, or possibly
other factors influencing death rates. Note also that this method can be
applied to quantities other than death rates, and that the adjustment process
can be extended to allow simultaneously for joint adjustment by 2 or more
factors [Jaffe (28)].

In contrast to the direct method, the indirect method of rate adjustment
requires as basic input:

1. The individual populations categorized by the specific variable or
   variables, such as age, for which we are adjusting.
2. A set of rates in a standard population.

Recalling that the direct method measured the force of the observed
specific rates on a standard population, by contrast, in the indirect method
we assess the effect of standard rates on observed specific populations. As
in the selection of the standard million, the choice of the proper standard
rates is critical to a sensible analysis. Frequently we know rates for a large
area, e.g., a state, and feel that these can be appropriately applied to some
particular subject areas, e.g., the counties within the state, where such rates
are unknown.

Table 13-5

POPULATION FOR DETROIT AND MUSKEGON, MICHIGAN
STANDARD METROPOLITAN STATISTICAL AREAS (SMSA), 1960
BY AGE AND SEX

	Age								
	0–9	10–19	20–29	30–39	40–49	50–59	60–69	70–79	80+	*Total*
Detroit SMSA										
Male	452,828	295,333	196,234	277,431	244,221	190,236	132,511	57,455	11,510	1,857,759
Female	436,420	298,514	230,183	290,361	252,426	184,503	129,529	64,174	18,491	1,904,601
Muskegon SMSA										
Male	19,362	13,475	7,627	9,679	9,258	7,043	4,638	2,312	666	74,060
Female	18,688	13,666	8,713	10,372	9,279	6,662	4,841	2,760	902	75,883

Consider the following illustration of this technique. A medical team is contemplating a large-scale study of malignant neoplasms. As part of their preliminary investigation 2 population centers are chosen for intensive scrutiny. The decision is made that one of these areas should center around a large industrial city and the other around a more isolated small city. Using a list of Standard Metropolitan Statistical Areas[1] (SMSA) as a guide, they select 2 such areas—one centered on Detroit and the other on Muskegon-Muskegon Heights, Michigan (Table 13-5).

Initial estimates of cancer death rates are desired. However, data are not available by age and sex for these areas. Since it is felt that some control of these variables is needed, adjusted rates are computed, data available for the state of Michigan (Table 13-6) being used to provide standard rates. Thus the investigators obtain rates for the 2 areas as if differences in age and sex composition between the 2 were eliminated.

The computation of such an adjusted death rate by the indirect method is straightforward; i.e., for each individual population multiply the age-sex specific population by the appropriate standard rate to obtain the number of deaths expected if that specific standard rate were in effect in the population. For example, there were 7,043 males aged 50–59 in the Muskegon SMSA in 1960, and the appropriate mortality rate was 312.4 per 100,000. The resulting multiplication yields $7,043 \times (312.4/100,000) = 22.00$, which is the expected number of deaths in the Muskegon SMSA if the standard rate is applied. All such calculations are listed in Table 13-7.

At this point in the calculations there are 2 options, depending on the additional information available. Let us call these techniques I and II. Technique I is appropriate when an investigator has data concerning the total number of deaths even though these deaths are not subcategorized by the particular variables being adjusted. For example, the total number of deaths resulting from malignant neoplasms have been tabulated for the SMSA's of Detroit and Muskegon for 1960. However, the age-sex specific deaths are not available. Using these totals, 5,416 for Detroit and 192 for Muskegon, and the expected age-sex totals computed above, we can develop adjusted death rates. Noting from Table 13-7 that these expected values are 5108.84 and 200.12 for Detroit and Muskegon respectively, we compute the ratios of observed to expected deaths for each area:

$$\text{Detroit:} \quad \frac{5,416}{5108.84} = 1.06; \qquad \text{Muskegon:} \quad \frac{192}{200.12} = 0.96.$$

[1] A Standard Metropolitan Statistical Area (SMSA) is a subdivision used by the Federal Bureau of the Census for demographic data. An SMSA is defined as a county or group of contiguous counties containing at least one city of 50,000 persons or more or "twin" cities having a combined population of at least 50,000.

Table 13-6

DEATH RATES PER 100,000 FOR MALIGNANT NEOPLASMS—MICHIGAN, 1960—
BY AGE AND SEX

	Age									
	0–9	10–19	20–29	30–39	40–49	50–59	60–69	70–79	80+	Total
Male	11.1	9.2	11.6	27.5	88.6	312.4	686.5	1,254.7	1,819.3	161.4
Female	6.7	6.6	7.7	34.7	122.6	253.5	473.5	780.2	1,206.4	127.7

Crude death rate   144.4

These ratios indicate the proportion of deaths due to cancer for observed deaths to deaths expected if the individual population had the same age-sex rates as the whole state.

Since the ratio for Detroit's SMSA is greater than 1 (1.06), we conclude that there are more observed deaths in this area than would be expected if the state rates applied. Similarly, the Muskegon ratio of 0.96 indicates that fewer deaths were observed than expected. Finally, to find the adjusted mortality rates we multiply each such ratio by the crude malignant neoplasm death rate for the state of Michigan. The crude death rate for cancer in Michigan in 1960 was $(11,298/7,823,194) \times 100,000$ or 144.4 per 100,000.

Table 13-7

EXPECTED NUMBERS OF DEATHS FROM MALIGNANT NEOPLASMA—
SMSA's FOR DETROIT AND MUSKEGON, 1960, USING MICHIGAN
MALIGNANT NEOPLASM DEATH RATES, 1960, BY AGE AND SEX

	Detroit		Muskegon	
Age	*Male*	*Female*	*Male*	*Female*
0–9	50.26	29.24	2.15	1.25
10–19	27.17	19.70	1.24	0.90
20–29	22.76	17.72	0.88	0.67
30–39	76.29	100.76	2.66	3.60
40–49	216.38	309.47	8.20	11.38
50–59	594.30	467.72	22.00	16.89
60–69	909.69	613.32	31.84	22.92
70–79	720.89	500.69	29.01	21.53
80+	209.40	223.08	12.12	10.88
Totals	2,827.14	2,281.70	110.10	90.02
Grand total	5,108.84		200.12	

Since the observed deaths in the Detroit SMSA exceeded the expected deaths, we must increase the statewide crude rate proportionately. Our adjusted rate becomes 1.06×144.4 per 100,000 or 153.1 per 100,000. Similarly, the adjusted rate for the Muskegon SMSA is 0.96×144.4 per 100,000 or 138.6 per 100,000.

It is interesting to compare these adjusted rates with the crude observed rates for these areas. In the Detroit SMSA the 5,416 observed deaths occurred in a population of 3,762,360 for a rate of $(5,416/3,762,360) \times 100,000 = 144.0$ per 100,000. In the Muskegon SMSA the 192 observed deaths occurred in a population of 149,943, for a rate of $(192/149,943) \times 100,000 = 128.0$ per 100,000. Thus our rates standardized by age and sex represent increases over the crude rates in both areas: 144.0 to 153.1 in Detroit and 128.0 to 138.6 in Muskegon. This indicates that if the age-sex specific rates were available for each of these population centers, many of them (but not necessarily all) would be less than the corresponding state rates. Overall, we can conclude that fewer persons in these areas are dying because of malignant neoplasms, on the basis of their age-sex distributions.

Technique II applies when no observed totals are available. In this case adjusted rates are found by computing expected numbers of deaths per population at risk times 100,000, i.e., for

$$\text{Detroit:} \quad \frac{5108.84}{3,762,360} \times 100,000 = 135.8 \text{ per } 100,000;$$

and

$$\text{Muskegon:} \quad \frac{200.12}{149,943} \times 100,000 = 133.5 \text{ per } 100,000.$$

We should not lose sight of the main purpose for adjusting rates—to allow rates to be compared when the underlying populations differ with respect to important characteristics such as age or sex. Both the direct and indirect methods bring rates from different populations to a common ground, by adjusting for the influence of forces that are out of balance in the individual populations.

Which method should we use—direct or indirect? Usually this is a moot question with the answer being dictated by the available data, but sometimes we have a choice. If we have specific rates for each of the populations involved, and if these rates are based on a large amount of data, the direct method is preferred. However, if specific rates are available, but based on very small numbers, the sampling variability associated with them may be so large as to preclude their use. The indirect method is applicable when each population's specific rates are unavailable (or of low quality), but the investigator feels that these rates, if available, would be reasonably similar to the standard rates used.

13-3-3 FURTHER REMARKS ABOUT RATES

Although we have dwelled on death rates, not all rates are quite so deadly. As we have noted, rates can measure such events as births, marriages, divorces, migration, etc., and adjustment methods apply just as well for such rates.

Two kinds of rates are used so often and with such ambiguity that special emphasis is needed. The tricky duo are the incidence rate and the prevalence rate. They are most frequently used as measures of morbidity—the amount of illness in a population. They differ in 2 respects:

1. Type of cases involved—either the total number of cases (prevalence) or the number of new cases (incidence).

2. Time of enumeration of the events—either at a point in time (prevalence) or for a period of time (incidence).

Definition

A *prevalence rate*[2] is defined as the total number of cases at a point in time divided by the population at risk at that point, whereas an *incidence rate* is the number of new cases occurring during a given period of time divided by the population at risk at the beginning of that period. Usually these rates are expressed per some convenient base in accord with the rule mentioned in Sec. 13-3-1.

Although the words "incidence" and "prevalence" are often used incorrectly in the literature, we hope that you will be conscious of the advantages of maintaining the distinction between them. One way to conceptualize this distinction is to note that an incidence rate is a dynamic measure, since it relates how the burden of disease is shifting over time. A prevalence rate, on the other hand, is a static measure, since it conveys amount of disease existing at a point in time.

As an example, suppose that a veterinarian is studying an epidemic of disease X which runs a clinical course ending either in recovery with permanent immunity or death. On July 1, 1967, a herd of animals in which the outbreak has already begun is investigated. Again on July 1, 1968, the same herd is reinvestigated. The following data summarize the findings:

1. Total herd size—July 1, 1967   600

2. Total number clinically ill (but alive) July 1, 1967   100

3. Total number becoming clinically ill between July 2, 1967 and July 1, 1968 inclusive   200

4. Total number of animals dying of the disease (none died before July 1, 1967 or after July 1, 1968)   120

The prevalence rate as of July 1, 1967 is $\frac{100}{600}$, or 16.7 per cent, while the incidence rate (July 2, 1967 to July 1, 1968) is $\frac{200}{500}$, or 40.0 per cent. It is crucial to note that the denominator for the incidence rate is 500, not 600, since we must first remove the 100 animals that were clinically ill on July 1, 1967. This is necessary, since these 100 animals are not at risk of becoming ill (they already were).

[2] More precisely, we here define "point" prevalence as opposed to "period" prevalence. In the formulation of a point prevalence rate we consider the total number of cases at a specified *point* in time. For a period prevalence rate we would consider the total number of cases that have occurred at *any time* in a specified period [see Hill (25)].

Another rate commonly used to measure disease severity is the case fatality rate, defined as the number of subjects dying of a disease divided by the number of cases of the disease. In our example there were 120 deaths and a total of 300 cases (100 as of July 1, 1967, plus the 200 from July 2, 1967 to July 1, 1968). Therefore, the case fatality rate is $\frac{120}{300}$ or 40.0 per cent.

13-4 Life Table

Space does not permit a full treatment of the life table, but we shall attempt to provide an overview.

The uses of the life table for the biologically oriented investigator are several. All relate to the notion of formalizing death data in a consistent probability pattern. In particular, the life table allows us to organize the mortality experience of a population. As pointed out in Sec. 13-3-1, we are frequently interested in comparing 2 populations. The life table is appropriate when comparison of the force or pattern of deaths by age is desired. The following are some situations in which the life table could prove beneficial:

1. A demographer may wish to relate life expectancy at birth for United States males born in 1900 with those born in 1960.

2. The federal government may want to estimate the social security or medical care case loads 5 or 10 years hence.

3. An investigator of cancer of the lung may want to map survival patterns for a set of patients.

4. A medical researcher might be interested in discerning what a population's age distribution would be if automobile accidents could be eliminated as a source of mortality.

Although the uses of the life table are varied, the tables themselves are only of 3 main types: current, cohort, and follow-up tables.

13-4-1 CURRENT AND COHORT LIFE TABLES

The current life table, the most common form of table, uses mortality rates observed at a given time. For example, age specific death rates for 1960 would be employed to generate a current life table giving the force of mortality as of 1960.

The cohort life table, on the other hand, follows a population born in the same year, the cohort, over time, applying mortality rates appropriate to that population at each age. For example, a cohort born in 1900 has the

1900 infant mortality rates, but that same population at age 10 has the 1910 age 10 mortality rates, etc.

There are advantages and disadvantages to both the current and the cohort life tables. The advantage of the current table is that it allows us to assess the force of mortality as it is at present and permits a glimpse at the future, if present rates carry forward. The advantage of the cohort table is that it gives us a way to assess mortality as it is experienced. The current life table is at a disadvantage if the current rates are atypical of the general time period, or if future rates fluctuate greatly, thus destroying the predictive value of the table.

It is also important to bear in mind that the mortality rates used in the current life table reflect, for different age groups, forces of mortality over different time periods. That is, 10-year-olds in 1970 have lived over the period 1960–1970, while 70-year-olds have lived from 1900 to 1970. Thus, the survivorship data relate to populations exposed to quantitatively different risks during their lifetimes.

The cohort life table is disadvantaged in that it may often be only of historical value. To generate this table we need the death rates appropriate to each time period the cohort has traveled through, none of which may be applicable at present. For example, if we wish to develop a life table for a cohort born in 1900, we need 1900 infant mortality rates, 1905 age 5 mortality rates, 1920 age 20 mortality rates, etc. Although these rates may be of historical interest to a biostatistician, they might be irrelevant in considering present mortality. Also, a cohort life table must stop at the present, since we are using mortality rates experienced by the cohort; c.g., a table constructed in 1970 for a cohort born in 1900 would have to end with the cohort at age 70, since any older ages would require death rates beyond 1970.

13-4-2 Life Table Terminology

We shall use the current life table to define and illustrate the life table functions, since these quantities are the same for any form of life table.

Table 13-8, a life table for United States males, 1964, will be used as an illustration.

A life table is called complete if every year of age is represented. If, however, age intervals, such as 5-year intervals, are used, we have an abridged table. A right-hand subscript denotes the age at the beginning of the interval, while a left-hand subscript shows the length of the interval. In a complete table no subscript is placed to the left, although a tacit 1 is assumed.

The most basic information we need to generate the life table consists of probabilities $_nq_x$, relating the probability that an individual alive at a stated age x will die before age $x + n$. These rates differ from the conventional age specific death rates, frequently termed central death rates, discussed

Table 13-8

ABRIDGED LIFE TABLE FOR MALE POPULATION: UNITED STATES 1964

Age Interval	$_nq_x$	l_x	$_nd_x$	$_nL_x$	T_x	$\overset{\circ}{e}_x$
<1	0.0278	100,000	2,779	97,527	6,685,013	66.9
1–5	0.0042	97,221	407	387,914	6,587,486	67.8
5–10	0.0026	96,814	249	483,405	6,199,572	64.0
10–15	0.0026	96,565	254	482,268	5,716,167	59.2
15–20	0.0066	96,311	637	480,110	5,233,899	54.3
20–25	0.0092	95,674	877	476,205	4,753,789	49.7
25–30	0.0090	94,797	850	471,855	4,277,584	45.1
30–35	0.0107	93,947	1,010	467,312	3,805,729	40.5
35–40	0.0149	92,937	1,386	461,453	3,338,417	35.9
40–45	0.0231	91,551	2,111	452,868	2,876,964	31.4
45–50	0.0366	89,440	3,270	439,649	2,424,096	27.1
50–55	0.0586	86,170	5,045	418,995	1,984,447	23.0
55–60	0.0897	81,125	7,280	388,368	1,565,452	19.3
60–65	0.1309	73,845	9,668	346,060	1,177,084	15.9
65–70	0.1924	64,177	12,347	290,784	831,024	12.9
70–75	0.2579	51,830	13,369	226,098	540,240	10.4
75–80	0.3454	38,461	13,283	159,016	314,142	8.2
80–85	0.4699	25,178	11,830	95,171	155,126	6.2
85+	1.0000	13,348	13,348	59,955	59,955	4.5

in Sec. 13-3-1. The $_nq_x$ rates are based on a population at risk at the start of the age interval, whereas the central rates, denoted by $_nm_x$, are based on mid-year populations. Many techniques have been developed to determine $_nq_x$ from $_nm_x$. It will suffice here to note that this is a complex problem for which a number of alternative solutions have been proposed. We will not discuss these methods. However, once an investigator has obtained the $_nq_x$ values, the rest of the quantities in the table are computationally straight-forward.

From Table 13-8 we note that $_5q_{10}$, meaning that the probability that an individual age 10 will die in the next 5 years was 0.0026 for United States males in 1964. The $_nq_x$ values are consistent with the fact that death rates are high in infancy, are low through age 30, and increase throughout the remainder of life.

Once we have obtained the $_nq_x$ rates, we multiply them by that portion of a hypothetical population surviving to age x. This population is usually taken to number 100,000 at birth. The number surviving to age x is denoted by l_x and is found by successively multiplying the $_nq_x$ to that portion of the original population still alive at age x. For example, the infant mortality rate q_0 in Table 13-8 is 0.0278 (0.02779), and l_0 is 100,000, so 100,000 $\times$ 0.02779 = 2,779 babies who would die before age 1. Thus of the original group, $100,000 - 2,779 = 97,221$ would be alive at age 1. This value is l_1. Multiplying this value by $0.0042 = {}_5q_1$, we obtain 407 deaths between ages 1 and 5. Subtracting 407 from 97,221 yields 96,814, which is l_5. Continuing in this fashion, we note that 96,311 will survive to age 15; 95,674 to age 20; and 13,348 to age 85.

Notice that ${}_nd_x$, the number dying in the interval x to $x + n$, is the difference between successive l_x values; e.g., ${}_5d_{20} = 877$, found from: $l_{20} - l_{25} = 95,674 - 94,797 = 877$. This means that 877 males in our population of 100,000 births will die between their 20th and 25th birthdays.

A particularly useful life table function is the average life expectancy at a given age. To find this entry, we must develop several preliminary quantities. They are, first, the total number of years in a given interval lived by individuals alive at the start of that interval, denoted by ${}_nL_x$.[3] For example, the 95,674 males alive at age 20 will live a total of 476,205 years between ages 20 and 25. We shall not consider methods used to compute ${}_nL_x$. Note, though, if no one died in this interval, ${}_5L_{20}$ would be simply 5 times the population extant at age 20: $95,674 \times 5 = 478,370$. The stated value of 476,205 for ${}_5L_{20}$, being smaller than the product $95,674 \times 5$, includes the necessary reduction resulting from those dying in the interval, none of whom accumulates a full 5 years of additional life.

Second, using ${}_nL_x$, we can accumulate the total number of years lived after attaining age x by those l_x individuals surviving to age x. This quantity T_x is thus the sum of all ${}_nL_x$ for age intervals greater than or equal to the specified age x. For example, $T_{20} = 4,753,789$ represents that the 95,674 men surviving to age 20 will live a total of 4,753,789 additional years. This value is found by summing ${}_5L_{20}, {}_5L_{25}, \ldots, L_{85+}$ or $476,205 + 471,855 + \cdots + 59,955$.

Having obtained T_x, we are now in a position to find the average life expectancy at age x, denoted by $\overset{\circ}{e}_x$.[4] Since T_x is the total number of years of life remaining to those alive at age x and l_x is the number of individuals at age x, $\overset{\circ}{e}_x = T_x/l_x$. Frequently $\overset{\circ}{e}_x$ is called the expectation of life at age x. This term, expectation, is dangerous, since it connotes life remaining for an individual, whereas $\overset{\circ}{e}_x$ is actually an arithmetic mean and as such refers to the group. In Table 13-8, $\overset{\circ}{e}_0$, the average length of life for United States males in 1964 at birth, was $66.9 = 6,685,013/100,000$. It is interesting to note that $\overset{\circ}{e}_1$ is larger than $\overset{\circ}{e}_0$, since the death rate q_0 is so large.

---

[3] There is an alternative explanation of ${}_nL_x$ and T_x predicted on a stationary population. See *Vital Statistics of the United States*, 1964, Sec. 5-2 (57).

[4] The notation including the small zero over e is used to distinguish $\overset{\circ}{e}_x$, the so-called complete expectation of life, from e_x, the curtate expectation of life. For an explanation of the difference between these terms, see Dublin, Lotka, and Spiegelman [(11), p. 21, footnote 16].

Summarizing the terms developed in this section, we have:

1. $_nq_x$ = the probability of death between ages x and $x + n$.

2. l_x = the number of individuals surviving to age x.

3. $_nd_x$ = the number of deaths between ages x and $x + n$.

4. $_nL_x$ = the number of years lived between ages x and $x + n$ by l_x.

5. T_x = the number of years lived in toto from age x onward by the l_x survivors to age x.

6. $\overset{\circ}{e}_x$ = the average years of life remaining at age x.

13-4-3 FOLLOW-UP LIFE TABLE

Frequently a medical investigator is interested in following patients for a long time after diagnosis. This type of follow-up is particularly useful in studying chronic diseases. In trying to summarize the course of the disease in these patients, the investigator is confronted with a number of problems, including patients dying, being followed for varying lengths of time from inception of the study or their disease, or being lost to follow-up by moving, changing physicians, or refusing to continue in the study. In spite of these difficulties, the investigator hopes to map a prognosis pattern of the disease that will be useful in predicting the probability of survival for x years. The follow-up life table technique affords a reasonably simple method of approaching this problem.

We shall develop this technique by considering an example presented by Merrell and Shulman (33) concerning 99 patients with systemic lupus erythematosus (SLE). Summarizing remarks by Merrell and Shulman, we see that there are 2 possibly complicating issues:

1. Patients entering the study may be observed at different states and duration of the disease, thus making it difficult to state the initial point from which prognosis should be estimated.

   It is meaningful to speak about the prognosis of a disease only if there is a well-defined point from which prognosis is assessed. This point may be quite simply noted, such as date of surgery or first treatment. However, in other situations precise dates may be more tenuous, such as date of first symptoms or of complete diagnosis. In any case, the key is that there must be a stated clear-cut reference point. Note that there is a unique chronological reference point for each patient, but the definition of the nature of this point is the same for all patients.

2. Patients may be followed for varying lengths of time because of different calendar entry and exit points.

Seldom will a physican have available a complete set of patients all of whom started a specific disease simultaneously and have been followed for the same length of time. Usually the population of patients builds gradually from the investigator's practice. Anxious to present informative findings, this same researcher is unwilling to wait until all patients have reached a specific endpoint, e.g., death. Perforce, he will have a motley set of subjects, some of whom are recent entries, others for whom extensive follow-up is available, and still others who have quit the study or died at various times from their reference point.

To illustrate these 2 possible sources of difficulty, consider the following 6 patients (numbered 1 through 6 for convenience) selected from the Merrell and Shulman data:

Patient	Date of Diagnosis		Date of Most Recent Observation and Status		
	Month	Year	Month	Year	Status
1	5	49	1	54	living
2	2	50	2	53	living
3	10	50	7	52	dead
4	12	51	4	52	living
5	6	52	1	54	living
6	12	53	2	54	dead

We can chart the course of follow-up for each patient by calendar years, as presented in Fig. 13-1. Note that from this presentation, first, reference points—date of diagnosis—vary considerably, covering 4 years and 7 months (from May, 1949, to December, 1953). Second, the length of follow-up varies from 2 months for patient number 6 to 4 years and 8 months for patient number 1. Third, the statuses of these patients are of 3 types: living and under study, living when last seen but no longer under study, and dead. From this presentation it is very difficult to discern any meaningful pattern for the course of SLE. However, if these same data are graphed to represent years of follow-up as in Fig. 13-2, a basis for prognosis is more apparent. For example, using these 6 cases, an investigator might ask about the prognostic pattern for SLE one year after diagnosis. Cutting the lines in Fig. 13-2 at one year (illustrated by the dotted line), he is able to observe the data in the form needed for a follow-up life table. Note that at this time (one year after diagnosis) 4 patients are alive and on study, one is out of the study but alive when last seen, and one is dead.

Figure 13-1

SAMPLE OF FOLLOW-UP PATTERNS FOR PATIENTS WITH
SLE BY CALENDAR YEARS

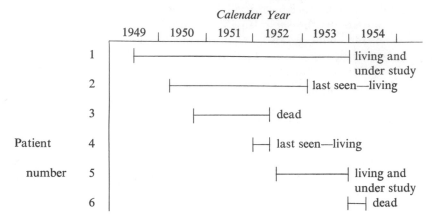

Figure 13-2

SAMPLE OF FOLLOW-UP PATTERNS FOR PATIENTS WITH
SLE BY YEAR OF FOLLOW-UP

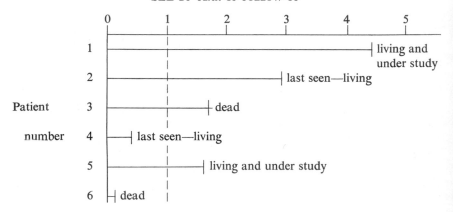

Thus in developing a follow-up life table we should consider the following items:

1. An exact reference point.

2. Duration of follow-up for patients who are alive and being studied, lost (refused to continue, moved, cannot be found), and dead.

Merrell and Shulman's data are presented in follow-up life table format in Table 13-9. In this example the reference point was diagnosis of SLE. This involved multiple clinical evidence or the agreement of several observers that disease existed or both. Let us consider the definition and computation of the values in each column.

1. $x =$ years after reference point. Note that this value may be a portion of a year, e.g., $\frac{1}{4}$, $\frac{1}{2}$, 1, $1\frac{1}{2}$ years.

2. $O_x =$ number of subjects observed at least x years after reference point.

3. $_nW_x =$ number of subjects last observed and alive in the interval x to $x + n$. This figure includes subjects who were observed only between x and $x + n$ years because that is all the time from their reference point to the present (time of the study) or because they withdrew (the original reason for the choice of the letter W) after that length of time. It includes those who moved, refused examination, or were lost to further study, but does not include those who died in the interval.

4. $_nd_x =$ number of subjects dying between x and $x + n$ years from their reference point.

5. $_nq_x =$ the probability of dying in the interval x to $x + n$ for those alive at time x. Computation of $_nq_x$ in the follow-up table is complicated by $_nW_x$, since these patients are not seen again. We do not know if they are alive or dead at time $x + n$, only that they were alive at time x. The most frequently employed estimate of $_nq_x$ is

$$_nq_x = \frac{_nd_x}{O_x - \frac{1}{2}\,_nW_x},$$

which assumes that the $_nW_x$ contribute on the average one-half of an interval of experience each. Further discussion of this formula is presented by Berkson and Gage (5).

6. $_nP_x =$ the probability an individual surviving x years will survive $x + n$ years. Thus $_nP_x = 1 - \,_nq_x$. To illustrate $_nq_x$ and $_nP_x$, consider the data in Table 13-9 for $x = 0.75$ years. Note that $_{0.25}d_{0.75} = 3$, $O_{0.75} = 64$ and $_{0.25}W_{0.75} = 8$, so

$$_{0.25}q_{0.75} = \frac{3}{64 - \frac{8}{2}} = \frac{3}{60} = 0.05$$

and

$$_{0.25}P_{0.75} = 1 - 0.05 = 0.95.$$

7. $P_x =$ the proportion of patients surviving x years. To survive x years, a person must survive the first interval, the second

interval, and the third, etc., and finally the interval ending at x. To calculate P_x multiply the $_np_x$'s for all intervals less than or equal to x. For example, $P_{0.50}$, the proportion surviving $x = 0.50$ years, is

$$P_{0.50} = 1.00 \times {}_{0.25}p_{0.00} \times {}_{0.25}p_{0.25}$$

$$= 1.00 \times 0.867 \times 0.975 = 0.845,$$

meaning that 84.5 per cent of the original group of patients with SLE will survive one-half year after diagnosis. We start with 1.00, since all of the population is alive at the time of diagnosis.

Table 13-9

FOLLOW-UP DATA AND CALCULATION OF SURVIVORSHIP
FOR 99 PATIENTS WITH SLE

Years after Diagnosis x	O_x	$_nW_x$	$_nd_x$	$_nq_x$	$_np_x$	P_x
0	99	3	13	0.133	0.867	100.0
0.25	83	9	2	0.025	0.975	86.7
0.50	72	6	2	0.029	0.971	84.5
0.75	64	8	3	0.050	0.950	82.1
1.00	53	18	6	0.136	0.864	78.0
2.00	29	9	2	0.082	0.918	67.4
3.00	18	12	2	0.167	0.833	61.9
4.00	4					51.6

We should note that the $_np_x$ are similar to binomial probabilities because in any time interval x to $x + n$, $_np_x$ represents the proportion or probability of surviving, while $_nq_x = 1 - {}_np_x$ is the proportion dying. Greenwood (23) derived a formula for the standard error of P_x:

$$\text{Standard error of } P_x = P_x \sqrt{\sum \frac{_nq_x}{(O_x - \frac{1}{2}{}_nW_x - {}_nd_x)}}$$

$$= P_x \sqrt{\sum \left(\frac{_nd_x}{O_x - \frac{1}{2}{}_nW_x}\right)\left(\frac{1}{O_x - \frac{1}{2}{}_nW_x - {}_nd_x}\right)},$$

where $_nq_x$ is as presented in the definitions, and where the sum extends from $x = 0$ to $x = n$, i.e., for the smallest x through the interval prior to that containing P_x. For example (expressing all P's as per cents), we have

Standard error of $P_{1.00}$

$$= 78.0\sqrt{\frac{13}{(97.5)(84.5)} + \frac{2}{(78.5)(76.5)} + \frac{2}{(69)(67)} + \frac{3}{(60)(57)}}$$

$$= 78.0\sqrt{0.00158 + 0.00033 + 0.00043 + 0.00088}$$

$$= 78.0\sqrt{0.00322}$$

$$= 78.0 \times 0.057 = 4.45 \text{ per cent.}$$

With this standard error one is in a position to test various hypotheses or to compute confidence intervals for the parameter corresponding to P_x.

EXERCISES

DATA FOR NEVADA, 1960:

Population for Age, Sex, and Race

Age	Male		Female	
	White	Nonwhite	White	Nonwhite
1	3,042	387	3,055	420
1–25	54,310	5,704	52,509	5,638
25–45	39,163	3,048	38,473	2,887
45–65	30,419	1,651	25,116	1,283
65–85	9,007	406	7,607	351
85+	357	27	385	33
Totals	136,298	11,223	127,145	10,612

Deaths by Age, Sex, and Race

Age	Male		Female	
	White	Nonwhite	White	Nonwhite
1	100	19	87	13
1–25	74	15	39	3
25–45	138	13	89	15
45–65	553	33	216	20
65–85	610	23	298	19
85+	84	2	71	4
Totals	1,559	105	800	74

Births by Sex and Race

	Male	Female
White	3,216	3,110
Nonwhite	490	454

13-1 Using the above data for Nevada, 1960, compute:

    a.   Crude death rate.
    b.   Male and female death rates.
    c.   Age specific death rates for all age groups.
    d.   White and nonwhite infant mortality rates.
    e.   Crude birth rate.
    f.   Vital indices for males, females, and overall.

13-2 Making use of the standard million population for the United States as presented in Table 13-3, compute by the direct method the age adjusted death rate for Nevada, 1960.

13-3 Suppose that death rates by sex for diseases of the cardiovascular system are unavailable for Las Vegas, Nevada, 1960. However, the following information is known: Total deaths for such diseases in Las Vegas are 409, while the statewide death rates for such deaths by sex are males 5.28 per 1,000 and females 2.94 per 1,000. Compute by the indirect method a sex-adjusted death rate for diseases of the cardiovascular system for the population of Las Vegas for 1960.

13-4 Consider the following data based on a study by Aanes (1):

    1.   23 pingue-fed sheep   4.   6 clinically ill
         11 males   2 males
         12 females   4 females

    2.   13 died   5.   4 not clinically ill
         8 males   1 male
         5 females   3 females

    3.   10 lived
         3 males
         7 females

Compute:

    a.   Crude death rate.
    b.   Sex specific death rates.
    c.   Crude case fatality rate.
    d.   Sex specific case fatality rates.

As a measure of clinical illness consider the attack rate, defined as the number of clinically apparent cases divided by the population at risk. Find:

e. Crude attack rate.
f. Sex specific attack rates.

13-5 Consider the following information for an isolated island. Assume that persons of all ages not previously infected are susceptible of being infected with infectious hepatitis (I.H.).

1. Population as of July 1, 1967   20,000
2. Number of cases of I.H. as of July 1, 1967   500
3. Number of cases of I.H. between July 1, 1967
    and June 30, 1968   1,000
4. Number of cases of I.H. extant as of June 30, 1968   1,300
5. Total number of deaths due to all causes between
    July 1, 1967 and June 30, 1968   400
6. Number of deaths due to I.H. between July 1, 1967
    and June 30, 1968   200
7. Total number of births between July 1, 1967 and
    June 30, 1968   250

a. Compute prevalence rate of I.H. as of July 1, 1967.
b. Compute the prevalence rate of I.H. as of June 30, 1968.
c. Estimate the population as of December 31, 1967 for use as denominator for parts d, e, and f. Explain your estimation.

Compute the following:

d. Incidence rate of I.H. for the interval July 1, 1967 to June 30, 1968.
e. Crude death rate July 1, 1967 to June 30, 1968.
f. Crude birth rate July 1, 1967 to June 30, 1968.
g. Case fatality rate for I.H. July 1, 1967 to June 30, 1968.

13-6 The following represent 112 patients with multiple myeloma. The time periods represent time from final laboratory diagnosis of disease.

X	O_x	$_nW_x$	$_nd_x$
0 months	112	13	18
3 months	81	9	10
6 months	62	6	8
9 months	48	7	5
12 months	36	3	4
18 months	29	2	7
24 to			
30 months	20	4	9

a. Compute the probability that a patient is still alive two years from time of diagnosis.
b. Find the standard error of this estimate.

References

1. Aanes, W. A., "Pingue (Hymenoxys richardsonii) Poisoning in Sheep," *American Journal of Veterinary Research*, **22**, 47–52, 1961.

2. Bartlett, M. S., "The Use of Transformations," *Biometrics*, **3**, 39–52, 1947.

3. Bayes, T., "An Essay Towards Solving a Problem in the Doctrine of Chances," *Philosophical Transaction*, **53**, 370–418, 1763.

4. Berkson, J., "Some Difficulties of Interpretation Encountered in the Application of the Chi-Square Test," *Journal of the American Statistical Association*, **33**, 526–536, 1938.

5. ———, and R. P. Gage, "Calculation of Survival Rates for Cancer," *Proceedings of the Staff Meeting Mayo Clinic*, **25**, 270, 1950.

6. Chernoff, H., and L. E. Moses, *Elementary Decision Theory*. New York: John Wiley & Sons, Inc., 1959.

7. Cochran, W. G., "Some Methods for Strengthening the Common Chi-Square Tests," *Biometrics*, **10**, 417–450, 1954.

8. David, F. N., *Games, Gods and Gambling*. London: Charles Griffin and Co., Ltd., 1962.

9. Dixon, W. J., and F. J. Massey, *Introduction to Statistical Analysis*, 3rd ed. New York: McGraw-Hill Book Company, 1969.

10. *Documenta Geigy Scientific Tables*, 6th Ed. New York: Ardsley, Geigy Pharmaceuticals, 1962.

11. Dublin, L. I., A. J. Lotka, and M. Spiegelman, *Length of Life*, rev. ed. New York: The Ronald Press Co., 1949.

12. Duncan, D. B., "Multiple Range and Multiple F-Tests," *Biometrics*, **11**, 1–42, 1955.

13. Dunnett, C. W., "A Multiple Comparison Procedure for Comparing Several Treatments with a Control," *Journal of the American Statistical Association*, **50**, 1096–1121, 1955.

14. Festinger, L., and D. Katz, *Research Methods in the Behavioral Sciences.* New York: The Dryden Press, 1953.

15. Fisher, R. A., *Statistical Methods for Research Workers*, 11th ed. London: Oliver and Boyd Ltd., 1950.

16. ———, "Frequency-distribution of the Values of the Correlation Coefficient in Samples from an Indefinitely Large Population," *Biometrika*, **10**, 507–521, 1915.

17. ———, "The Conditions Under Which χ^2 Measures the Discrepancy Between Observation and Hypothesis," *Journal of the Royal Statistical Society*, **87**, 442–450, 1924.

18. ———, and F. Yates, *Statistical Tables for Biological, Agricultural, and Medical Research*, 6th ed. New York: Hafner Publishing Co. Inc., 1963.

19. Freund, J. E., *Modern Elementary Statistics*, 3rd ed. Englewood Cliffs, N.J.: Prentice-Hall, Inc., 1967.

20. Galton, F., *Natural Inheritance.* London: Macmillan and Co., 1889.

21. Geary, R. C., "Moments of the Ratio of the Mean Deviation to the Standard Deviation for Normal Samples," *Biometrika*, **28**, 295–307, 1936.

22. Gosset, W. S. ("Student"), "The Probable Error of the Mean," *Biometrika*, **6**, 1–25, 1908.

23. Greenwood, M., "Reports on Public Health and Statistical Subjects, No. 33. A Report on the Natural Duration of Cancer. Appendix 1. The Errors of Sampling of the Survivorship Tables." London: His Majesty's Stationery Office, 1926.

24. Grizzle, J. E., "Continuity Correction in the χ^2 Test for 2×2 Tables," *The American Statistician*, **21**, 28–32, 1967.

25. Hill, A. B., *Principles of Medical Statistics*, 8th ed. New York: Oxford University Press, 1966.

26. Hoel, P. G., *Introduction to Mathematical Statistics*, 3rd ed. New York: John Wiley & Sons, Inc., 1962.

27. Huff, D., *How to Lie with Statistics.* New York: W. W. Norton and Co., Inc., 1954.

28. Jaffe, A. J., *Handbook of Statistical Methods for Demographers.* Washington D.C., U.S. Government Printing Office, 1960.

29. Johnson, B. C., and R. D. Remington, "A Sampling Study of Blood Pressure Levels in White and Negro Residents of Nassau, Bahamas," *Journal of Chronic Diseases*, **13**, 39–51, 1961.

30. Kaufman, K., "Serum Cholinesterase Activity in the Normal Individual and in People with Liver Disease," *Annals of Internal Medicine*, **41**, 533–545, 1954.

31. Mann, H. B., and D. R. Whitney, "On a Test of Whether One of Two Random Variables is Stochastically Larger than the Other," *Annals of Mathematical Statistics*, **18**, 50–61, 1947.

32. McNemar, A., *Psychological Statistics.* New York: John Wiley & Sons, Inc., 1955.

33. Merrell, Margaret, and Lawrence Shulman, "Determination of Prognosis in Chronic Disease, Illustrated by Systemic Lupus Erythematosus," *Journal of Chronic Diseases*, **1**, 12–32, 1955.

34. *Michigan Health Statistics Annual Statistical Report*, 1961. Statistical Methods Section, Michigan Department of Health, Lansing, Michigan, 1961.

35. Milam, D. F., and H. Muench, "Hemoglobin Levels in Specific Race, Age and Sex Groups of a Normal North Carolina Population," *Journal Lab. and Clinical Medicine*, **31**, 878–885, 1946.

36. Miller, R. G., *Simultaneous Statistical Inference.* New York: McGraw-Hill Book Company, 1966.

37. Mood, A. M., *Introduction to the Theory of Statistics.* New York: McGraw-Hill Book Company, 1950.

38. ——, and A. Graybill, *Introduction to the Theory of Statistics*, 2nd ed. New York: McGraw-Hill Book Company, 1963.

39. Ore, O., *Cardano: The Gambling Scholar.* Princeton, N.J.: Princeton University Press, 1953.

40. O'Rourke, R. A., D. P. Fischer, E. E. Escabor, V. L. Bishop, and E. Rapaport, "Effect of Acute Pericardial Tamponade on Coronary Blood Flow," *American Journal of Physiology*, **212**, 549–552, 1967.

41. Pearson, K., "On the Criterion that a Given System of Deviations from the Probable in the Case of a Correlated System of Variables is such that It Can Be Reasonably Supposed to Have Arisen from Random Sampling," *Philosophical Magazine*, 5th Series, **50**, 157–175, 1900.

42. ——, "I. Historical Note on The Origin of the Normal Curve of Errors," *Biometrika*, **16**, 402–404, 1924.

43. Pearson, E. S., and H. O. Hartley, *Biometrika Tables for Statisticians*, Volume 1, 3rd ed. Cambridge: Cambridge University Press, 1966.

44. Pearson, K., and A. Lee, "On the Laws of Inheritance in Man. I. Inheritance of Physical Characters," *Biometrika*, **2**, 357–462, 1902.

45. Pircher, F. J., E. A. Carr, and M. E. Patno, "Evaluation of Quantitative Aspects of the Radioisotope Renogram," *Journal of Nuclear Medicine*, **4**, 117–131, 1963.

46. Putter, J., "The Treatment of Ties in Some Nonparametric Tests," *Annals of Mathematical Statistics*, **26**, 368–386, 1955.

47. Scheffe, H., *The Analysis of Variance*. New York: John Wiley & Sons, Inc., 1959.

48. Swed, F. S., and C. Eisenhart, "Tables for Testing Randomness of Grouping in A Sequence of Alternatives," *Annals of Mathematical Statistics*, **14**, 66–87, 1943.

49. Taylor, H. L., Jr., personal communication, 1967.

50. Todhunter, I., *A History of the Mathematical Theory of Probability*. Cambridge and London: Macmillan and Co., 1865.

51. Tukey, J. W., "The Problem of Multiple Comparisons," *Princeton University Dittoed Statistical Series*, Princeton University, 1953.

52. U.S. Bureau of the Census, U.S. Census of Population: 1960, General Population Characteristics of Alaska—Final Report PC(1)-3B. U.S. Government Printing Office, Washington, D.C., 1961.

53. U.S. Bureau of the Census, U.S. Census of Population: 1960 General Population Characteristics of Hawaii—Final Report PC(1)-13B. U.S. Government Printing Office, Washington, D.C., 1961.

54. U.S. Bureau of the Census, U.S. Census of Population: 1960 General Population Characteristics of Michigan—Final Report PC(1)-24B. U.S. Government Printing Office, Washington, D.C., 1961.

55. U.S. Bureau of the Census, U.S. Census of Population: 1960, General Social and Economic Characteristics, Final Report PC(1)-1C. U.S. Government Printing Office, Washington, D.C., 1962.

56. *Vital Statistics of the United States, 1960*. U.S. Department of Health, Education, and Welfare, Washington, D.C., U.S. Government Printing Office, 1963.

57. *Vital Statistics of the United States, 1964*, Vol. II, Mortality, Sec. 5-2, U.S. Department of Health, Education, and Welfare, Washington, D.C., U.S. Government Printing Office, 1966.

58. Welch, B. L., "The Generalization of Student's Problem When Several Different Population Variances are Involved," *Biometrika*, **34**, 28–35, 1947.

59. Wilcoxon, F., *Some Rapid Approximate Statistical Procedures*. Stamford, Conn., American Cyanamid Co., 1949.

60. Wilcoxon, F., "Individual Comparisons by Ranking Methods," *Biometrics*, **1**, 50–83, 1945.

61. Wilk, M. B., and O. Kempthorne, "Non-Additivities in a Latin Square Design," *Journal of the American Statistical Association*, **52**, No. 278, 218–236, 1957.

62. Yates, F., "Contingency Tables Involving Small Numbers and the χ^2 Test," *Journal of the Royal Statistical Society Supplement*, **1**, 217–235, 1934.

Appendix

Tables

Table A-1

```
53872 34774   19087 81775   71440 12082   75092 34608   75448 13148
04226 62404   71577 00984   56056 32404   87641 53392   92561 33388
28666 44190   75524 62038   21423 46281   92238 96306   72606 80601
63817 30279   14088 86434   16183 06401   90586 80292   54555 47371
22359 16442   83879 47486   19838 32252   39560 95851   36758 36141

50968 28728   83525 16031   77583 65578   84794 51367   32535 83834
39652 24248   96617 91200   10769 52386   39559 75921   49375 22847
35493 00529   69632 29684   80284 87828   72418 80950   86311 34016
75687 53919   80439 20534   96185 72345   96391 52625   50866 45132
31509 93521   10681 44124   88345 84969   88768 48819   22311 41235

40389 76282   37506 60661   23295 67357   95419 10864   87833 09152
59244 54664   63424 97899   44153 69251   08781 18604   02312 21658
99876 17075   40934 08912   96196 58503   63613 24486   98092 45672
06457 50072   18060 71023   84349 40984   59487 77782   32107 53770
14297 07687   05517 10362   35783 62236   63764 45542   68889 03862

51661 57130   97442 29590   21634 79772   73801 70122   46467 47152
53455 41788   16117 09698   24409 05079   76603 57563   33461 46791
48086 31512   62819 27689   63744 11023   11184 87679   22218 70139
19108 01602   96950 41536   39974 88287   83546 69187   45539 78263
39001 77727   33095 58785   29179 45421   71416 20418   38558 78700

72346 55617   14714 21930   14851 38209   52202 03979   05970 74483
19094 64359   89829 10942   53101 37758   29583 26792   42840 45872
82247 77127   01652 50774   04970 83300   33760 22172   67516 62135
75968 18386   31874 52249   21015 20365   57475 32756   58268 75739
01963 38095   99960 91307   99654 74279   80145 53303   11870 50485

64828 15817   80923 55226   51893 93362   15757 47430   84855 95822
64347 61578   44160 06266   35118 52558   56436 96155   10293 67506
54746 52337   84826 39012   59118 19851   10156 78167   41473 99025
22241 41501   02993 99340   91044 67268   51088 12751   74008 33773
11906 20043   10415 44425   31712 54831   85591 62237   88797 14382

76637 07609   95378 95580   86909 50609   99008 99042   50364 36664
93896 47120   98926 30636   28136 49458   84145 79205   79517 93446
75292 88232   14360 12455   13656 65736   70428 66917   64412 38502
98792 29828   10577 48184   29433 98278   22543 76155   82107 22066
65751 91049   94127 47558   99880 79667   86254 72797   67117 44699

72064 62102   39155 79462   82975 02638   00302 79476   72656 84003
01227 35821   80607 61734   02600 45564   72344 71034   48370 96826
44768 56504   13993 59701   88238 92483   09497 66058   36651 37927
69838 91226   85736 72247   64099 86305   49877 76215   66980 30228
01800 39313   57730 84410   47637 81369   51830 43536   58937 91901

11756 45441   59948 57975   92422 70057   50210 30345   55912 31638
39056 86614   53643 62909   27198 04454   33789 86463   66603 48083
88086 93172   68311 39164   42012 10447   45933 28844   36844 57684
12648 27948   76750 19915   66815 34015   43011 27150   94264 89516
16254 87661   66181 68609   58626 58428   75051 27558   49463 66646

69682 19109   94189 94626   09299 10649   55405 54571   57855 54921
61336 86663   13010 40412   50139 30769   13048 61407   41056 60510
65727 66488   12304 70011   93324 58764   87274 43103   96002 06984
55705 34418   99410 32635   42984 40981   91750 27431   05142 77950
95402 51746   98184 38830   97590 00066   82770 42325   28778 83571

79228 94510   57711 64366   89040 43278   69072 22003   89465 61483
48103 56760   82564 33649   35176 32278   51357 05489   47462 55931
70969 27677   99621 63065   73194 70462   19316 77945   45004 39895
69931 20237   75246 59124   12484 22012   79731 82435   56301 99752
37208 22741   41946 74109   03760 24094   40210 76617   52317 50643

60151 92327   85150 27728   64813 47667   66078 03628   95240 03808
46210 47674   53747 95354   67757 75477   26396 09592   96239 50854
55399 48142   12284 95298   56399 61358   87541 12998   79639 63633
23677 64950   97041 43088   80143 34294   91468 01066   90350 78891
41947 70066   90311 17133   11674 00826   75760 37586   33621 14199
```

* Abstracted with kind permission from *A Million Random Digits with* 100,000 *Normal Deviates* by the RAND Corporation, The Free Press of Glencoe, New York, 1955.

RANDOM NUMBERS

```
16972 42181   87945 94104   95701 00743   75411 51930   54869 98991
74938 79042   38473 89672   45752 35715   89537 78155   09851 24983
78075 53671   81047 92759   94519 59473   91679 90536   41676 35230
76744 26190   21649 79753   21287 17698   39490 00533   34823 08134
82273 69293   23383 59365   18258 54530   47274 69686   55081 28731

30239 23081   09526 26055   87099 41372   55542 32754   87317 94638
41177 77163   38252 10349   49511 17540   61781 32769   51662 55606
07715 88600   69730 78912   19642 39764   47146 19472   84012 08887
16855 47454   98638 15189   87345 80509   33392 50866   17629 28208
27985 61979   02979 98092   41184 73815   57939 91057   04860 66667

77411 98433   42302 86602   26596 64175   64359 97570   64437 55592
19453 18731   01039 18933   92188 83767   56148 56261   79920 78514
03381 35119   30355 08287   00448 32800   24106 04054   70572 71063
11659 27315   09204 26213   57325 51470   56108 23141   16121 53925
35032 14283   20642 15311   36238 12079   67596 00017   51789 90737

32061 51250   39825 08554   88716 40945   68579 33784   62025 32535
81855 16888   24630 15077   47256 08529   54837 24161   95621 53483
48422 09247   43406 16093   01168 28523   31406 49360   99243 85090
86190 56195   31409 88248   52436 70161   98500 74702   99546 74570
90627 37048   50285 69189   97489 83007   31477 13908   97472 74448

60103 76739   57644 56746   63005 08804   47081 65928   65045 58629
09606 69465   16536 94055   86328 56533   16670 57295   26249 18524
62479 29610   03235 51050   15855 66828   08115 16166   32854 74206
40232 52840   02512 99258   09327 55073   86030 29933   00528 67359
10690 55550   81275 78369   33658 47000   89425 60573   81137 25474

73958 38949   99568 72713   22665 03244   17399 83950   66820 08704
56554 57926   41529 00619   51972 09442   60298 81066   28362 41165
35676 20333   77622 93718   57255 09780   26798 60083   58959 45691
01383 85677   96572 16401   31379 88519   41325 33938   36342 03327
29448 88487   05814 82402   42132 85708   89754 57495   57655 78644

56863 94737   68661 43498   33376 81659   07422 58435   24855 15523
20269 34456   48608 11787   86056 88290   17463 66628   03033 80771
06790 99803   86439 94235   48560 62912   82302 43198   97087 97104
73690 79726   06492 77431   49864 69775   46450 02122   09083 92746
76222 20006   98660 88690   01190 05588   76651 03461   11987 80756

18434 21893   80472 19499   80423 58643   27088 66458   78358 56606
20463 75133   41713 84279   56045 79079   20212 91560   60548 95128
27105 77095   72016 23683   01386 40381   74673 11811   36625 62958
47736 56338   07546 36084   73126 33364   78730 47282   76795 95719
60938 13970   90288 79457   50343 92054   12541 93216   58624 37392

02743 59982   92806 62853   39755 42550   31081 38860   35712 78632
74802 59354   91213 26293   18112 93831   01473 10798   18229 18642
06933 78651   45636 77509   28610 34307   68045 15107   62935 34149
40345 80092   50587 18535   19001 82179   12572 77589   33459 35130
70055 98685   10244 11760   21952 73985   68903 66934   42442 07608

34552 76373   40928 93696   97711 15818   31004 03263   05626 07460
45253 86947   42417 28778   14936 94099   90775 42001   86675 62770
71558 21692   84077 17814   33316 49494   31817 90127   39485 92302
95474 76468   12019 04274   01893 23930   88771 31142   65859 28948
34619 91898   28499 00279   35351 87736   83909 43736   19258 95068

44546 75524   68535 77434   18543 15479   58850 73802   10636 82735
22917 96024   04784 05809   52788 83577   02269 68632   23310 46261
33043 31433   47833 75234   74539 38529   57893 45997   71749 28666
99357 54593   21688 64216   85938 51742   12898 09737   61504 18946
01072 31679   80961 34029   56463 09594   11939 51777   64796 52452

90838 50179   42064 62987   13072 84227   24060 59438   05695 38136
35914 39441   90149 67957   16955 39960   26142 45600   75486 74103
87047 77284   12753 45644   47843 55781   06672 57548   84706 25453
93727 46613   48045 49685   28385 37200   98473 56808   86774 07305
37439 50362   44171 18495   57370 77691   28006 55318   39723 25299
```

Table A-1 (*cont.*)

98892 53633	33909 81674	91956 84531	60422 55574	31670 61059
95398 77381	21912 24873	26372 12044	43234 08503	86716 08095
28982 24589	88896 31137	87512 33216	29665 26014	02919 17639
31303 70209	42174 10757	98531 35725	68208 61239	26705 43916
08457 10085	35741 79416	72457 59502	46986 09051	70963 19759
30698 80818	90073 78320	83675 78361	49929 70495	92247 04318
27142 41186	52273 81087	67396 16795	98542 83820	48765 24164
36775 63628	70856 43164	88426 51415	37514 24870	55665 05311
02560 51679	79600 23297	36434 17174	00109 02731	05909 58959
36744 66697	08331 50201	56303 09171	55995 60232	31305 30689
66482 04302	29770 46201	04588 42575	99318 84406	83405 21186
76375 41539	65940 57820	29283 94564	96598 00619	60468 97375
95772 72925	19454 63712	21401 96665	77750 21218	02990 50796
59013 81632	85000 39180	99975 73253	46534 59083	60243 27664
52392 04440	45628 34976	92012 16596	28596 15493	80754 48760
08027 07629	04339 77570	47155 77128	24498 67455	06320 82004
27284 39416	57313 03508	71443 42543	73335 68620	87559 77927
20513 38581	82309 69951	82658 60958	18290 60534	30741 89647
36076 12821	68723 37934	62818 64157	54590 98263	70109 06755
60679 43862	43675 03653	21060 81096	71332 28930	44207 08354
49416 58370	63738 87515	39290 87656	36130 23490	30963 57350
65757 39149	11780 92494	41335 35835	69882 56431	08091 01981
17379 77731	65133 44979	90939 29184	76634 58007	34873 83816
00757 13129	09648 07644	81689 68088	34882 04971	27565 66577
68276 79035	78273 83412	97328 81003	65938 85510	78367 29316
64716 91696	45448 92281	73854 67452	52145 41582	81549 82434
83695 11496	57066 48153	74754 56383	09253 65456	32438 96357
58275 66797	35380 41155	44389 94860	42074 31178	27967 12666
58005 84170	29999 23631	93032 41592	55688 78599	59902 21568
99993 80083	08810 07244	42067 76669	19686 64064	67141 80520
31692 51607	89056 74472	91284 20263	16039 94491	33767 73915
82997 58320	04852 52595	95514 56543	06636 61291	67504 57205
05043 40582	46051 60261	04996 82256	47375 87507	05112 88489
75781 38768	70475 00601	18378 32077	36523 30843	07057 78326
21033 15175	30741 45814	92222 16704	00197 51267	33224 40276
99092 60991	12571 71753	65214 33885	82939 50723	88987 69761
07204 93373	85112 29610	30375 64836	18459 08235	67650 72930
88859 97254	07771 21393	64657 42013	12753 03028	24224 24918
30497 91407	72900 15699	58653 38063	25072 48698	88083 48040
09726 18075	45852 54968	43743 82050	78412 79456	95032 10984
95330 01985	24128 60514	42539 91907	25694 37097	39566 24043
09760 32388	05601 49923	66126 54146	67213 52234	48381 89442
01534 81967	15337 95831	84643 40792	47562 95494	62087 18064
11234 59350	48368 57195	36287 03046	87136 36057	93913 70080
71056 48762	80221 59683	27504 21121	94711 11807	80882 48359
34208 05374	60304 43178	97247 24875	26259 67622	14657 80354
47132 62839	82198 92445	60650 76219	02772 48651	66449 89213
55685 93302	43019 45861	95493 16106	12783 37248	83533 25440
17803 18184	10510 27159	83008 20544	41665 99439	70606 28974
55045 17219	66737 59080	78489 12626	60661 53733	70062 14289
01923 33647	98442 59293	83318 33425	76412 87062	01295 11083
07202 76476	71888 54845	17468 41964	68694 59662	55905 26898
68825 68242	95750 11033	58634 78411	08523 19313	29327 47526
68525 06496	17446 41378	32368 82019	66101 56733	43308 82641
80819 33515	97373 43064	16221 99697	37951 07947	12935 49391
64200 96929	26044 49283	56545 67200	21325 85056	51345 06309
30156 29121	75874 42399	41121 90643	19585 06364	47203 19679
50467 14282	89098 66717	14753 73356	47781 34165	82842 00121
53764 83212	26675 64184	64455 29023	03181 13674	08838 83829
81727 35572	95469 36825	81882 95083	68323 14965	34166 32351

RANDOM NUMBERS

30807	55558	96026	97398	21723	86560	52617	07771	61886	48234
75104	23682	78756	72728	85940	57290	75507	78715	01426	02310
06180	62724	36835	80288	25075	32609	33312	21348	87710	55457
22098	34834	66117	36252	82717	50585	43639	79999	07414	84003
13173	64783	20984	11929	18849	26211	77375	49561	96747	67007
75273	36108	55265	15653	82270	99216	27805	60088	06056	97377
89849	65756	44454	04602	14292	74458	57777	35934	05160	26359
91108	43562	18883	16569	49599	73871	67101	12054	56492	15981
51843	01542	17881	12954	94913	39583	94969	61146	35907	72184
02644	23564	85464	62947	92571	89377	85004	84654	20465	86212
38608	83374	74032	62183	08740	05279	30455	31032	71512	16476
43164	28909	88624	14992	85359	10193	32491	14769	63694	92640
80933	52950	45646	36636	05085	28053	27596	54873	68476	65823
67690	96766	69250	19344	47855	43489	77479	62418	54079	40069
68579	17014	25362	15114	30982	27250	29052	71115	83369	46776
46353	39733	44677	50133	26623	15979	10651	04263	34087	67005
30039	09532	52215	09164	20930	88230	43403	63230	83525	93550
89200	92772	42195	91634	39272	46462	76835	27755	03151	75692
58118	57942	14807	68214	76093	47484	24468	91764	52907	16675
97230	33027	70166	43232	98802	70715	30216	35586	18909	79658

Table A-2
BINOMIAL PROBABILITIES

n	x	0.01	0.05	0.10	0.15	0.20	0.25	0.30	1/3	0.35	0.40	0.45	0.50
1	0	0.9900	0.9500	0.9000	0.8500	0.8000	0.7500	0.7000	0.6667	0.6500	0.6000	0.5500	0.5000
	1	0.0100	0.0500	0.1000	0.1500	0.2000	0.2500	0.3000	0.3333	0.3500	0.4000	0.4500	0.5000
2	0	0.9801	0.9025	0.8100	0.7225	0.6400	0.5625	0.4900	0.4444	0.4225	0.3600	0.3025	0.2500
	1	0.0198	0.0950	0.1800	0.2550	0.3200	0.3750	0.4200	0.4444	0.4550	0.4800	0.4950	0.5000
	2	0.0001	0.0025	0.0100	0.0225	0.0400	0.0625	0.0900	0.1111	0.1225	0.1600	0.2025	0.2500
3	0	0.9703	0.8574	0.7290	0.6141	0.5120	0.4219	0.3430	0.2963	0.2746	0.2160	0.1664	0.1250
	1	0.0294	0.1354	0.2430	0.3251	0.3840	0.4219	0.4410	0.4444	0.4436	0.4320	0.4084	0.3750
	2	0.0003	0.0071	0.0270	0.0574	0.0960	0.1406	0.1890	0.2222	0.2389	0.2880	0.3341	0.3750
	3	0.0000	0.0001	0.0010	0.0034	0.0080	0.0156	0.0270	0.0370	0.0429	0.0640	0.0911	0.1250
4	0	0.9605	0.8145	0.6561	0.5220	0.4096	0.3164	0.2401	0.1975	0.1785	0.1296	0.0915	0.0625
	1	0.0388	0.1715	0.2916	0.3685	0.4096	0.4219	0.4116	0.3951	0.3845	0.3456	0.2995	0.2500
	2	0.0006	0.0135	0.0486	0.0975	0.1536	0.2109	0.2646	0.2963	0.3105	0.3456	0.3675	0.3750
	3	0.0000	0.0005	0.0036	0.0115	0.0256	0.0469	0.0756	0.0988	0.1115	0.1536	0.2005	0.2500
	4	0.0000	0.0000	0.0001	0.0005	0.0016	0.0039	0.0081	0.0123	0.0150	0.0256	0.0410	0.0625
5	0	0.9510	0.7738	0.5905	0.4437	0.3277	0.2373	0.1681	0.1317	0.1160	0.0778	0.0503	0.0312
	1	0.0480	0.2036	0.3280	0.3915	0.4096	0.3955	0.3601	0.3292	0.3124	0.2592	0.2059	0.1563
	2	0.0010	0.0214	0.0729	0.1382	0.2048	0.2637	0.3087	0.3292	0.3364	0.3456	0.3369	0.3125
	3	0.0000	0.0012	0.0081	0.0244	0.0512	0.0879	0.1323	0.1646	0.1812	0.2304	0.2757	0.3125
	4	0.0000	0.0000	0.0005	0.0021	0.0064	0.0146	0.0284	0.0412	0.0487	0.0768	0.1127	0.1563
	5	0.0000	0.0000	0.0000	0.0001	0.0003	0.0010	0.0024	0.0041	0.0053	0.0102	0.0185	0.0312
6	0	0.9415	0.7351	0.5314	0.3771	0.2621	0.1780	0.1176	0.0878	0.0754	0.0467	0.0277	0.0156
	1	0.0570	0.2321	0.3543	0.3994	0.3932	0.3559	0.3026	0.2634	0.2437	0.1866	0.1359	0.0938
	2	0.0015	0.0306	0.0984	0.1762	0.2458	0.2967	0.3241	0.3292	0.3280	0.3110	0.2779	0.2344
	3	0.0000	0.0021	0.0146	0.0414	0.0819	0.1318	0.1852	0.2195	0.2355	0.2765	0.3032	0.3125
	4	0.0000	0.0001	0.0012	0.0055	0.0154	0.0330	0.0596	0.0823	0.0951	0.1382	0.1861	0.2344
	5	0.0000	0.0000	0.0001	0.0004	0.0015	0.0044	0.0102	0.0165	0.0205	0.0369	0.0609	0.0938
	6	0.0000	0.0000	0.0000	0.0000	0.0001	0.0002	0.0007	0.0014	0.0018	0.0041	0.0083	0.0156

n	x							p					
		0.01	0.05	0.10	0.15	0.20	0.25	0.30	1/3	0.35	0.40	0.45	0.50
7	0	0.9321	0.6983	0.4783	0.3206	0.2097	0.1335	0.0824	0.0585	0.0490	0.0280	0.0152	0.0078
	1	0.0659	0.2573	0.3720	0.3960	0.3670	0.3114	0.2470	0.2048	0.1848	0.1306	0.0872	0.0547
	2	0.0020	0.0406	0.1240	0.2096	0.2753	0.3115	0.3177	0.3073	0.2985	0.2613	0.2140	0.1641
	3	0.0000	0.0036	0.0230	0.0617	0.1147	0.1730	0.2269	0.2561	0.2679	0.2903	0.2919	0.2734
	4	0.0000	0.0002	0.0025	0.0109	0.0286	0.0577	0.0972	0.1280	0.1442	0.1935	0.2388	0.2734
	5	0.0000	0.0000	0.0002	0.0011	0.0043	0.0116	0.0250	0.0384	0.0466	0.0775	0.1172	0.1641
	6	0.0000	0.0000	0.0000	0.0001	0.0004	0.0012	0.0036	0.0064	0.0084	0.0172	0.0320	0.0547
	7	0.0000	0.0000	0.0000	0.0000	0.0000	0.0001	0.0002	0.0005	0.0006	0.0016	0.0037	0.0078
8	0	0.9227	0.6634	0.4305	0.2725	0.1678	0.1001	0.0576	0.0390	0.0319	0.0168	0.0084	0.0039
	1	0.0746	0.2794	0.3826	0.3847	0.3355	0.2670	0.1977	0.1561	0.1372	0.0896	0.0548	0.0313
	2	0.0026	0.0514	0.1488	0.2376	0.2936	0.3114	0.2965	0.2731	0.2587	0.2090	0.1570	0.1093
	3	0.0001	0.0054	0.0331	0.0838	0.1468	0.2077	0.2541	0.2731	0.2786	0.2787	0.2569	0.2188
	4	0.0000	0.0004	0.0046	0.0185	0.0459	0.0865	0.1361	0.1707	0.1875	0.2322	0.2626	0.2734
	5	0.0000	0.0000	0.0004	0.0027	0.0092	0.0231	0.0467	0.0683	0.0808	0.1239	0.1718	0.2188
	6	0.0000	0.0000	0.0000	0.0002	0.0011	0.0038	0.0100	0.0171	0.0217	0.0413	0.0704	0.1093
	7	0.0000	0.0000	0.0000	0.0000	0.0001	0.0004	0.0012	0.0024	0.0034	0.0078	0.0164	0.0313
	8	0.0000	0.0000	0.0000	0.0000	0.0000	0.0000	0.0001	0.0002	0.0002	0.0007	0.0017	0.0039
9	0	0.9135	0.6302	0.3874	0.2316	0.1342	0.0751	0.0404	0.0260	0.0207	0.0101	0.0046	0.0020
	1	0.0831	0.2986	0.3874	0.3678	0.3020	0.2252	0.1556	0.1171	0.1004	0.0604	0.0339	0.0175
	2	0.0033	0.0628	0.1722	0.2597	0.3020	0.3004	0.2668	0.2341	0.2162	0.1613	0.1110	0.0703
	3	0.0001	0.0078	0.0447	0.1070	0.1762	0.2336	0.2669	0.2731	0.2716	0.2508	0.2119	0.1641
	4	0.0000	0.0006	0.0074	0.0283	0.0660	0.1168	0.1715	0.2048	0.2194	0.2508	0.2600	0.2461
	5	0.0000	0.0000	0.0008	0.0050	0.0165	0.0389	0.0735	0.1024	0.1181	0.1672	0.2128	0.2461
	6	0.0000	0.0000	0.0001	0.0006	0.0028	0.0087	0.0210	0.0341	0.0424	0.0744	0.1160	0.1641
	7	0.0000	0.0000	0.0000	0.0000	0.0003	0.0012	0.0039	0.0073	0.0098	0.0212	0.0407	0.0703
	8	0.0000	0.0000	0.0000	0.0000	0.0000	0.0001	0.0004	0.0009	0.0013	0.0035	0.0083	0.0175
	9	0.0000	0.0000	0.0000	0.0000	0.0000	0.0000	0.0000	0.0001	0.0001	0.0003	0.0008	0.0020

Table A-2 (cont.)
BINOMIAL PROBABILITIES

n	x	0.01	0.05	0.10	0.15	0.20	0.25	0.30	1/3	0.35	0.40	0.45	0.50
10	0	0.9044	0.5987	0.3487	0.1969	0.1074	0.0563	0.0282	0.0173	0.0135	0.0060	0.0025	0.0010
	1	0.0913	0.3152	0.3874	0.3474	0.2684	0.1877	0.1211	0.0867	0.0725	0.0404	0.0208	0.0097
	2	0.0042	0.0746	0.1937	0.2759	0.3020	0.2816	0.2335	0.1951	0.1756	0.1209	0.0763	0.0440
	3	0.0001	0.0105	0.0574	0.1298	0.2013	0.2503	0.2668	0.2601	0.2522	0.2150	0.1664	0.1172
	4	0.0000	0.0009	0.0112	0.0401	0.0881	0.1460	0.2001	0.2276	0.2377	0.2508	0.2384	0.2051
	5	0.0000	0.0001	0.0015	0.0085	0.0264	0.0584	0.1030	0.1366	0.1536	0.2007	0.2340	0.2460
	6	0.0000	0.0000	0.0001	0.0013	0.0055	0.0162	0.0367	0.0569	0.0689	0.1114	0.1596	0.2051
	7	0.0000		0.0000	0.0001	0.0008	0.0031	0.0090	0.0163	0.0212	0.0425	0.0746	0.1172
	8			0.0000	0.0000	0.0001	0.0004	0.0015	0.0030	0.0043	0.0106	0.0229	0.0440
	9				0.0000	0.0000	0.0000	0.0001	0.0003	0.0005	0.0016	0.0042	0.0097
	10	0.0000				0.0000	0.0000	0.0000	0.0000	0.0000	0.0001	0.0003	0.0010
11	0	0.8953	0.5688	0.3138	0.1673	0.0859	0.0422	0.0198	0.0116	0.0088	0.0036	0.0014	0.0005
	1	0.0995	0.3293	0.3835	0.3248	0.2362	0.1549	0.0932	0.0636	0.0518	0.0266	0.0125	0.0054
	2	0.0050	0.0867	0.2130	0.2866	0.2953	0.2581	0.1998	0.1590	0.1395	0.0887	0.0513	0.0268
	3	0.0002	0.0136	0.0711	0.1518	0.2215	0.2581	0.2568	0.2384	0.2255	0.1774	0.1259	0.0806
	4	0.0000	0.0015	0.0157	0.0535	0.1107	0.1721	0.2201	0.2384	0.2427	0.2365	0.2060	0.1611
	5	0.0000	0.0001	0.0025	0.0132	0.0387	0.0803	0.1321	0.1669	0.1830	0.2207	0.2360	0.2256
	6		0.0000	0.0003	0.0024	0.0097	0.0267	0.0566	0.0835	0.0986	0.1471	0.1931	0.2256
	7			0.0000	0.0003	0.0018	0.0064	0.0173	0.0298	0.0379	0.0701	0.1128	0.1611
	8			0.0000	0.0000	0.0002	0.0011	0.0037	0.0075	0.0102	0.0234	0.0462	0.0806
	9				0.0000	0.0000	0.0001	0.0005	0.0012	0.0018	0.0052	0.0126	0.0268
	10					0.0000	0.0000	0.0000	0.0001	0.0002	0.0007	0.0020	0.0054
	11	0.0000				0.0000	0.0000	0.0000	0.0000	0.0000	0.0000	0.0002	0.0005
12	0	0.8864	0.5404	0.2824	0.1422	0.0687	0.0317	0.0138	0.0077	0.0057	0.0022	0.0008	0.0002
	1	0.1074	0.3412	0.3766	0.3012	0.2062	0.1267	0.0712	0.0462	0.0367	0.0174	0.0075	0.0030
	2	0.0060	0.0988	0.2301	0.2924	0.2834	0.2323	0.1678	0.1272	0.1089	0.0638	0.0338	0.0161
	3	0.0002	0.0174	0.0853	0.1720	0.2363	0.2581	0.2397	0.2120	0.1954	0.1419	0.0924	0.0537
	4	0.0000	0.0020	0.0213	0.0683	0.1328	0.1936	0.2312	0.2384	0.2366	0.2129	0.1700	0.1208
	5	0.0000	0.0002	0.0038	0.0193	0.0532	0.1032	0.1585	0.1908	0.2040	0.2270	0.2225	0.1934
	6		0.0000	0.0004	0.0040	0.0155	0.0401	0.0792	0.1113	0.1281	0.1766	0.2124	0.2256
	7			0.0000	0.0006	0.0033	0.0115	0.0291	0.0477	0.0591	0.1009	0.1489	0.1934
	8			0.0000	0.0001	0.0005	0.0024	0.0078	0.0149	0.0199	0.0420	0.0762	0.1208
	9				0.0000	0.0001	0.0004	0.0015	0.0033	0.0048	0.0125	0.0277	0.0537
	10					0.0000	0.0000	0.0002	0.0005	0.0007	0.0025	0.0068	0.0161
	11					0.0000	0.0000	0.0000	0.0000	0.0001	0.0003	0.0010	0.0030
	12	0.0000					0.0000	0.0000	0.0000	0.0000	0.0000	0.0001	0.0002

Table A-2 (cont.)
BINOMIAL PROBABILITIES

p

n	x	0.01	0.05	0.10	0.15	0.20	0.25	0.30	1/3	0.35	0.40	0.45	0.50
15	0	0.8601	0.4633	0.2059	0.0874	0.0352	0.0134	0.0047	0.0023	0.0016	0.0005	0.0001	0.0000
	1	0.1301	0.3667	0.3431	0.2312	0.1319	0.0668	0.0306	0.0171	0.0126	0.0047	0.0016	0.0005
	2	0.0092	0.1348	0.2669	0.2856	0.2309	0.1559	0.0915	0.0599	0.0475	0.0219	0.0090	0.0032
	3	0.0004	0.0307	0.1285	0.2185	0.2502	0.2252	0.1701	0.1299	0.1110	0.0634	0.0317	0.0139
	4	0.0000	0.0049	0.0429	0.1156	0.1876	0.2252	0.2186	0.1948	0.1792	0.1268	0.0780	0.0416
	5		0.0005	0.0105	0.0449	0.1031	0.1651	0.2061	0.2143	0.2124	0.1859	0.1404	0.0917
	6		0.0001	0.0019	0.0132	0.0430	0.0918	0.1473	0.1786	0.1905	0.2066	0.1914	0.1527
	7		0.0000	0.0003	0.0030	0.0139	0.0393	0.0811	0.1148	0.1320	0.1771	0.2013	0.1964
	8			0.0000	0.0005	0.0034	0.0131	0.0348	0.0574	0.0710	0.1181	0.1657	0.1964
	9				0.0001	0.0007	0.0034	0.0115	0.0223	0.0298	0.0612	0.1049	0.1527
	10				0.0000	0.0001	0.0007	0.0030	0.0067	0.0096	0.0245	0.0514	0.0917
	11					0.0000	0.0001	0.0006	0.0015	0.0023	0.0074	0.0192	0.0416
	12						0.0000	0.0001	0.0003	0.0004	0.0016	0.0052	0.0139
	13							0.0000	0.0000	0.0001	0.0003	0.0010	0.0032
	14							0.0000	0.0000	0.0000	0.0000	0.0001	0.0005
	15	0.0000	0.0000	0.0000	0.0000	0.0000	0.0000	0.0000	0.0000	0.0000	0.0000	0.0000	0.0000
20	0	0.8179	0.3585	0.1216	0.0388	0.0115	0.0032	0.0008	0.0003	0.0002	0.0000	0.0000	0.0000
	1	0.1652	0.3773	0.2701	0.1368	0.0577	0.0211	0.0068	0.0030	0.0019	0.0005	0.0001	0.0000
	2	0.0159	0.1887	0.2852	0.2293	0.1369	0.0670	0.0279	0.0143	0.0100	0.0031	0.0008	0.0002
	3	0.0010	0.0596	0.1901	0.2428	0.2054	0.1339	0.0716	0.0429	0.0323	0.0124	0.0040	0.0011
	4	0.0000	0.0133	0.0898	0.1821	0.2182	0.1896	0.1304	0.0911	0.0738	0.0350	0.0139	0.0046
	5		0.0022	0.0319	0.1028	0.1746	0.2023	0.1789	0.1457	0.1272	0.0746	0.0364	0.0148
	6		0.0003	0.0089	0.0454	0.1091	0.1686	0.1916	0.1821	0.1714	0.1244	0.0746	0.0370
	7		0.0000	0.0020	0.0160	0.0545	0.1124	0.1643	0.1821	0.1844	0.1659	0.1221	0.0739
	8			0.0004	0.0046	0.0222	0.0609	0.1144	0.1480	0.1614	0.1797	0.1623	0.1201
	9			0.0001	0.0011	0.0074	0.0271	0.0654	0.0987	0.1158	0.1597	0.1771	0.1602
	10			0.0000	0.0002	0.0020	0.0099	0.0308	0.0543	0.0686	0.1171	0.1593	0.1762
	11				0.0000	0.0005	0.0030	0.0120	0.0247	0.0336	0.0710	0.1185	0.1602
	12					0.0001	0.0008	0.0039	0.0092	0.0136	0.0355	0.0728	0.1201
	13					0.0000	0.0002	0.0010	0.0028	0.0045	0.0145	0.0366	0.0739
	14						0.0000	0.0002	0.0007	0.0012	0.0049	0.0150	0.0370
	15						0.0000	0.0000	0.0001	0.0003	0.0013	0.0049	0.0148
	16								0.0000	0.0000	0.0003	0.0013	0.0046
	17										0.0000	0.0002	0.0011
	18											0.0000	0.0002
	19												0.0000
	20	0.0000	0.0000	0.0000	0.0000	0.0000	0.0000	0.0000	0.0000	0.0000	0.0000	0.0000	0.0000

Table A-3

INDIVIDUAL PROBABILITIES $e^{-\mu}\mu^x/x!$ FOR THE POISSON DISTRIBUTION*

x	0.2	0.4	0.6	0.8	1.0	1.2	1.4	1.6	x
0	0.818731	0.670320	0.548812	0.449329	0.367879	0.301194	0.246597	0.201897	0
1	0.163746	0.268128	0.329287	0.359463	0.367879	0.361433	0.345236	0.323034	1
2	0.016375	0.053626	0.098786	0.143785	0.183940	0.216860	0.241665	0.258428	2
3	0.001092	0.007150	0.019757	0.038343	0.061313	0.086744	0.112777	0.137828	3
4	0.000055	0.000715	0.002964	0.007669	0.015328	0.026023	0.039472	0.055131	4
5	0.000002	0.000057	0.000356	0.001227	0.003066	0.006246	0.011052	0.017642	5
6		0.000004	0.000036	0.000164	0.000511	0.001249	0.002579	0.004705	6
7			0.000003	0.000019	0.000073	0.000214	0.000516	0.001075	7
8				0.000002	0.000009	0.000032	0.000090	0.000215	8
9					0.000001	0.000004	0.000014	0.000038	9
10						0.000001	0.000002	0.000006	10
11								0.000001	11

x	1.8	2.0	2.5	3.0	3.5	4.0	4.5	5.0	x
0	0.165299	0.135335	0.082085	0.049787	0.030197	0.018316	0.011109	0.006738	0
1	0.297538	0.270671	0.205212	0.149361	0.105691	0.073263	0.049990	0.033690	1
2	0.267784	0.270671	0.256516	0.224042	0.184959	0.146525	0.112479	0.084224	2
3	0.160671	0.180447	0.213763	0.224042	0.215785	0.195367	0.168718	0.140374	3
4	0.072302	0.090224	0.133602	0.168031	0.188812	0.195367	0.189808	0.175467	4
5	0.026029	0.036089	0.066801	0.100819	0.132169	0.156293	0.170827	0.175467	5
6	0.007809	0.012030	0.027834	0.050409	0.077098	0.104196	0.128120	0.146223	6
7	0.002008	0.003437	0.009941	0.021604	0.038549	0.059540	0.082363	0.104445	7
8	0.000452	0.000859	0.003106	0.008102	0.016865	0.029770	0.046329	0.065278	8
9	0.000090	0.000191	0.000863	0.002701	0.006559	0.013231	0.023165	0.036266	9
10	0.000016	0.000038	0.000216	0.000810	0.002296	0.005292	0.010424	0.018138	10
11	0.000003	0.000007	0.000049	0.000221	0.000730	0.001925	0.004264	0.008242	11
12		0.000001	0.000010	0.000055	0.000213	0.000642	0.001599	0.003434	12
13			0.000002	0.000013	0.000057	0.000197	0.000554	0.001321	13
14				0.000003	0.000014	0.000056	0.000178	0.000472	14
15				0.000001	0.000003	0.000015	0.000053	0.000157	15
16					0.000001	0.000004	0.000015	0.000049	16
17						0.000001	0.000004	0.000014	17
18							0.000001	0.000004	18
19								0.000001	19

x	5.5	6.0	6.5	7.0	7.5	8.0	9.0	10.0	x
0	0.004087	0.002479	0.001503	0.000912	0.000553	0.000335	0.000123	0.000045	0
1	0.022477	0.014873	0.009772	0.006383	0.004148	0.002684	0.001111	0.000454	1
2	0.061812	0.044618	0.031760	0.022341	0.015555	0.010735	0.004998	0.002270	2
3	0.113323	0.089235	0.068814	0.052129	0.038889	0.028626	0.014994	0.007567	3

* The data of this table are extracted with kind permission from: *Biometrika Tables for Statisticians*, 3rd Ed. Vol. I, Table 39, 1966. London: Bentley House.

INDIVIDUAL PROBABILITIES $e^{-\mu}\mu^x/x!$ FOR THE POISSON DISTRIBUTION

x				μ					x
	5.5	6.0	6.5	7.0	7.5	8.0	9.0	10.0	
4	0.155819	0.133853	0.111822	0.091226	0.072916	0.057252	0.033737	0.018917	4
5	0.171401	0.160623	0.145369	0.127717	0.109375	0.091604	0.060727	0.037833	5
6	0.157117	0.160623	0.157483	0.149003	0.136718	0.122138	0.091090	0.063055	6
7	0.123449	0.137677	0.146234	0.149003	0.146484	0.139587	0.117116	0.090079	7
8	0.084871	0.103258	0.118815	0.130377	0.137329	0.139587	0.131756	0.112599	8
9	0.051866	0.068838	0.085811	0.101405	0.114440	0.124077	0.131756	0.125110	9
10	0.028526	0.041303	0.055777	0.070983	0.085830	0.099262	0.118580	0.125110	10
11	0.014263	0.022529	0.032959	0.045171	0.058521	0.072190	0.097020	0.113736	11
12	0.006537	0.011264	0.017853	0.026350	0.036575	0.048127	0.072765	0.094780	12
13	0.002766	0.005199	0.008926	0.014188	0.021101	0.029616	0.050376	0.072908	13
14	0.001087	0.002228	0.004144	0.007094	0.011304	0.016924	0.032384	0.052077	14
15	0.000398	0.000891	0.001796	0.003311	0.005652	0.009026	0.019431	0.034718	15
16	0.000137	0.000334	0.000730	0.001448	0.002649	0.004513	0.010930	0.021699	16
17	0.000044	0.000118	0.000279	0.000596	0.001169	0.002124	0.005786	0.012764	17
18	0.000014	0.000039	0.000101	0.000232	0.000487	0.000944	0.002893	0.007091	18
19	0.000004	0.000012	0.000034	0.000085	0.000192	0.000397	0.001370	0.003732	19
20	0.000001	0.000004	0.000011	0.000030	0.000072	0.000159	0.000617	0.001866	20
21		0.000001	0.000003	0.000010	0.000026	0.000061	0.000264	0.000899	21
22			0.000001	0.000003	0.000009	0.000022	0.000108	0.000404	22
23				0.000001	0.000003	0.000008	0.000042	0.000176	23
24					0.000001	0.000003	0.000016	0.000073	24
25						0.000001	0.000006	0.000029	25
26							0.000002	0.000011	26
27							0.000001	0.000004	27
28								0.000001	28
29								0.000001	29

	11.0	12.0	13.0	14.0	15.0				
0	0.000017	0.000006	0.000002	0.000001					0
1	0.000184	0.000074	0.000029	0.000012	0.000005				1
2	0.001010	0.000442	0.000191	0.000081	0.000034				2
3	0.003705	0.001770	0.000828	0.000380	0.000172				3
4	0.010189	0.005309	0.002690	0.001331	0.000645				4
5	0.022415	0.012741	0.006994	0.003727	0.001936				5
6	0.041095	0.025481	0.015153	0.008696	0.004839				6
7	0.064577	0.043682	0.028141	0.017392	0.010370				7
8	0.088794	0.065523	0.045730	0.030435	0.019444				8
9	0.108526	0.087364	0.066054	0.047344	0.032407				9
10	0.119378	0.104837	0.085870	0.066282	0.048611				10
11	0.119378	0.114368	0.101483	0.084359	0.066287				11
12	0.109430	0.114363	0.109940	0.098418	0.082859				12

INDIVIDUAL PROBABILITIES $e^{-\mu}\mu^x/x!$ FOR THE POISSON DISTRIBUTION

x	μ					x
	11.0	12.0	13.0	14.0	15.0	
13	0.092595	0.105570	0.109940	0.105989	0.095607	13
14	0.072753	0.090489	0.102087	0.105989	0.102436	14
15	0.053352	0.072391	0.088475	0.098923	0.102436	15
16	0.036680	0.054293	0.071886	0.086558	0.096034	16
17	0.023734	0.038325	0.054972	0.071283	0.084736	17
18	0.014504	0.025550	0.039702	0.055442	0.070613	18
19	0.008397	0.016137	0.027164	0.040852	0.055747	19
20	0.004618	0.009682	0.017657	0.028597	0.041810	20
21	0.002419	0.005533	0.010930	0.019064	0.029865	21
22	0.001210	0.003018	0.006459	0.012132	0.020362	22
23	0.000578	0.001575	0.003651	0.007385	0.013280	23
24	0.000265	0.000787	0.001977	0.004308	0.008300	24
25	0.000117	0.000378	0.001028	0.002412	0.004980	25
26	0.000049	0.000174	0.000514	0.001299	0.002873	26
27	0.000020	0.000078	0.000248	0.000674	0.001596	27
28	0.000008	0.000033	0.000115	0.000337	0.000855	28
29	0.000003	0.000014	0.000052	0.000163	0.000442	29
30	0.000001	0.000005	0.000022	0.000076	0.000221	30
31		0.000002	0.000009	0.000034	0.000107	31
32		0.000001	0.000004	0.000015	0.000052	32
33			0.000002	0.000006	0.000032	33
34			0.000001	0.000003	0.000010	34
35				0.000001	0.000004	35
36					0.000002	36
37					0.000001	37

Table A-4

THE NORMAL DISTRIBUTION*

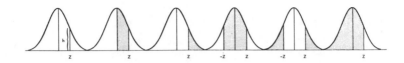

z	h	A†	B	C	D	E
0.00	0.3989	0.0000	0.5000	0.0000	1.0000	0.5000
0.01	0.3989	0.0040	0.4960	0.0080	0.9920	0.5040
0.02	0.3989	0.0080	0.4920	0.0160	0.9840	0.5080
0.0251	0.3988	0.01	0.49	0.02	0.98	0.51
0.03	0.3988	0.0120	0.4880	0.0239	0.9761	0.5120
0.04	0.3986	0.0160	0.4840	0.0319	0.9681	0.5160
0.05	0.3984	0.0199	0.4801	0.0399	0.9601	0.5199
0.0502	0.3984	0.02	0.48	0.04	0.96	0.52
0.06	0.3982	0.0239	0.4761	0.0478	0.9522	0.5239
0.07	0.3980	0.0279	0.4721	0.0558	0.9442	0.5279
0.0753	0.3978	0.03	0.47	0.06	0.94	0.53
0.08	0.3977	0.0319	0.4681	0.0638	0.9362	0.5319
0.09	0.3973	0.0359	0.4641	0.0717	0.9283	0.5359
0.10	0.3970	0.0398	0.4602	0.0797	0.9203	0.5398
0.1004	0.3969	0.04	0.46	0.08	0.92	0.54
0.11	0.3965	0.0438	0.4562	0.0876	0.9124	0.5438
0.12	0.3961	0.0478	0.4522	0.0955	0.9045	0.5478
0.1257	0.3958	0.05	0.45	0.10	0.9	0.55
0.13	0.3956	0.0517	0.4483	0.1034	0.8966	0.5517
0.14	0.3951	0.0557	0.4443	0.1113	0.8887	0.5557
0.15	0.3945	0.0596	0.4404	0.1192	0.8808	0.5596
0.1510	0.3944	0.06	0.44	0.12	0.88	0.56
0.16	0.3939	0.0636	0.4364	0.1271	0.8729	0.5636
0.17	0.3932	0.0675	0.4325	0.1350	0.8650	0.5675
0.1764	0.3928	0.07	0.43	0.14	0.86	0.57
0.18	0.3925	0.0714	0.4286	0.1429	0.8571	0.5714
0.19	0.3918	0.0753	0.4247	0.1507	0.8493	0.5753
0.20	0.3910	0.0793	0.4207	0.1585	0.8415	0.5793
0.2019	0.3909	0.08	0.42	0.16	0.84	0.58
0.21	0.3902	0.0832	0.4168	0.1663	0.8337	0.5832
0.22	0.3894	0.0871	0.4129	0.1741	0.8259	0.5871
0.2275	0.3888	0.09	0.41	0.18	0.82	0.59
0.23	0.3885	0.0910	0.4090	0.1819	0.8181	0.5910
0.24	0.3876	0.0948	0.4052	0.1897	0.8103	0.5948
0.25	0.3867	0.0987	0.4013	0.1974	0.8026	0.5987
0.2533	0.3863	0.10	0.40	0.20	0.80	0.60
0.26	0.3857	0.1026	0.3974	0.2051	0.7949	0.6026
0.27	0.3847	0.1064	0.3936	0.2128	0.7872	0.6064
0.2793	0.3837	0.11	0.39	0.22	0.78	0.61
0.28	0.3836	0.1103	0.3897	0.2205	0.7795	0.6103
0.29	0.3825	0.1141	0.3859	0.2282	0.7718	0.6141
0.30	0.3814	0.1179	0.3821	0.2358	0.7642	0.6179
0.3055	0.3808	0.12	0.38	0.24	0.76	0.62
0.31	0.3802	0.1217	0.3783	0.2434	0.7566	0.6217
0.32	0.3790	0.1255	0.3745	0.2510	0.7490	0.6255

* Abstracted from *National Bureau of Standards—Applied Mathematics Series*—23, U.S. Govt. Printing Office, Washington, D.C., 1953.

† Letters *A–E* refer to the shaded areas under the corresponding normal curves.

THE NORMAL DISTRIBUTION

z	h	A	B	C	D	E
0.33	0.3778	0.1293	0.3707	0.2586	0.7414	0.6293
0.3319	0.3776	0.13	0.37	0.26	0.74	0.63
0.34	0.3765	0.1331	0.3669	0.2661	0.7339	0.6331
0.35	0.3752	0.1368	0.3632	0.2737	0.7263	0.6368
0.3585	0.3741	0.14	0.36	0.28	0.72	0.64
0.36	0.3739	0.1406	0.3594	0.2812	0.7188	0.6406
0.37	0.3725	0.1443	0.3557	0.2886	0.7114	0.6443
0.38	0.3712	0.1480	0.3520	0.2961	0.7039	0.6480
0.3853	0.3704	0.15	0.35	0.30	0.70	0.65
0.39	0.3697	0.1517	0.3483	0.3035	0.6965	0.6517
0.40	0.3683	0.1554	0.3446	0.3108	0.6892	0.6554
0.41	0.3668	0.1591	0.3409	0.3182	0.6818	0.6591
0.4125	0.3664	0.16	0.34	0.32	0.68	0.66
0.42	0.3653	0.1628	0.3372	0.3255	0.6745	0.6628
0.43	0.3637	0.1664	0.3336	0.3328	0.6672	0.6664
0.4399	0.3622	0.17	0.33	0.34	0.66	0.67
0.44	0.3621	0.1700	0.3300	0.3401	0.6599	0.6700
0.45	0.3605	0.1736	0.3264	0.3473	0.6527	0.6736
0.46	0.3589	0.1772	0.3228	0.3545	0.6455	0.6772
0.4677	0.3576	0.18	0.32	0.36	0.64	0.68
0.47	0.3572	0.1808	0.3192	0.3616	0.6384	0.6808
0.48	0.3555	0.1844	0.3156	0.3688	0.6312	0.6844
0.49	0.3538	0.1879	0.3121	0.3759	0.6241	0.6879
0.4959	0.3528	0.19	0.31	0.38	0.62	0.69
0.50	0.3521	0.1915	0.3085	0.3829	0.6171	0.6915
0.51	0.3503	0.1950	0.3050	0.3899	0.6101	0.6950
0.52	0.3485	0.1985	0.3015	0.3969	0.6031	0.6985
0.5244	0.3477	0.2	0.3	0.40	0.6	0.70
0.53	0.3467	0.2019	0.2981	0.4039	0.5961	0.7019
0.54	0.3448	0.2054	0.2946	0.4108	0.5892	0.7054
0.55	0.3429	0.2088	0.2912	0.4177	0.5823	0.7088
0.5534	0.3423	0.21	0.29	0.42	0.58	0.71
0.56	0.3410	0.2123	0.2877	0.4245	0.5755	0.7123
0.57	0.3391	0.2157	0.2843	0.4313	0.5687	0.7157
0.58	0.3372	0.2190	0.2810	0.4381	0.5619	0.7190
0.5828	0.3366	0.22	0.28	0.44	0.56	0.72
0.59	0.3352	0.224	0.2776	0.4448	0.5552	0.7224
0.60	0.3332	0.2257	0.2743	0.4515	0.5485	0.7257
0.61	0.3312	0.2291	0.2709	0.4581	0.5419	0.7291
0.6128	0.3306	0.23	0.27	0.46	0.54	0.73
0.62	0.3292	0.2324	0.2676	0.4647	0.5353	0.7324
0.63	0.3271	0.2357	0.2643	0.4713	0.5287	0.7357
0.64	0.3251	0.2389	0.2611	0.4778	0.5222	0.7389
0.6433	0.3244	0.24	0.26	0.48	0.52	0.74
0.65	0.3230	0.2422	0.2578	0.4843	0.5157	0.7422
0.66	0.3209	0.2454	0.2546	0.4907	0.5093	0.7454
0.67	0.3187	0.2486	0.2514	0.4971	0.5029	0.7486
0.6745	0.3178	0.25	0.25	0.50	0.50	0.75
0.68	0.3166	0.2517	0.2483	0.5035	0.4965	0.7517
0.69	0.3144	0.2549	0.2451	0.5098	0.4902	0.7549
0.70	0.3123	0.2580	0.2420	0.5161	0.4839	0.7580
0.7063	0.3109	0.26	0.24	0.52	0.48	0.76
0.71	0.3101	0.2611	0.2389	0.5223	0.4777	0.7611
0.72	0.3079	0.2642	0.2358	0.5285	0.4715	0.7642
0.73	0.3056	0.2673	0.2327	0.5346	0.4654	0.7673
0.7388	0.3037	0.27	0.23	0.54	0.46	0.77
0.74	0.3034	0.2704	0.2296	0.5407	0.4593	0.7704
0.75	0.3011	0.2734	0.2266	0.5467	0.4533	0.7734
0.76	0.2989	0.2764	0.2236	0.5527	0.4473	0.7764
0.77	0.2966	0.2794	0.2206	0.5587	0.4413	0.7794

THE NORMAL DISTRIBUTION

z	h	A	B	C	D	E
0.7722	0.2961	0.28	0.22	0.56	0.44	0.78
0.78	0.2943	0.2823	0.2177	0.5646	0.4354	0.7823
0.79	0.2920	0.2852	0.2148	0.5705	0.4295	0.7852
0.80	0.2897	0.2881	0.2119	0.5763	0.4237	0.7881
0.8064	0.2882	0.29	0.21	0.58	0.42	0.79
0.81	0.2874	0.2910	0.2090	0.5821	0.4179	0.7910
0.82	0.2850	0.2939	0.2061	0.5878	0.4122	0.7939
0.83	0.2827	0.2967	0.2033	0.5935	0.4065	0.7967
0.84	0.2803	0.2995	0.2005	0.5991	0.4009	0.7995
0.8416	0.2800	0.30	0.20	0.60	0.40	0.80
0.85	0.2780	0.3023	0.1977	0.6047	0.3953	0.8023
0.86	0.2756	0.3051	0.1949	0.6102	0.3898	0.8051
0.87	0.2732	0.3078	0.1922	0.6157	0.3843	0.8078
0.8779	0.2714	0.31	0.19	0.62	0.38	0.81
0.88	0.2709	0.3106	0.1894	0.6211	0.3789	0.8106
0.89	0.2685	0.3133	0.1867	0.6265	0.3735	0.8133
0.90	0.2661	0.3159	0.1841	0.6319	0.3681	0.8159
0.91	0.2637	0.3186	0.1814	0.6372	0.3628	0.8186
0.9154	0.2624	0.32	0.18	0.64	0.36	0.82
0.92	0.2613	0.3212	0.1788	0.6424	0.3576	0.8212
0.93	0.2589	0.3238	0.1762	0.6476	0.3524	0.8238
0.94	0.2565	0.3264	0.1736	0.6528	0.3472	0.8264
0.95	0.2541	0.3289	0.1711	0.6579	0.3421	0.8289
0.9542	0.2531	0.33	0.17	0.66	0.34	0.83
0.96	0.2516	0.3315	0.1685	0.6629	0.3371	0.8315
0.97	0.2492	0.3340	0.1660	0.6680	0.3320	0.8340
0.98	0.2468	0.3365	0.1635	0.6729	0.3271	0.8365
0.99	0.2444	0.3389	0.1611	0.6778	0.3222	0.8389
0.9945	0.2433	0.34	0.16	0.68	0.32	0.84
1.00	0.2420	0.3413	0.1587	0.6827	0.3173	0.8413
1.01	0.2396	0.3438	0.1562	0.6875	0.3125	0.8438
1.02	0.2371	0.3461	0.1539	0.6923	0.3077	0.8461
1.03	0.2347	0.3485	0.1515	0.6970	0.3030	0.8485
1.036	0.2332	0.35	0.15	0.70	0.3	0.85
1.04	0.2323	0.3508	0.1492	0.7017	0.2983	0.8508
1.05	0.2299	0.3531	0.1469	0.7063	0.2937	0.8531
1.06	0.2275	0.3554	0.1446	0.7109	0.2891	0.8554
1.07	0.2251	0.3577	0.1423	0.7154	0.2846	0.8577
1.08	0.2227	0.3599	0.1401	0.7199	0.2801	0.8599
1.080	0.2226	0.36	0.14	0.72	0.28	0.86
1.09	0.2203	0.3621	0.1379	0.7243	0.2757	0.8621
1.10	0.2179	0.3643	0.1357	0.7287	0.2713	0.8643
1.11	0.2155	0.3665	0.1335	0.7330	0.2670	0.8665
1.12	0.2131	0.3686	0.1314	0.7373	0.2627	0.8686
1.1264	0.2115	0.37	0.13	0.74	0.26	0.87
1.13	0.2107	0.3708	0.1292	0.7415	0.2585	0.8708
1.14	0.2083	0.3729	0.1271	0.7457	0.2543	0.8729
1.15	0.2059	0.3749	0.1251	0.7499	0.2501	0.8749
1.16	0.2036	0.3770	0.1230	0.7540	0.2460	0.8770
1.17	0.2012	0.3790	0.1210	0.7580	0.2420	0.8790
1.175	0.2000	0.38	0.12	0.76	0.24	0.88
1.18	0.1989	0.3810	0.1190	0.7620	0.2380	0.8810
1.19	0.1965	0.3830	0.1170	0.7660	0.2340	0.8830
1.20	0.1942	0.3849	0.1151	0.7699	0.2301	0.8849
1.21	0.1919	0.3869	0.1131	0.7737	0.2263	0.8869
1.22	0.1895	0.3888	0.1112	0.7775	0.2225	0.8888
1.227	0.1880	0.39	0.11	0.78	0.22	0.89
1.23	0.1872	0.3907	0.1093	0.7813	0.2187	0.8907
1.24	0.1849	0.3925	0.1075	0.7850	0.2150	0.8925
1.25	0.1826	0.3944	0.1056	0.7887	0.2113	0.8944

THE NORMAL DISTRIBUTION

z	h	A	B	C	D	E
1.26	0.1804	0.3962	0.1038	0.7923	0.2077	0.8962
1.27	0.1781	0.3980	0.1020	0.7959	0.2041	0.8980
1.28	0.1758	0.3997	0.1003	0.7995	0.2005	0.8997
1.282	0.1755	0.40	0.10	0.80	0.20	0.90
1.29	0.1736	0.4015	0.0985	0.8029	0.1971	0.9015
1.30	0.1714	0.4032	0.0968	0.8064	0.1936	0.9032
1.31	0.1691	0.4049	0.0951	0.8098	0.1902	0.9049
1.32	0.1669	0.4066	0.0934	0.8132	0.1868	0.9066
1.33	0.1647	0.4082	0.0918	0.8165	0.1835	0.9082
1.34	0.1626	0.4099	0.0901	0.8198	0.1802	0.9099
1.341	0.1624	0.41	0.09	0.82	0.18	0.91
1.35	0.1604	0.4115	0.0885	0.8230	0.1770	0.9115
1.36	0.1582	0.4131	0.0869	0.8262	0.1738	0.9131
1.37	0.1561	0.4147	0.0853	0.8293	0.1707	0.9147
1.38	0.1539	0.4162	0.0838	0.8324	0.1676	0.9162
1.39	0.1518	0.4177	0.0823	0.8355	0.1645	0.9177
1.40	0.1497	0.4192	0.0808	0.8385	0.1615	0.9192
1.405	0.1487	0.42	0.08	0.84	0.16	0.92
1.41	0.1476	0.4207	0.0793	0.8415	0.1585	0.9207
1.42	0.1456	0.4222	0.0778	0.8444	0.1556	0.9222
1.43	0.1435	0.4236	0.0764	0.8473	0.1527	0.9236
1.44	0.1415	0.4251	0.0749	0.8501	0.1499	0.9251
1.45	0.1394	0.4265	0.0735	0.8529	0.1471	0.9265
1.46	0.1374	0.4279	0.0721	0.8557	0.1443	0.9279
1.47	0.1354	0.4292	0.0708	0.8584	0.1416	0.9292
1.476	0.1343	0.43	0.07	0.86	0.14	0.93
1.48	0.1334	0.4306	0.0694	0.8611	0.1389	0.9306
1.49	0.1315	0.4319	0.0681	0.8638	0.1362	0.9319
1.50	0.1295	0.4332	0.0668	0.8664	0.1336	0.9332
1.51	0.1276	0.4345	0.0655	0.8690	0.1310	0.9345
1.52	0.1257	0.4357	0.0643	0.8715	0.1285	0.9357
1.53	0.1238	0.4370	0.0630	0.8740	0.1260	0.9370
1.54	0.1219	0.4382	0.0618	0.8764	0.1236	0.9382
1.55	0.1200	0.4394	0.0606	0.8789	0.1211	0.9394
1.555	0.1191	0.44	0.06	0.88	0.12	0.94
1.56	0.1182	0.4406	0.0594	0.8812	0.1188	0.9406
1.57	0.1163	0.4418	0.0582	0.8836	0.1164	0.9418
1.58	0.1145	0.4429	0.0571	0.8859	0.1141	0.9429
1.59	0.1127	0.4441	0.0559	0.8882	0.1118	0.9441
1.60	0.1109	0.4452	0.0548	0.8904	0.1096	0.9452
1.61	0.1092	0.4463	0.0537	0.8926	0.1074	0.9463
1.62	0.1074	0.4474	0.0526	0.8948	0.1052	0.9474
1.63	0.1057	0.4484	0.0516	0.8969	0.1031	0.9484
1.64	0.1040	0.4495	0.0505	0.8990	0.1010	0.9495
1.645	0.1031	0.45	0.05	0.90	0.10	0.95
1.65	0.1023	0.4505	0.0495	0.9011	0.0989	0.9505
1.66	0.1006	0.4515	0.0485	0.9031	0.0969	0.9515
1.67	0.0989	0.4525	0.0475	0.9051	0.0949	0.9525
1.68	0.0973	0.4535	0.0465	0.9070	0.0930	0.9535
1.69	0.0957	0.4545	0.0455	0.9090	0.0910	0.9545
1.70	0.0940	0.4554	0.0446	0.9109	0.0891	0.9554
1.71	0.0925	0.4564	0.0436	0.9127	0.0873	0.9564
1.72	0.0909	0.4573	0.0427	0.9146	0.0854	0.9573
1.73	0.0893	0.4582	0.0418	0.9164	0.0836	0.9582
1.74	0.0878	0.4591	0.0409	0.9181	0.0819	0.9591
1.75	0.0863	0.4599	0.0401	0.9199	0.0801	0.9599
1.751	0.0862	0.46	0.04	0.92	0.08	0.96
1.76	0.0848	0.4608	0.0392	0.9216	0.0784	0.9608
1.77	0.0833	0.4616	0.0384	0.9233	0.0767	0.9616
1.78	0.0818	0.4625	0.0375	0.9249	0.0751	0.9625

THE NORMAL DISTRIBUTION

z	h	A	B	C	D	E
1.79	0.0804	0.4633	0.0367	0.9266	0.0734	0.9633
1.80	0.0790	0.4641	0.0359	0.9281	0.0719	0.9641
1.81	0.0775	0.4649	0.0352	0.9297	0.0703	0.9649
1.82	0.0761	0.4656	0.0344	0.9312	0.0688	0.9656
1.83	0.0748	0.4664	0.0336	0.9328	0.0672	0.9664
1.84	0.0734	0.4671	0.0329	0.9342	0.0658	0.9671
1.85	0.0721	0.4678	0.0322	0.9357	0.0643	0.9678
1.86	0.0707	0.4686	0.0314	0.9371	0.0629	0.9686
1.87	0.0694	0.4693	0.0307	0.9385	0.0615	0.9693
1.88	0.0681	0.4699	0.0301	0.9399	0.0601	0.9699
1.881	0.0680	0.47	0.03	0.94	0.06	0.97
1.89	0.0669	0.4706	0.0294	0.9412	0.0588	0.9706
1.90	0.0656	0.4713	0.0287	0.9426	0.0574	0.9713
1.91	0.0644	0.4719	0.0281	0.9439	0.0561	0.9719
1.92	0.0632	0.4726	0.0274	0.9451	0.0549	0.9726
1.93	0.0620	0.4732	0.0268	0.9464	0.0536	0.9732
1.94	0.0608	0.4738	0.0262	0.9476	0.0524	0.9738
1.95	0.0596	0.4744	0.0256	0.9488	0.0512	0.9744
1.960	0.0585	0.475	0.025	0.95	0.05	0.975
1.97	0.0573	0.4756	0.0244	0.9512	0.0488	0.9756
1.98	0.0562	0.4761	0.0239	0.9523	0.0477	0.9761
1.99	0.0551	0.4767	0.0233	0.9534	0.0466	0.9767
2.00	0.0540	0.4772	0.0228	0.9545	0.0455	0.9772
2.01	0.0529	0.4778	0.0222	0.9556	0.0444	0.9778
2.02	0.0519	0.4783	0.0217	0.9566	0.0434	0.9783
2.03	0.0508	0.4788	0.0212	0.9576	0.0424	0.9788
2.04	0.0498	0.4793	0.0207	0.9586	0.0414	0.9793
2.05	0.0488	0.4798	0.0202	0.9596	0.0404	0.9798
2.054	0.0484	0.48	0.02	0.96	0.04	0.98
2.06	0.0478	0.4803	0.0197	0.9606	0.0394	0.9803
2.07	0.0468	0.4808	0.0192	0.9615	0.0385	0.9808
2.08	0.0459	0.4812	0.0188	0.9625	0.0375	0.9812
2.09	0.0449	0.4817	0.0183	0.9634	0.0366	0.9817
2.10	0.0440	0.4821	0.0179	0.9643	0.0357	0.9821
2.11	0.0431	0.4826	0.0174	0.9651	0.0349	0.9826
2.12	0.0422	0.4830	0.0170	0.9660	0.0340	0.9830
2.13	0.0413	0.4834	0.0166	0.9668	0.0332	0.9834
2.14	0.0404	0.4838	0.0162	0.9676	0.0324	0.9838
2.15	0.0396	0.4842	0.0158	0.9684	0.0316	0.9842
2.16	0.0387	0.4846	0.0154	0.9692	0.0308	0.9846
2.17	0.0379	0.4850	0.0150	0.9700	0.0300	0.9850
2.18	0.0371	0.4854	0.0146	0.9707	0.0293	0.9854
2.19	0.0363	0.4857	0.0143	0.9715	0.0285	0.9857
2.20	0.0355	0.4861	0.0139	0.9722	0.0278	0.9861
2.21	0.0347	0.4864	0.0136	0.9729	0.0271	0.9864
2.22	0.0339	0.4868	0.0132	0.9736	0.0264	0.9868
2.23	0.0332	0.4871	0.0129	0.9743	0.0257	0.9871
2.24	0.0325	0.4875	0.0125	0.9749	0.0251	0.9875
2.25	0.0317	0.4878	0.0122	0.9756	0.0244	0.9878
2.26	0.0310	0.4881	0.0119	0.9762	0.0238	0.9881
2.27	0.0303	0.4884	0.0116	0.9768	0.0232	0.9884
2.28	0.0297	0.4887	0.0113	0.9774	0.0226	0.9887
2.29	0.0290	0.4890	0.0110	0.9780	0.0220	0.9890
2.30	0.0283	0.4893	0.0107	0.9786	0.0214	0.9893
2.31	0.0277	0.4896	0.0104	0.9791	0.0209	0.9896
2.32	0.0270	0.4898	0.0102	0.9797	0.0203	0.9898
2.326	0.0267	0.49	0.01	0.98	0.02	0.99
2.33	0.0264	0.4901	0.0099	0.9802	0.0198	0.9901
2.34	0.0258	0.4904	0.0096	0.9807	0.0193	0.9904
2.35	0.0252	0.4906	0.0094	0.9812	0.0188	0.9906

THE NORMAL DISTRIBUTION

z	h	A	B	C	D	E
2.36	0.0246	0.4909	0.0091	0.9817	0.0183	0.9909
2.37	0.0241	0.4911	0.0089	0.9822	0.0178	0.9911
2.38	0.0235	0.4913	0.0087	0.9827	0.0173	0.9913
2.39	0.0229	0.4916	0.0084	0.9832	0.0168	0.9916
2.40	0.0224	0.4918	0.0082	0.9836	0.0164	0.9918
2.41	0.0219	0.4920	0.0080	0.9840	0.0160	0.9920
2.42	0.0213	0.4922	0.0078	0.9845	0.0155	0.9922
2.43	0.0208	0.4925	0.0075	0.9849	0.0151	0.9925
2.44	0.0203	0.4927	0.0073	0.9853	0.0147	0.9927
2.45	0.0198	0.4929	0.0071	0.9857	0.0143	0.9929
2.46	0.0194	0.4931	0.0069	0.9861	0.0139	0.9931
2.47	0.0189	0.4932	0.0068	0.9865	0.0135	0.9932
2.48	0.0184	0.4934	0.0066	0.9869	0.0131	0.9934
2.49	0.0180	0.4936	0.0064	0.9872	0.0128	0.9936
2.50	0.0175	0.4938	0.0062	0.9876	0.0124	0.9938
2.51	0.0171	0.4940	0.0060	0.9879	0.0121	0.9940
2.52	0.0167	0.4941	0.0059	0.9883	0.0117	0.9941
2.53	0.0163	0.4943	0.0057	0.9886	0.0114	0.9943
2.54	0.0158	0.4945	0.0055	0.9889	0.0111	0.9945
2.55	0.0154	0.4946	0.0054	0.9892	0.0108	0.9946
2.56	0.0151	0.4948	0.0052	0.9895	0.0105	0.9948
2.57	0.0147	0.4949	0.0051	0.9898	0.0102	0.9949
2.576	0.0145	0.495	0.005	0.99	0.01	0.995
2.58	0.0143	0.4951	0.0049	0.9901	0.0099	0.9951
2.59	0.0139	0.4952	0.0048	0.9904	0.0096	0.9952
2.60	0.0136	0.4953	0.0047	0.9907	0.0093	0.9953
2.61	0.0132	0.4955	0.0045	0.9909	0.0091	0.9955
2.62	0.0129	0.4956	0.0044	0.9912	0.0088	0.9956
2.63	0.0126	0.4957	0.0043	0.9915	0.0085	0.9957
2.64	0.0122	0.4959	0.0041	0.9917	0.0083	0.9959
2.65	0.0119	0.4960	0.0040	0.9920	0.0080	0.9960
2.70	0.0104	0.4965	0.0035	0.9931	0.0069	0.9965
2.75	0.0091	0.4970	0.0030	0.9940	0.0060	0.9970
2.80	0.0079	0.4974	0.0026	0.9949	0.0051	0.9974
2.85	0.0069	0.4978	0.0022	0.9956	0.0044	0.9978
2.90	0.0060	0.4981	0.0019	0.9963	0.0037	0.9981
2.95	0.0051	0.4984	0.0016	0.9968	0.0032	0.9984
3.00	0.0044	0.4987	0.0013	0.9973	0.0027	0.9987
3.05	0.0038	0.4989	0.0011	0.9977	0.0023	0.9989
3.090	0.0034	0.499	0.001	0.998	0.002	0.999
3.10	0.0033	0.4990	0.0010	0.9981	0.0019	0.9990
3.15	0.0028	0.4992	0.0008	0.9984	0.0016	0.9992
3.20	0.0024	0.4993	0.0007	0.9986	0.0014	0.9993
3.25	0.0020	0.4994	0.0006	0.9988	0.0012	0.9994
3.291	0.0018	0.4995	0.0005	0.999	0.001	0.9995
3.30	0.0017	0.4995	0.0005	0.9990	0.0010	0.9995
3.35	0.0015	0.4996	0.0004	0.9992	0.0008	0.9996
3.40	0.0012	0.4997	0.0003	0.9993	0.0007	0.9997
3.45	0.0010	0.4997	0.0003	0.9994	0.0006	0.9997
3.50	0.0009	0.4998	0.0002	0.9995	0.0005	0.9998
3.55	0.0007	0.4998	0.0002	0.9996	0.0004	0.9998
3.60	0.0006	0.4998	0.0002	0.9997	0.0003	0.9998
3.65	0.0005	0.4999	0.0001	0.9997	0.0003	0.9999
3.70	0.0004	0.4999	0.0001	0.9998	0.0002	0.9999
3.75	0.0004	0.4999	0.0001	0.9998	0.0002	0.9999
3.80	0.0003	0.4999	0.0001	0.9999	0.0001	0.9999

Table A-5

PERCENTILES OF THE t DISTRIBUTION*

d.f.	$t_{0.60}$	$t_{0.70}$	$t_{0.80}$	$t_{0.90}$	$t_{0.95}$	$t_{0.975}$	$t_{0.99}$	$t_{0.995}$	$t_{0.9995}$
1	0.3250	0.7270	1.376	3.078	6.3138	12.706	31.821	63.657	636.619
2	0.2885	0.6172	1.061	1.886	2.9200	4.3027	6.965	9.9248	31.598
3	0.2766	0.5840	0.978	1.638	2.3534	3.1825	4.541	5.8409	12.924
4	0.2707	0.5692	0.941	1.533	2.1318	2.7764	3.747	4.6041	8.610
5	0.2672	0.5598	0.920	1.476	2.0150	2.5706	3.365	4.0321	6.869
6	0.2648	0.5536	0.906	1.440	1.9432	2.4469	3.143	3.7074	5.959
7	0.2632	0.5493	0.896	1.415	1.8946	2.3646	2.998	3.4995	5.408
8	0.2619	0.5461	0.889	1.397	1.8595	2.3060	2.896	3.3554	5.041
9	0.2610	0.5436	0.883	1.383	1.8331	2.2622	2.821	3.2498	4.781
10	0.2602	0.5416	0.879	1.372	1.8125	2.2281	2.764	3.1693	4.587
11	0.2596	0.5400	0.876	1.363	1.7939	2.2010	2.718	3.1058	4.437
12	0.2590	0.5387	0.873	1.356	1.7823	2.1788	2.681	3.0545	4.318
13	0.2586	0.5375	0.870	1.350	1.7709	2.1604	2.650	3.0123	4.221
14	0.2582	0.5366	0.868	1.345	1.7613	2.1448	2.624	2.9768	4.140
15	0.2579	0.5358	0.866	1.341	1.7530	2.1315	2.602	2.9467	4.073
16	0.2576	0.5358	0.865	1.337	1.7459	2.1199	2.583	2.9208	4.015
17	0.2574	0.5344	0.863	1.333	1.7396	2.1098	2.567	2.8982	3.965
18	0.2571	0.5338	0.862	1.330	1.7341	2.1009	2.552	2.8784	3.922
19	0.2569	0.5333	0.861	1.328	1.7291	2.0930	2.539	2.8609	3.883
20	0.2567	0.5329	0.860	1.325	1.7247	2.0860	2.528	2.8453	3.850
21	0.2566	0.5325	0.859	1.323	1.7207	2.0796	2.518	2.8314	3.819
22	0.2564	0.5321	0.858	1.321	1.7171	2.0739	2.508	2.8188	3.792
23	0.2563	0.5318	0.858	1.319	1.7139	2.0687	2.500	2.9073	3.767
24	0.2562	0.5315	0.857	1.318	1.7109	2.0639	2.492	2.7969	3.745
25	0.2561	0.5312	0.856	1.316	1.7081	2.0595	2.485	2.7874	3.725
26	0.2560	0.5309	0.856	1.315	1.7056	2.0555	2.479	2.7787	3.707
27	0.2559	0.5307	0.855	1.314	1.7033	2.0518	2.473	2.7707	3.690
28	0.2558	0.5304	0.855	1.313	1.7011	2.0484	2.467	2.7633	3.674
29	0.2557	0.5302	0.854	1.311	1.6991	2.0452	2.462	2.7564	3.659
30	0.2556	0.5300	0.854	1.310	1.6973	2.0423	2.457	2.7500	3.616
35	0.2553	0.5292	0.8521	1.3062	1.6896	2.0301	2.438	2.7239	3.5919
40	0.2550	0.5286	0.8507	1.3031	1.6839	2.0211	2.423	2.7045	3.5511
45	0.2549	0.5281	0.8497	1.3007	1.6794	2.0141	2.412	2.6896	3.5207
50	0.2547	0.5278	0.8489	1.2987	1.6759	2.0086	2.403	2.6778	3.4965
60	0.2545	0.5272	0.8477	1.2959	1.6707	2.0003	2.390	2.6603	3.4606
70	0.2543	0.5268	0.8468	1.2938	1.6669	1.9945	2.381	2.6480	3.4355
80	0.2542	0.5265	0.8462	1.2922	1.6641	1.9901	2.374	2.6388	3.4169
90	0.2541	0.5263	0.8457	1.2910	1.6620	1.9867	2.368	2.6316	3.4022
100	0.2540	0.5261	0.8452	1.2901	1.6602	1.9840	2.364	2.6260	3.3909
120	0.2539	0.5258	0.8446	1.2887	1.6577	1.9799	2.358	2.6175	3.3736
140	0.2538	0.5256	0.8442	1.2876	1.6558	1.9771	2.353	2.6114	3.3615
160	0.2538	0.5255	0.8439	1.2869	1.6545	1.9749	2.350	2.6070	3.3527
180	0.2537	0.5253	0.8436	1.2863	1.6534	1.9733	2.347	2.6035	3.3456
200	0.2537	0.5252	0.8434	1.2858	1.6525	1.9719	2.345	2.6006	3.3400
∞	0.2533	0.5244	0.8416	1.2816	1.6449	1.9600	2.326	2.5758	3.2905

* The data of this table are extracted with kind permission from *Documenta Geigy Scientific Tables*, 6th Ed., pp. 32–35, Geigy Pharmaceuticals, Division of Geigy Chemical Corporation, Ardsley, N.Y.

Table A-6

PERCENTILES OF THE CHI-SQUARE DISTRIBUTION*

d.f.	$\chi^2_{0.0005}$	$\chi^2_{0.005}$	$\chi^2_{0.01}$	$\chi^2_{0.025}$	$\chi^2_{0.05}$	$\chi^2_{0.10}$	$\chi_{0.20}$	$\chi^2_{0.30}$	$\chi^2_{0.40}$
1	0.000000393	0.0000393	0.000157	0.000982	0.00393	0.0158	0.0642	0.148	0.275
2	0.00100	0.0100	0.0201	0.0506	0.103	0.211	0.446	0.713	1.022
3	0.0153	0.0717	0.115	0.216	0.352	0.584	1.005	1.424	1.869
4	0.0639	0.207	0.297	0.484	0.711	1.004	1.649	2.195	2.753
5	0.158	0.412	0.554	0.831	1.145	1.610	2.343	3.000	3.655
6	0.299	0.676	0.872	1.237	1.635	2.204	3.070	3.828	4.570
7	0.485	0.989	1.239	1.690	2.167	2.833	3.822	4.671	5.493
8	0.710	1.344	1.646	2.180	2.733	3.490	4.594	5.527	6.423
9	0.972	1.735	2.088	2.700	3.325	4.168	5.380	6.393	7.357
10	1.265	2.156	2.558	3.247	3.940	4.865	6.179	7.267	8.295
11	1.587	2.603	3.053	3.816	4.575	5.578	6.989	8.148	9.237
12	1.934	3.074	3.571	4.404	5.226	6.304	7.807	9.034	10.182
13	2.305	3.565	4.107	5.009	5.892	7.042	8.634	9.926	11.129
14	2.697	4.075	4.660	5.629	6.571	7.790	9.467	10.821	12.079
15	3.108	4.601	5.229	6.262	7.261	8.547	10.307	11.721	13.030
16	3.536	5.142	5.812	6.908	7.962	9.312	11.152	12.624	13.983
17	3.980	5.697	6.408	7.564	8.672	10.085	12.002	13.531	14.937
18	4.439	6.265	7.015	8.231	9.390	10.865	12.857	14.440	15.893
19	4.912	6.844	7.633	8.907	10.117	11.651	13.716	15.352	16.850
20	5.398	7.434	8.260	9.591	10.851	12.443	14.578	16.266	17.809
21	5.896	8.034	8.897	10.283	11.591	13.240	15.445	17.182	18.768
22	6.405	8.643	9.542	10.982	12.338	14.041	16.314	18.101	19.729
23	6.924	9.260	10.196	11.688	13.091	14.848	17.187	19.021	20.690
24	7.453	9.886	10.856	12.401	13.848	15.659	18.062	19.943	21.652
25	7.991	10.520	11.524	13.120	14.611	16.473	18.940	20.867	22.616
26	8.538	11.160	12.198	13.844	15.379	17.292	19.820	21.792	23.579
27	9.093	11.808	12.879	14.573	16.151	18.114	20.703	22.719	24.544
28	9.656	12.461	13.565	15.308	16.928	18.939	21.588	23.647	25.509
29	10.227	13.121	14.256	16.047	17.708	19.768	22.475	24.577	26.475
30	10.804	13.787	14.953	16.791	18.493	20.599	23.364	25.508	27.442
35	13.788	17.192	18.509	20.569	22.465	24.797	27.836	30.178	32.282
40	16.906	20.707	22.164	24.433	26.509	29.051	32.345	34.872	37.134
45	20.136	24.311	25.901	28.366	30.612	33.350	36.884	39.585	41.995
50	23.461	27.991	29.707	32.357	34.764	37.689	41.449	44.313	46.864
60	30.340	35.535	37.485	40.482	43.188	46.459	50.641	53.809	56.620
70	37.467	43.275	45.442	48.758	51.739	55.329	59.898	63.346	66.396
80	44.791	51.172	53.540	57.153	60.391	64.278	69.207	72.915	76.188
90	52.276	59.196	61.754	65.647	69.126	73.291	78.558	82.511	85.993
100	59.897	67.328	70.065	74.222	77.930	82.358	87.945	92.129	95.808
120	75.468	83.852	86.924	91.573	95.705	100.624	106.806	111.419	115.465
140	91.393	100.655	104.035	109.137	113.659	119.029	125.758	130.766	135.149
160	107.598	117.680	121.346	126.870	131.756	137.546	144.783	150.158	154.856
180	124.033	134.885	138.821	144.741	149.969	156.153	163.868	169.588	174.580
200	140.661	152.241	156.432	162.728	168.279	174.835	183.003	189.049	194.319

PERCENTILES OF THE CHI-SQUARE DISTRIBUTION

d.f.	$\chi^2_{0.50}$	$\chi^2_{0.60}$	$\chi^2_{0.70}$	$\chi^2_{0.80}$	$\chi^2_{0.90}$	$\chi^2_{0.95}$	$\chi^2_{0.975}$	$\chi^2_{0.99}$	$\chi^2_{0.995}$	$\chi^2_{0.9995}$
1	0.455	0.708	1.074	1.642	2.706	3.841	5.024	6.635	7.879	12.116
2	1.386	1.833	2.408	3.219	4.605	5.991	7.378	9.210	10.597	15.202
3	2.366	2.946	3.665	4.642	6.251	7.815	9.348	11.345	12.838	17.730
4	3.357	4.045	4.878	5.989	7.779	9.488	11.143	13.277	14.860	19.998
5	4.351	5.132	6.064	7.289	9.236	11.070	12.832	15.086	16.750	22.105
6	5.348	6.211	7.231	8.558	10.645	12.592	14.449	16.812	18.548	24.103
7	6.346	7.283	8.383	9.803	12.017	14.067	16.013	18.475	20.278	26.018
8	7.344	8.351	9.524	11.030	13.362	15.507	17.535	20.090	21.955	27.868
9	8.343	9.414	10.656	12.242	14.684	16.919	19.023	21.666	23.589	29.666
10	9.342	10.473	11.781	13.442	15.987	18.307	20.483	23.209	25.188	31.419
11	10.341	11.530	12.899	14.631	17.275	19.675	21.920	24.725	26.757	33.136
12	11.340	12.584	14.011	15.812	18.549	21.026	23.336	26.217	28.300	34.821
13	12.340	13.636	15.119	16.985	19.812	22.362	24.736	27.688	29.819	36.478
14	13.339	14.685	16.222	18.151	21.064	23.685	26.119	29.141	31.319	38.109
15	14.339	15.733	17.322	19.311	22.307	24.996	27.488	30.578	32.601	39.719
16	15.338	16.780	18.418	20.465	23.542	26.296	28.845	32.000	34.267	41.308
17	16.338	17.824	19.511	21.615	24.769	27.587	30.191	33.409	35.718	42.879
18	17.338	18.868	20.601	22.760	25.989	28.869	31.526	34.805	37.156	44.434
19	18.338	19.910	21.689	23.900	27.204	30.144	32.852	36.191	38.582	45.973
20	19.337	20.951	22.775	25.038	28.412	31.410	34.170	37.566	39.997	47.498
21	20.337	21.991	23.858	26.171	29.615	32.671	35.479	38.932	41.401	49.010
22	21.337	23.031	24.939	27.301	30.813	33.924	36.781	40.289	42.796	50.511
23	22.337	24.069	26.018	28.429	32.007	35.172	38.076	41.638	44.181	52.000
24	23.337	25.106	27.096	29.553	33.196	36.415	39.364	42.980	45.558	53.479
25	24.337	26.143	28.172	30.675	34.382	37.652	40.646	44.314	46.928	54.947
26	25.336	27.179	29.246	31.795	35.563	38.885	41.923	45.642	48.290	56.407
27	26.336	28.214	30.319	32.912	36.741	40.113	43.194	46.963	49.645	57.858
28	27.336	29.249	31.391	34.027	37.916	41.337	44.461	48.278	50.993	59.300
29	28.336	30.283	32.461	35.139	39.087	42.537	45.722	49.588	52.336	60.734
30	29.336	31.316	33.530	36.250	40.256	43.773	46.979	50.892	53.672	62.161
35	34.336	36.475	38.859	41.778	46.059	49.802	53.203	57.342	60.275	69.198
40	39.335	41.622	44.165	47.269	51.805	55.758	59.342	63.691	66.766	76.095
45	44.335	46.761	49.452	52.729	57.505	61.656	65.410	69.957	73.166	82.876
50	49.335	51.892	54.723	58.164	63.167	67.505	71.420	76.154	79.490	89.561
60	59.335	62.135	65.226	68.972	74.397	79.082	83.298	88.379	91.952	102.695
70	69.334	72.358	75.689	79.715	85.527	90.531	95.023	100.425	104.215	115.577
80	79.334	82.566	86.120	90.405	96.578	101.879	106.629	112.329	116.321	128.261
90	89.334	92.761	96.524	101.054	107.565	113.145	118.136	124.116	128.299	140.783
100	99.334	102.946	106.906	111.667	118.498	124.342	129.561	135.806	140.169	153.165
120	119.334	123.289	127.616	132.806	140.233	146.567	152.211	158.950	163.648	177.602
140	139.334	143.604	148.269	153.854	161.827	168.613	174.648	181.840	186.846	201.682
160	159.334	163.898	168.876	174.828	183.311	190.516	196.915	204.530	209.824	225.480
180	179.334	184.173	189.446	195.743	204.704	212.304	219.044	227.056	232.620	249.048
200	199.334	204.434	209.985	216.609	226.021	233.994	241.058	249.445	255.264	272.422

* The data of this table are extracted with kind permission from *Documenta Geigy Scientific Tables*, *6th Ed.*, pp. 36–39, Geigy Pharmaceuticals, Division of Geigy Chemical Corporation, Ardsley, N.Y. and Hald, A. and Sinkbaek, S. A., "A Table of Percentage Points of the χ^2-Distribution," *Skandinavisk Aktuarietidskrift*, 33, 168–175, 1950.

Table A-7

PERCENTILES OF THE F DISTRIBUTION

$$F_{0.999}$$

f_2 \ f_1	1	2	3	4	5	6	7	8	9
1	4053*	5000*	5404*	5625*	5764*	5859*	5929*	5981*	6023*
2	998.5	999.0	999.2	999.2	999.3	999.3	999.4	999.4	999.4
3	167.0	148.5	141.1	137.1	134.6	132.8	131.6	130.6	129.9
4	74.14	61.25	56.18	53.44	51.71	50.53	49.66	49.00	48.47
5	47.18	37.12	33.20	31.09	29.75	28.82	28.16	27.64	27.24
6	35.51	27.00	23.70	21.92	20.81	20.03	19.46	19.03	18.69
7	29.25	21.69	18.77	17.19	16.21	15.52	15.02	14.63	14.33
8	25.42	18.49	15.83	14.39	13.49	12.86	12.40	12.04	11.77
9	22.86	16.39	13.90	12.56	11.71	11.13	10.70	10.37	10.11
10	21.04	14.91	12.55	11.28	10.48	9.92	9.52	9.20	8.96
11	19.69	13.81	11.56	10.35	9.58	9.05	8.66	8.35	8.12
12	18.64	12.97	10.80	9.63	8.89	8.38	8.00	7.71	7.48
13	17.81	12.31	10.21	9.07	8.35	7.86	7.49	7.21	6.98
14	17.14	11.78	9.73	8.62	7.92	7.43	7.08	6.80	6.58
15	16.59	11.34	9.34	8.25	7.57	7.09	6.74	6.47	6.26
16	16.12	10.97	9.00	7.94	7.27	6.81	6.46	6.19	5.98
17	15.72	10.66	8.73	7.68	7.02	6.56	6.22	5.96	5.75
18	15.38	10.39	8.49	7.46	6.81	6.35	6.02	5.76	5.56
19	15.08	10.16	8.28	7.26	6.62	6.18	5.85	5.59	5.39
20	14.82	9.95	8.10	7.10	6.46	6.02	5.69	5.44	5.24
21	14.59	9.77	7.94	6.95	6.32	5.88	5.56	5.31	5.11
22	14.38	9.61	7.80	6.81	6.19	5.76	5.44	5.19	4.99
23	14.19	9.47	7.67	6.69	6.08	5.65	5.33	5.09	4.89
24	14.03	9.34	7.55	6.59	5.98	5.55	5.23	4.99	4.80
25	13.88	9.22	7.45	6.49	5.88	5.46	5.15	4.91	4.71
26	13.74	9.12	7.36	6.41	5.80	5.38	5.07	4.83	4.64
27	13.61	9.02	7.27	6.33	5.73	5.31	5.00	4.76	4.57
28	13.50	8.93	7.19	6.25	5.66	5.24	4.93	4.69	4.50
29	13.39	8.85	7.12	6.19	5.59	5.18	4.87	4.64	4.45
30	13.29	8.77	7.05	6.12	5.53	5.12	4.82	4.58	4.39
40	12.61	8.25	6.60	5.70	5.13	4.73	4.44	4.21	4.02
60	11.97	7.76	6.17	5.31	4.76	4.37	4.09	3.87	3.69
120	11.38	7.32	5.79	4.95	4.42	4.04	3.77	3.55	3.38
∞	10.83	6.91	5.42	4.62	4.10	3.74	3.47	3.27	3.10

* Multiply these entries by 100.

f_1 is the number of degrees of freedom in the numerator.

f_2 is the number of degrees of freedom in the denominator.

The relation $F_{\alpha/2}(f_1, f_2) = \dfrac{1}{F_{1-(\alpha/2)}(f_2, f_1)}$ gives lower percentiles of the F distribution. Note the reversal of degrees of freedom.

$$F_{0.999}$$

f_2 \ f_1	10	12	15	20	24	30	40	60	120	∞
1	6056*	6107*	6158*	6209*	6235*	6261*	6287*	6313*	6340*	6366*
2	999.4	999.4	999.4	999.4	999.5	999.5	999.5	999.5	999.5	999.5
3	129.2	128.3	127.4	126.4	125.9	125.4	125.0	124.5	124.0	123.5
4	48.05	47.41	46.76	46.10	45.77	45.43	45.09	44.75	44.40	44.05
5	26.92	26.42	25.91	25.39	25.14	24.87	24.60	24.33	24.06	23.79
6	18.41	17.99	17.56	17.12	16.89	16.67	16.44	16.21	15.99	15.75
7	14.08	13.71	13.32	12.93	12.73	12.53	12.33	12.12	11.91	11.70
8	11.54	11.19	10.84	10.48	10.30	10.11	9.92	9.73	9.53	9.33
9	9.89	9.57	9.24	8.90	8.72	8.55	8.37	8.19	8.00	7.81
10	8.75	8.45	8.13	7.80	7.64	7.47	7.30	7.12	6.94	6.76
11	7.92	7.63	7.32	7.01	6.85	6.68	6.52	6.35	6.17	6.00
12	7.29	7.00	6.71	6.40	6.25	6.09	5.93	5.76	5.59	5.42
13	6.80	6.52	6.23	5.93	5.78	5.63	5.47	5.30	5.14	4.97
14	6.40	6.13	5.85	5.56	5.41	5.25	5.10	4.94	4.77	4.60
15	6.08	5.81	5.54	5.25	5.10	4.95	4.80	4.64	4.47	4.31
16	5.81	5.55	5.27	4.99	4.85	4.70	4.54	4.39	4.23	4.06
17	5.58	5.32	5.05	4.78	4.63	4.48	4.33	4.18	4.02	3.85
18	5.39	5.13	4.87	4.59	4.45	4.30	4.15	4.00	3.84	3.67
19	5.22	4.97	4.70	4.43	4.29	4.14	3.99	3.84	3.68	3.51
20	5.08	4.82	4.56	4.29	4.15	4.00	3.86	3.70	3.54	3.38
21	4.95	4.70	4.44	4.17	4.03	3.88	3.74	3.58	3.42	3.26
22	4.83	4.58	4.33	4.06	3.92	3.78	3.63	3.48	3.32	3.15
23	4.73	4.48	4.23	3.96	3.82	3.68	3.53	3.38	3.22	3.05
24	4.64	4.39	4.14	3.87	3.74	3.59	3.45	3.29	3.14	2.97
25	4.56	4.31	4.06	3.79	3.66	3.52	3.37	3.22	3.06	2.89
26	4.48	4.24	3.99	3.72	3.59	3.44	3.30	3.15	2.99	2.82
27	4.41	4.17	3.92	3.66	3.52	3.38	3.23	3.08	2.92	2.75
28	4.35	4.11	3.86	3.60	3.46	3.32	3.18	3.02	2.86	2.69
29	4.29	4.05	3.80	3.54	3.41	3.27	3.12	2.97	2.81	2.64
30	4.24	4.00	3.75	3.49	3.36	3.22	3.07	2.92	2.76	2.59
40	3.87	3.64	3.40	3.15	3.01	2.87	2.73	2.57	2.41	2.23
60	3.54	3.31	3.08	2.83	2.69	2.55	2.41	2.25	2.08	1.89
120	3.24	3.02	2.78	2.53	2.40	2.26	2.11	1.95	1.76	1.54
∞	2.96	2.74	2.51	2.27	2.13	1.99	1.84	1.66	1.45	1.00

Table A-7 (cont.)

PERCENTILES OF THE F DISTRIBUTION

$$F_{0.995}$$

f_1 is the number of degrees of freedom in the numerator.
f_2 is the number of degrees of freedom in the denominator.

$f_2 \backslash f_1$	1	2	3	4	5	6	7	8	9
1	16211	20000	21615	22500	23056	23437	23715	23925	24091
2	198.5	199.0	199.2	199.2	199.3	199.3	199.4	199.4	199.4
3	55.5	49.80	47.47	46.19	45.39	44.84	44.43	44.13	43.88
4	31.33	26.28	24.26	23.15	22.46	21.97	21.62	21.35	21.14
5	22.78	18.31	16.53	15.56	14.94	14.51	14.20	13.96	13.77
6	18.63	14.54	12.92	12.03	11.46	11.07	10.79	10.57	10.39
7	16.24	12.40	10.88	10.05	9.52	9.16	8.89	8.68	8.51
8	14.69	11.04	9.60	8.81	8.30	7.95	7.69	7.50	7.34
9	13.61	10.11	8.72	7.96	7.47	7.13	6.88	6.69	6.54
10	12.83	9.43	8.08	7.34	6.87	6.54	6.30	6.12	5.97
11	12.23	8.91	7.60	6.88	6.42	6.10	5.86	5.68	5.54
12	11.75	8.51	7.23	6.52	6.07	5.76	5.52	5.35	5.20
13	11.37	8.19	6.93	6.23	5.79	5.48	5.25	5.08	4.94
14	11.06	7.92	6.68	6.00	5.56	5.26	5.03	4.86	4.72
15	10.80	7.70	6.48	5.80	5.37	5.07	4.85	4.67	4.54
16	10.58	7.51	6.30	5.64	5.21	4.91	4.69	4.52	4.38
17	10.38	7.35	6.16	5.50	5.07	4.78	4.56	4.39	4.25
18	10.22	7.21	6.03	5.37	4.96	4.66	4.44	4.28	4.14
19	10.07	7.09	5.92	5.27	4.85	4.56	4.34	4.18	4.04
20	9.94	6.99	5.82	5.17	4.76	4.47	4.26	4.09	3.96
21	9.83	6.89	5.73	5.09	4.68	4.39	4.18	4.01	3.88
22	9.73	6.81	5.65	5.02	4.61	4.32	4.11	3.94	3.81
23	9.63	6.73	5.58	4.95	4.54	4.26	4.05	3.88	3.75
24	9.55	6.66	5.52	4.89	4.49	4.20	3.99	3.83	3.69
25	9.48	6.60	5.46	4.84	4.43	4.15	3.94	3.78	3.64
26	9.41	6.54	5.41	4.79	4.38	4.10	3.89	3.73	3.60
27	9.34	6.49	5.36	4.74	4.34	4.06	3.85	3.69	3.56
28	9.28	6.44	5.32	4.70	4.30	4.02	3.81	3.65	3.52
29	9.23	6.40	5.28	4.66	4.26	3.98	3.77	3.61	3.48
30	9.18	6.35	5.24	4.62	4.23	3.95	3.74	3.58	3.45
40	8.83	6.07	4.98	4.37	3.99	3.71	3.51	3.35	3.22
60	8.49	5.79	4.73	4.14	3.76	3.49	3.29	3.13	3.01
120	8.18	5.54	4.50	3.92	3.55	3.28	3.09	2.93	2.81
∞	7.88	5.30	4.28	3.72	3.35	3.09	2.90	2.74	2.62

$$F_{0.995}$$

f_2 \\ f_1	10	12	15	20	24	30	40	60	120	∞
1	24224	24425	24630	24836	24940	25044	25148	25253	25359	25465
2	199.4	199.4	199.4	199.4	199.5	199.5	199.5	199.5	199.5	199.5
3	43.69	43.39	43.08	42.78	42.62	42.47	42.31	42.15	41.99	41.83
4	20.97	20.70	20.44	20.17	20.03	19.89	19.75	19.61	19.47	19.32
5	13.62	13.38	13.15	12.90	12.78	12.66	12.53	12.40	12.27	12.14
6	10.25	10.03	9.81	9.59	9.47	9.36	9.24	9.12	9.00	8.88
7	8.38	8.18	7.97	7.75	7.65	7.53	7.42	7.31	7.19	7.08
8	7.21	7.01	6.81	6.61	6.50	6.40	6.29	6.18	6.06	5.95
9	6.42	6.23	6.03	5.83	5.73	5.62	5.52	5.41	5.30	5.19
10	5.85	5.66	5.47	5.27	5.17	5.07	4.97	4.86	4.75	4.64
11	5.42	5.24	5.05	4.86	4.76	4.65	4.55	4.44	4.34	4.23
12	5.09	4.91	4.72	4.53	4.43	4.33	4.23	4.12	4.01	3.90
13	4.82	4.64	4.46	4.27	4.17	4.07	3.97	3.87	3.76	3.65
14	4.60	4.43	4.25	4.06	3.96	3.86	3.76	3.66	3.55	3.44
15	4.42	4.25	4.07	3.88	3.79	3.69	3.58	3.48	3.37	3.26
16	4.27	4.10	3.92	3.73	3.64	3.54	3.44	3.33	3.22	3.11
17	4.14	3.97	3.79	3.61	3.51	3.41	3.31	3.21	3.10	2.98
18	4.03	3.86	3.68	3.50	3.40	3.30	3.20	3.10	2.99	2.87
19	3.93	3.76	3.59	3.40	3.31	3.21	3.11	3.00	2.89	2.78
20	3.85	3.68	3.50	3.32	3.22	3.12	3.02	2.92	2.81	2.69
21	3.77	3.60	3.43	3.24	3.15	3.05	2.95	2.84	2.73	2.61
22	3.70	3.54	3.36	3.18	3.08	2.98	2.88	2.77	2.66	2.55
23	3.64	3.47	3.30	3.12	3.02	2.92	2.82	2.71	2.60	2.48
24	3.59	3.42	3.25	3.06	2.97	2.87	2.77	2.66	2.55	2.43
25	3.54	3.37	3.20	3.01	2.92	2.82	2.72	2.61	2.50	2.38
26	3.49	3.33	3.15	2.97	2.87	2.77	2.67	2.56	2.45	2.33
27	3.45	3.28	3.11	2.93	2.83	2.73	2.63	2.52	2.41	2.29
28	3.41	3.25	3.07	2.89	2.79	2.69	2.59	2.48	2.37	2.25
29	3.38	3.21	3.04	2.86	2.76	2.66	2.56	2.45	2.33	2.21
30	3.34	3.18	3.01	2.82	2.73	2.63	2.52	2.42	2.30	2.18
40	3.12	2.95	2.78	2.60	2.50	2.40	2.30	2.18	2.06	1.93
60	2.90	2.74	2.57	2.39	2.29	2.19	2.08	1.96	1.83	1.69
120	2.71	2.54	2.37	2.19	2.09	1.98	1.87	1.75	1.61	1.43
∞	2.52	2.36	2.19	2.00	1.90	1.79	1.67	1.53	1.36	1.00

Table A-7 (cont.)

PERCENTILES OF THE F DISTRIBUTION

$$F_{0.99}$$

f_2 \ f_1	1	2	3	4	5	6	7	8	9
1	4052	4999.5	5403	5625	5764	5859	5928	5981	6022
2	98.50	99.00	99.17	99.25	99.30	99.33	99.36	99.37	99.39
3	34.12	30.82	29.46	28.71	28.24	27.91	27.67	27.49	27.35
4	21.20	18.00	16.69	15.98	15.52	15.21	14.98	14.80	14.55
5	16.26	13.27	12.06	11.39	10.97	10.67	10.46	10.29	10.16
6	13.75	10.92	9.78	9.15	8.75	8.47	8.26	8.10	7.98
7	12.25	9.55	8.45	7.85	7.46	7.19	6.99	6.84	6.72
8	11.26	8.65	7.59	7.01	6.63	6.37	6.18	6.03	5.91
9	10.56	8.02	6.99	6.42	6.06	5.80	5.61	5.47	5.35
10	10.04	7.56	6.55	5.99	5.64	5.39	5.20	5.06	4.94
11	9.65	7.21	6.22	5.67	5.32	5.07	4.89	4.74	4.63
12	9.33	6.93	5.95	5.41	5.06	4.82	4.64	4.50	4.39
13	9.07	6.70	5.74	5.21	4.86	4.62	4.44	4.30	4.19
14	8.86	6.51	5.56	5.04	4.69	4.46	4.28	4.14	4.03
15	8.68	6.36	5.42	4.89	4.56	4.32	4.14	4.00	3.89
16	8.53	6.23	5.29	4.77	4.44	4.20	4.03	3.89	3.78
17	8.40	6.11	5.18	4.67	4.34	4.10	3.93	3.79	3.68
18	8.29	6.01	5.09	4.58	4.25	4.01	3.84	3.71	3.60
19	8.18	5.93	5.01	4.50	4.17	3.94	3.77	3.63	3.52
20	8.10	5.85	4.94	4.43	4.10	3.87	3.70	3.56	3.46
21	8.02	5.78	4.87	4.37	4.04	3.81	3.64	3.51	3.40
22	7.95	5.72	4.82	4.31	3.99	3.76	3.59	3.45	3.35
23	7.88	5.66	4.76	4.26	3.94	3.71	3.54	3.41	3.30
24	7.82	5.61	4.72	4.22	3.90	3.67	3.50	3.36	3.26
25	7.77	5.57	4.68	4.18	3.85	3.63	3.46	3.32	3.22
26	7.72	5.53	4.64	4.14	3.82	3.59	3.42	3.29	3.18
27	7.68	5.49	4.60	4.11	3.78	3.56	3.39	3.26	3.15
28	7.64	5.45	4.57	4.07	3.75	3.53	3.36	3.23	3.12
29	7.60	5.42	4.54	4.04	3.73	3.50	3.33	3.20	3.09
30	7.56	5.39	4.51	4.02	3.70	3.47	3.30	3.17	3.07
40	7.31	5.18	4.31	3.83	3.51	3.29	3.12	2.99	2.89
60	7.08	4.98	4.13	3.65	3.34	3.12	2.95	2.82	2.72
120	6.85	4.79	3.95	3.48	3.17	2.96	2.79	2.66	2.56
∞	6.63	4.61	3.78	3.32	3.02	2.80	2.64	2.51	2.41

f_1 is the number of degrees of freedom in the numerator.
f_2 is the number of degrees of freedom in the denominator.

$F_{0.99}$

f_2 \ f_1	10	12	15	20	24	30	40	60	120	∞
1	6056	6106	6157	6209	6235	6261	6287	6313	6339	6366
2	99.40	99.42	99.43	99.45	99.46	99.47	99.47	99.48	99.49	99.50
3	27.23	27.05	26.87	26.69	26.60	26.50	26.41	26.32	26.22	26.13
4	14.55	14.37	14.20	14.02	13.93	13.84	13.75	13.65	13.56	13.46
5	10.05	9.89	9.72	9.55	9.47	9.38	9.29	9.20	9.11	9.02
6	7.87	7.72	7.56	7.40	7.31	7.23	7.14	7.06	6.97	6.88
7	6.62	6.47	6.31	6.16	6.07	5.99	5.91	5.82	5.74	5.65
8	5.81	5.67	5.52	5.36	5.28	5.20	5.12	5.03	4.95	4.86
9	5.26	5.11	4.96	4.81	4.73	4.65	4.57	4.48	4.40	4.31
10	4.85	4.71	4.56	4.41	4.33	4.25	4.17	4.08	4.00	3.91
11	4.54	4.40	4.25	4.10	4.02	3.94	3.86	3.78	3.69	3.60
12	4.30	4.16	4.01	3.86	3.78	3.70	3.62	3.54	3.45	3.36
13	4.10	3.96	3.82	3.66	3.59	3.51	3.43	3.34	3.25	3.17
14	3.94	3.80	3.66	3.51	3.43	3.35	3.27	3.18	3.09	3.00
15	3.80	3.67	3.52	3.37	3.29	3.21	3.13	3.05	2.96	2.87
16	3.69	3.55	3.41	3.26	3.18	3.10	3.02	2.93	2.84	2.75
17	3.59	3.46	3.31	3.16	3.08	3.00	2.92	2.83	2.75	2.65
18	3.51	3.37	3.23	3.08	3.00	2.92	2.84	2.75	2.66	2.57
19	3.43	3.30	3.15	3.00	2.92	2.84	2.76	2.67	2.58	2.49
20	3.37	3.23	3.09	2.94	2.86	2.78	2.69	2.61	2.52	2.42
21	3.31	3.17	3.03	2.88	2.80	2.72	2.64	2.55	2.46	2.36
22	3.26	3.12	2.98	2.83	2.75	2.67	2.58	2.50	2.40	2.31
23	3.21	3.07	2.93	2.78	2.70	2.62	2.54	2.45	2.35	2.26
24	3.17	3.03	2.89	2.74	2.66	2.58	2.49	2.40	2.31	2.21
25	3.13	2.99	2.85	2.70	2.62	2.54	2.45	2.36	2.27	2.17
26	3.09	2.96	2.81	2.66	2.58	2.50	2.42	2.33	2.23	2.13
27	3.06	2.93	2.78	2.63	2.55	2.47	2.38	2.29	2.20	2.10
28	3.03	2.90	2.75	2.60	2.52	2.44	2.35	2.26	2.17	2.06
29	3.00	2.87	2.73	2.57	2.49	2.41	2.33	2.23	2.14	2.03
30	2.98	2.84	2.70	2.55	2.47	2.39	2.30	2.21	2.11	2.01
40	2.80	2.66	2.52	2.37	2.29	2.20	2.11	2.02	1.92	1.80
60	2.63	2.50	2.35	2.20	2.12	2.03	1.94	1.84	1.73	1.60
120	2.47	2.34	2.19	2.03	1.95	1.86	1.76	1.66	1.53	1.38
∞	2.32	2.18	2.04	1.88	1.79	1.70	1.59	1.47	1.32	1.00

Table A-7 (cont.)

PERCENTILES OF THE F DISTRIBUTION
$F_{0.975}$

f_2 \ f_1	1	2	3	4	5	6	7	8	9
1	647.8	799.5	864.2	899.2	921.8	937.1	948.2	956.7	963.3
2	38.51	39.00	39.17	39.25	39.30	39.33	39.36	39.37	39.39
3	17.44	16.04	15.14	15.10	14.88	14.73	14.62	14.54	14.47
4	12.22	10.65	9.98	9.60	9.36	9.20	9.07	8.98	8.90
5	10.01	8.43	7.76	7.39	7.15	6.98	6.85	6.76	6.68
6	8.81	7.26	6.60	6.23	5.99	5.82	5.70	5.60	5.52
7	8.07	6.54	5.89	5.52	5.29	5.12	4.99	4.90	4.82
8	7.57	6.06	5.42	5.05	4.82	4.65	4.53	4.43	4.36
9	7.21	5.71	5.08	4.72	4.48	4.32	4.20	4.10	4.03
10	6.94	5.46	4.83	4.47	4.24	4.07	3.95	3.85	3.78
11	6.72	5.26	4.63	4.28	4.04	3.88	3.76	3.66	3.59
12	6.55	5.10	4.47	4.12	3.89	3.73	3.61	3.51	3.44
13	6.41	4.97	4.35	4.00	3.77	3.60	3.48	3.39	3.31
14	6.30	4.86	4.24	3.89	3.66	3.50	3.38	3.29	3.21
15	6.20	4.77	4.15	3.80	3.58	3.41	3.29	3.20	3.12
16	6.12	4.69	4.08	3.73	3.50	3.34	3.22	3.12	3.05
17	6.04	4.62	4.01	3.66	3.44	3.28	3.16	3.06	2.98
18	5.98	4.56	3.95	3.61	3.38	3.22	3.10	3.01	2.93
19	5.92	4.51	3.90	3.56	3.33	3.17	3.05	2.96	2.88
20	5.87	4.46	3.86	3.51	3.29	3.13	3.01	2.91	2.84
21	5.83	4.42	3.82	3.48	3.25	3.09	2.97	2.87	2.80
22	5.79	4.38	3.78	3.44	3.22	3.05	2.93	2.84	2.76
23	5.75	4.35	3.75	3.41	3.18	3.02	2.90	2.81	2.73
24	5.72	4.32	3.72	3.38	3.15	2.99	2.87	2.78	2.70
25	5.69	4.29	3.69	3.35	3.13	2.97	2.85	2.75	2.68
26	5.66	4.27	3.67	3.33	3.10	2.94	2.82	2.73	2.65
27	5.63	4.24	3.65	3.31	3.08	2.92	2.80	2.71	2.63
28	5.61	4.22	3.63	3.29	3.06	2.90	2.78	2.69	2.61
29	5.59	4.20	3.61	3.27	3.04	2.88	2.76	2.67	2.59
30	5.57	4.18	3.59	3.25	3.03	2.87	2.75	2.65	2.57
40	5.42	4.05	3.46	3.13	2.90	2.74	2.62	2.53	2.45
60	5.29	3.93	3.34	3.01	2.79	2.63	2.51	2.41	2.33
120	5.15	3.80	3.23	2.89	2.67	2.52	2.39	2.30	2.22
∞	5.02	3.69	3.12	2.79	2.57	2.41	2.29	2.19	2.11

f_1 is the number of degrees of freedom in the numerator.
f_2 is the number of degrees of freedom in the denominator.

$F_{0.975}$

f_2 \ f_1	10	12	15	20	24	30	40	60	120	∞
1	968.6	976.7	984.9	993.1	997.2	1001	1006	1010	1014	1018
2	39.40	39.41	39.43	39.45	39.46	39.46	39.47	39.48	39.49	39.50
3	14.42	14.34	14.25	14.17	14.12	14.08	14.04	13.99	13.95	13.90
4	8.84	8.75	8.66	8.56	8.51	8.46	8.41	8.36	8.31	8.26
5	6.62	6.52	6.43	6.33	6.28	6.23	6.18	6.12	6.07	6.02
6	5.46	5.37	5.27	5.17	5.12	5.07	5.01	4.96	4.90	4.85
7	4.76	4.67	4.57	4.47	4.42	4.36	4.31	4.25	4.20	4.14
8	4.30	4.20	4.10	4.00	3.95	3.89	3.84	3.78	3.73	3.67
9	3.96	3.87	3.77	3.67	3.61	3.56	3.51	3.45	3.39	3.33
10	3.72	3.62	3.52	3.42	3.37	3.31	3.26	3.20	3.14	3.08
11	3.53	3.43	3.33	3.23	3.17	3.12	3.06	3.00	2.94	2.88
12	3.37	3.28	3.18	3.07	3.02	2.96	2.91	2.85	2.79	2.72
13	3.25	3.15	3.05	2.95	2.89	2.84	2.78	2.72	2.66	2.60
14	3.15	3.05	2.95	2.84	2.79	2.73	2.67	2.61	2.55	2.49
15	3.06	2.96	2.86	2.76	2.70	2.64	2.59	2.52	2.46	2.40
16	2.99	2.89	2.79	2.68	2.63	2.57	2.51	2.45	2.38	2.32
17	2.92	2.82	2.72	2.62	2.56	2.50	2.44	2.38	2.32	2.25
18	2.87	2.77	2.67	2.56	2.50	2.44	2.38	2.32	2.26	2.19
19	2.82	2.72	2.62	2.51	2.45	2.39	2.33	2.27	2.20	2.13
20	2.77	2.68	2.57	2.46	2.41	2.35	2.29	2.22	2.16	2.09
21	2.73	2.64	2.53	2.42	2.37	2.31	2.25	2.18	2.11	2.04
22	2.70	2.60	2.50	2.39	2.33	2.27	2.21	2.14	2.08	2.00
23	2.67	2.57	2.47	2.36	2.30	2.24	2.18	2.11	2.04	1.97
24	2.64	2.54	2.44	2.33	2.27	2.21	2.15	2.08	2.01	1.94
25	2.61	2.51	2.41	2.30	2.24	2.18	2.12	2.05	1.98	1.91
26	2.59	2.49	2.39	2.28	2.22	2.16	2.09	2.03	1.95	1.88
27	2.57	2.47	2.36	2.25	2.19	2.13	2.07	2.00	1.93	1.85
28	2.55	2.45	2.34	2.23	2.17	2.11	2.05	1.98	1.91	1.83
29	2.53	2.43	2.32	2.21	2.15	2.09	2.03	1.96	1.89	1.81
30	2.51	2.41	2.31	2.20	2.14	2.07	2.01	1.94	1.87	1.79
40	2.39	2.29	2.18	2.07	2.01	1.94	1.88	1.80	1.72	1.64
60	2.27	2.17	2.06	1.94	1.88	1.82	1.74	1.67	1.58	1.48
120	2.16	2.05	1.94	1.82	1.76	1.69	1.61	1.53	1.43	1.31
∞	2.05	1.94	1.83	1.71	1.64	1.57	1.48	1.39	1.27	1.00

Table A-7 (*cont.*)

PERCENTILES OF THE F DISTRIBUTION
$F_{0.95}$

f_2 \ f_1	1	2	3	4	5	6	7	8	9
1	161.4	199.5	215.7	224.6	230.2	234.0	236.8	238.9	240.5
2	18.51	19.00	19.16	19.25	19.30	19.33	19.35	19.37	19.38
3	10.13	9.55	9.28	9.12	9.01	8.94	8.89	8.85	8.81
4	7.71	6.94	6.59	6.39	6.26	6.16	6.09	6.04	6.00
5	6.61	5.79	5.41	5.19	5.05	4.95	4.88	4.82	4.77
6	5.99	5.14	4.76	4.53	4.39	4.28	4.21	4.15	4.10
7	5.59	4.74	4.35	4.12	3.97	3.87	3.79	3.73	3.68
8	5.32	4.46	4.07	3.84	3.69	3.58	3.50	3.44	3.39
9	5.12	4.26	3.86	3.63	3.48	3.37	3.29	3.23	3.18
10	4.96	4.10	3.71	3.48	3.33	3.22	3.14	3.07	3.02
11	4.84	3.98	3.59	3.36	3.20	3.09	3.01	2.95	2.90
12	4.75	3.89	3.49	3.26	3.11	3.00	2.91	2.85	2.80
13	4.67	3.81	3.41	3.18	3.03	2.92	2.83	2.77	2.71
14	4.60	3.74	3.34	3.11	2.96	2.85	2.76	2.70	2.65
15	4.54	3.68	3.29	3.06	2.90	2.79	2.71	2.64	2.59
16	4.49	3.63	3.24	3.01	2.85	2.74	2.66	2.59	2.54
17	4.45	3.59	3.20	2.96	2.81	2.70	2.61	2.55	2.49
18	4.41	3.55	3.16	2.93	2.77	2.66	2.58	2.51	2.46
19	4.38	3.52	3.13	2.90	2.74	2.63	2.54	2.48	2.42
20	4.35	3.49	3.10	2.87	2.71	2.60	2.51	2.45	2.39
21	4.32	3.47	3.07	2.84	2.68	2.57	2.49	2.42	2.37
22	4.30	3.44	3.05	2.82	2.66	2.55	2.46	2.40	2.34
23	4.28	3.42	3.03	2.80	2.64	2.53	2.44	2.37	2.32
24	4.26	3.40	3.01	2.78	2.62	2.51	2.42	2.36	2.30
25	4.24	3.39	2.99	2.76	2.60	2.49	2.40	2.34	2.28
26	4.23	3.37	2.98	2.74	2.59	2.47	2.39	2.32	2.27
27	4.21	3.35	2.96	2.73	2.57	2.46	2.37	2.31	2.25
28	4.20	3.34	2.95	2.71	2.56	2.45	2.36	2.29	2.24
29	4.18	3.33	2.93	2.70	2.55	2.43	2.35	2.28	2.22
30	4.17	3.32	2.92	2.69	2.53	2.42	2.33	2.27	2.21
40	4.08	3.23	2.84	2.61	2.45	2.34	2.25	2.18	2.12
60	4.00	3.15	2.76	2.53	2.37	2.25	2.17	2.10	2.04
120	3.92	3.07	2.68	2.45	2.29	2.17	2.09	2.02	1.96
∞	3.84	3.00	2.60	2.37	2.21	2.10	2.01	1.94	1.88

f_1 is the number of degrees of freedom in the numerator.
f_2 is the number of degrees of freedom in the denominator.

$$F_{0.95}$$

f_2 \ f_1	10	12	15	20	24	30	40	60	120	∞
1	241.9	243.9	245.9	248.0	249.1	250.1	251.1	252.2	253.3	254.3
2	19.40	19.41	19.43	19.45	19.45	19.46	19.47	19.48	19.49	19.50
3	8.79	8.74	8.70	8.66	8.64	8.62	8.59	8.57	8.55	8.53
4	5.96	5.91	5.86	5.80	5.77	5.75	5.72	5.69	5.66	5.63
5	4.74	4.68	4.62	4.56	4.53	4.50	4.46	4.43	4.40	4.36
6	4.06	4.00	3.94	3.87	3.84	3.81	3.77	3.74	3.70	3.67
7	3.64	3.57	3.51	3.44	3.41	3.38	3.34	3.30	3.27	3.23
8	3.35	3.28	3.22	3.15	3.12	3.08	3.04	3.01	2.97	2.93
9	3.14	3.07	3.01	2.94	2.90	2.86	2.83	2.79	2.75	2.71
10	2.98	2.91	2.85	2.77	2.74	2.70	2.66	2.62	2.58	2.54
11	2.85	2.79	2.72	2.65	2.61	2.57	2.53	2.49	2.45	2.40
12	2.75	2.69	2.62	2.54	2.51	2.47	2.43	2.38	2.34	2.30
13	2.67	2.60	2.53	2.46	2.42	2.38	2.34	2.30	2.25	2.21
14	2.60	2.53	2.46	2.39	2.35	2.31	2.27	2.22	2.18	2.13
15	2.54	2.48	2.40	2.33	2.29	2.25	2.20	2.16	2.11	2.07
16	2.49	2.42	2.35	2.28	2.24	2.19	2.15	2.11	2.06	2.01
17	2.45	2.38	2.31	2.23	2.19	2.15	2.10	2.06	2.01	1.96
18	2.41	2.34	2.27	2.19	2.15	2.11	2.06	2.02	1.97	1.92
19	2.38	2.31	2.23	2.16	2.11	2.07	2.03	1.98	1.93	1.88
20	2.35	2.28	2.20	2.12	2.08	2.04	1.99	1.95	1.90	1.84
21	2.32	2.25	2.18	2.10	2.05	2.01	1.96	1.92	1.87	1.81
22	2.30	2.23	2.15	2.07	2.03	1.98	1.94	1.89	1.84	1.78
23	2.27	2.20	2.13	2.05	2.01	1.96	1.91	1.86	1.81	1.76
24	2.25	2.18	2.11	2.03	1.98	1.94	1.89	1.84	1.79	1.73
25	2.24	2.16	2.09	2.01	1.96	1.92	1.87	1.82	1.77	1.71
26	2.22	2.15	2.07	1.99	1.95	1.90	1.85	1.80	1.75	1.69
27	2.20	2.13	2.06	1.97	1.93	1.88	1.84	1.79	1.73	1.67
28	2.19	2.12	2.04	1.96	1.91	1.87	1.82	1.77	1.71	1.65
29	2.18	2.10	2.03	1.94	1.90	1.85	1.81	1.75	1.70	1.64
30	2.16	2.09	2.01	1.93	1.89	1.84	1.79	1.74	1.68	1.62
40	2.08	2.00	1.92	1.84	1.79	1.74	1.69	1.64	1.58	1.51
60	1.99	1.92	1.84	1.75	1.70	1.65	1.59	1.53	1.47	1.39
120	1.91	1.83	1.75	1.66	1.61	1.55	1.50	1.43	1.35	1.25
∞	1.83	1.75	1.67	1.57	1.52	1.46	1.39	1.32	1.22	1.00

Table A-7 (cont.)

PERCENTILES OF THE F DISTRIBUTION
$F_{0.90}$

f_2 \ f_1	1	2	3	4	5	6	7	8	9
1	39.86	49.50	53.39	55.83	57.24	58.20	58.91	59.44	59.86
2	8.53	9.00	9.16	9.24	9.29	9.33	9.35	9.37	9.38
3	5.54	5.46	5.39	5.34	5.31	5.28	5.27	5.25	5.24
4	4.54	4.32	4.19	4.11	4.05	4.01	3.98	3.95	3.94
5	4.06	3.78	3.62	3.52	3.45	3.40	3.37	3.34	3.32
6	3.78	3.46	3.29	3.18	3.11	3.05	3.01	2.98	2.96
7	3.59	3.26	3.07	2.96	2.88	2.83	2.78	2.75	2.72
8	3.46	3.11	2.92	2.81	2.73	2.67	2.62	2.59	2.56
9	3.36	3.01	2.81	2.69	2.61	2.55	2.51	2.47	2.44
10	3.29	2.92	2.73	2.61	2.52	2.46	2.41	2.38	2.35
11	3.23	2.86	2.66	2.54	2.45	2.39	2.34	2.30	2.27
12	3.18	2.81	2.61	2.48	2.39	2.33	2.28	2.24	2.21
13	3.14	2.76	2.56	2.43	2.35	2.28	2.23	2.20	2.16
14	3.10	2.73	2.52	2.39	2.31	2.24	2.19	2.15	2.12
15	3.07	2.70	2.49	2.36	2.27	2.21	2.16	2.12	2.09
16	3.05	2.67	2.46	2.33	2.24	2.18	2.13	2.09	2.06
17	3.03	2.64	2.44	2.31	2.22	2.15	2.10	2.06	2.03
18	3.01	2.62	2.42	2.29	2.20	2.13	2.08	2.04	2.00
19	2.99	2.61	2.40	2.27	2.18	2.11	2.06	2.02	1.98
20	2.97	2.59	2.38	2.25	2.16	2.09	2.04	2.00	1.96
21	2.96	2.57	2.36	2.23	2.14	2.08	2.02	1.98	1.95
22	2.95	2.56	2.35	2.22	2.13	2.06	2.01	1.97	1.93
23	2.94	2.55	2.34	2.21	2.11	2.05	1.99	1.95	1.92
24	2.93	2.54	2.33	2.19	2.10	2.04	1.98	1.94	1.91
25	2.92	2.53	2.32	2.18	2.09	2.02	1.97	1.93	1.89
26	2.91	2.52	2.31	2.17	2.08	2.01	1.96	1.92	1.88
27	2.90	2.51	2.30	2.17	2.07	2.00	1.95	1.91	1.87
28	2.89	2.50	2.29	2.16	2.06	2.00	1.94	1.90	1.87
29	2.89	2.50	2.28	2.15	2.06	1.99	1.93	1.89	1.86
30	2.88	2.49	2.28	2.14	2.05	1.98	1.93	1.88	1.85
40	2.84	2.44	2.23	2.09	2.00	1.93	1.87	1.83	1.79
60	2.79	2.39	2.18	2.04	1.95	1.87	1.82	1.77	1.74
120	2.75	2.35	2.13	1.99	1.90	1.82	1.77	1.72	1.68
∞	2.71	2.30	2.08	1.94	1.85	1.77	1.72	1.67	1.63

f_1 is the number of degrees of freedom in the numerator.
f_2 is the number of degrees of freedom in the denominator.

$F_{0.90}$

f_1 / f_2	10	12	15	20	24	30	40	60	120	∞
1	60.19	60.71	61.22	61.74	62.00	62.26	62.53	62.79	63.06	63.33
2	9.39	9.41	9.42	9.44	9.45	9.46	9.47	9.47	9.48	9.49
3	5.23	5.22	5.20	5.18	5.18	5.17	5.16	5.15	5.14	5.13
4	3.92	3.90	3.87	3.84	3.83	3.82	3.80	3.79	3.78	3.76
5	3.30	3.27	3.24	3.21	3.19	3.17	3.16	3.14	3.12	3.10
6	2.94	2.90	2.87	2.84	2.82	2.80	2.78	2.76	2.74	2.72
7	2.70	2.67	2.63	2.59	2.58	2.56	2.54	2.51	2.49	2.47
8	2.54	2.50	2.46	2.42	2.40	2.38	2.36	2.34	2.32	2.29
9	2.42	2.38	2.34	2.30	2.28	2.25	2.23	2.21	2.18	2.16
10	2.32	2.28	2.24	2.20	2.18	2.16	2.13	2.11	2.08	2.06
11	2.25	2.21	2.17	2.12	2.10	2.08	2.05	2.03	2.00	1.97
12	2.19	2.15	2.10	2.06	2.04	2.01	1.99	1.96	1.93	1.90
13	2.14	2.10	2.05	2.01	1.98	1.96	1.93	1.90	1.88	1.85
14	2.10	2.05	2.01	1.96	1.94	1.91	1.89	1.86	1.83	1.80
15	2.06	2.02	1.97	1.92	1.90	1.87	1.85	1.82	1.79	1.76
16	2.03	1.99	1.94	1.89	1.87	1.84	1.81	1.78	1.75	1.72
17	2.00	1.96	1.91	1.86	1.84	1.81	1.78	1.75	1.72	1.69
18	1.98	1.93	1.89	1.84	1.81	1.78	1.75	1.72	1.69	1.66
19	1.96	1.91	1.86	1.81	1.79	1.76	1.73	1.70	1.67	1.63
20	1.94	1.89	1.84	1.79	1.77	1.74	1.71	1.68	1.64	1.61
21	1.92	1.87	1.83	1.78	1.75	1.72	1.69	1.66	1.62	1.59
22	1.90	1.86	1.81	1.76	1.73	1.70	1.67	1.64	1.60	1.57
23	1.89	1.84	1.80	1.74	1.72	1.69	1.66	1.62	1.59	1.55
24	1.88	1.83	1.78	1.73	1.70	1.67	1.64	1.61	1.57	1.53
25	1.87	1.82	1.77	1.72	1.69	1.66	1.63	1.59	1.56	1.52
26	1.86	1.81	1.76	1.71	1.68	1.65	1.61	1.58	1.54	1.50
27	1.85	1.80	1.75	1.70	1.67	1.64	1.60	1.57	1.53	1.49
28	1.84	1.79	1.74	1.69	1.66	1.63	1.59	1.56	1.52	1.48
29	1.83	1.78	1.73	1.68	1.65	1.62	1.58	1.55	1.51	1.47
30	1.82	1.77	1.72	1.67	1.64	1.61	1.57	1.54	1.50	1.46
40	1.76	1.71	1.66	1.61	1.57	1.54	1.51	1.47	1.42	1.38
60	1.71	1.66	1.60	1.54	1.51	1.48	1.44	1.40	1.35	1.29
120	1.65	1.60	1.55	1.48	1.45	1.41	1.37	1.32	1.26	1.19
∞	1.60	1.55	1.49	1.42	1.38	1.34	1.30	1.24	1.17	1.00

Table A-7 (cont.)

PERCENTILES' OF THE F DISTRIBUTION[1]

$F_{0.75}$

f_2 \ f_1	1	2	3	4	5	6	7	8	9
1	5.83	7.50	8.20	8.58	8.82	8.98	9.10	9.19	9.26
2	2.57	3.00	3.15	3.23	3.28	3.31	3.34	3.35	3.37
3	2.02	2.28	2.36	2.39	2.41	2.42	2.43	2.44	2.44
4	1.81	2.00	2.05	2.06	2.07	2.08	2.08	2.08	2.08
5	1.69	1.85	1.88	1.89	1.89	1.89	1.89	1.89	1.89
6	1.62	1.76	1.78	1.79	1.79	1.78	1.78	1.78	1.77
7	1.57	1.70	1.72	1.72	1.71	1.71	1.70	1.70	1.69
8	1.54	1.66	1.67	1.66	1.66	1.65	1.64	1.64	1.63
9	1.51	1.62	1.63	1.63	1.62	1.61	1.60	1.60	1.59
10	1.49	1.60	1.60	1.59	1.59	1.58	1.57	1.56	1.56
11	1.47	1.58	1.58	1.57	1.56	1.55	1.54	1.53	1.53
12	1.46	1.56	1.56	1.55	1.54	1.53	1.52	1.51	1.51
13	1.45	1.55	1.55	1.53	1.52	1.51	1.50	1.49	1.49
14	1.44	1.53	1.53	1.52	1.51	1.50	1.49	1.48	1.47
15	1.43	1.52	1.52	1.51	1.49	1.48	1.47	1.46	1.46
16	1.42	1.51	1.51	1.50	1.48	1.47	1.46	1.46	1.45
17	1.42	1.51	1.50	1.49	1.47	1.46	1.45	1.44	1.43
18	1.41	1.50	1.49	1.48	1.46	1.45	1.44	1.43	1.42
19	1.41	1.49	1.49	1.47	1.46	1.44	1.43	1.42	1.41
20	1.40	1.49	1.48	1.47	1.45	1.44	1.43	1.42	1.41
21	1.40	1.48	1.48	1.46	1.44	1.43	1.42	1.41	1.40
22	1.40	1.48	1.47	1.45	1.44	1.42	1.41	1.40	1.39
23	1.39	1.47	1.47	1.45	1.43	1.42	1.41	1.40	1.39
24	1.39	1.47	1.46	1.44	1.43	1.41	1.40	1.39	1.38
25	1.39	1.47	1.46	1.44	1.42	1.41	1.40	1.39	1.38
26	1.38	1.46	1.45	1.44	1.42	1.41	1.39	1.38	1.37
27	1.38	1.46	1.45	1.43	1.42	1.40	1.39	1.38	1.37
28	1.38	1.46	1.45	1.43	1.41	1.40	1.39	1.38	1.37
29	1.38	1.45	1.45	1.43	1.41	1.40	1.38	1.37	1.36
30	1.38	1.45	1.44	1.42	1.41	1.39	1.38	1.37	1.36
40	1.36	1.44	1.42	1.40	1.39	1.37	1.36	1.35	1.34
60	1.35	1.42	1.41	1.38	1.37	1.35	1.33	1.32	1.31
120	1.34	1.40	1.39	1.37	1.35	1.33	1.31	1.30	1.29
∞	1.32	1.39	1.37	1.35	1.33	1.31	1.29	1.28	1.27

f_1 is the number of degrees of freedom in the numerator.
f_2 is the number of degrees of freedom in the denominator.
[1] The data of this table are extracted with kind permission from *Biometrika Tables for Statisticians*, 3rd Ed., Vol. I, Table 18, 1966. London: Bentley House.

$F_{0.75}$

f_2 \ f_1	10	12	15	20	24	30	40	60	120	∞
1	9.32	9.41	9.49	9.58	9.63	9.67	9.71	9.76	9.80	9.85
2	3.38	3.39	3.41	3.43	3.43	3.44	3.45	3.46	3.47	3.48
3	2.44	2.45	2.46	2.46	2.46	2.47	2.47	2.47	2.47	2.47
4	2.08	2.08	2.08	2.08	2.08	2.08	2.08	2.08	2.08	2.08
5	1.89	1.89	1.89	1.88	1.88	1.88	1.88	1.87	1.87	1.87
6	1.77	1.77	1.76	1.76	1.75	1.75	1.75	1.74	1.74	1.74
7	1.69	1.68	1.68	1.67	1.67	1.66	1.66	1.65	1.65	1.65
8	1.63	1.62	1.62	1.61	1.60	1.60	1.59	1.59	1.58	1.58
9	1.59	1.58	1.57	1.56	1.56	1.55	1.54	1.54	1.53	1.53
10	1.55	1.54	1.53	1.52	1.52	1.51	1.51	1.50	1.49	1.48
11	1.52	1.51	1.50	1.49	1.49	1.48	1.47	1.47	1.46	1.45
12	1.50	1.49	1.48	1.47	1.46	1.45	1.45	1.44	1.43	1.42
13	1.48	1.47	1.46	1.45	1.44	1.43	1.42	1.42	1.41	1.40
14	1.46	1.45	1.44	1.43	1.42	1.41	1.41	1.40	1.39	1.38
15	1.45	1.44	1.43	1.41	1.41	1.40	1.39	1.38	1.37	1.36
16	1.44	1.43	1.41	1.40	1.39	1.38	1.37	1.36	1.35	1.34
17	1.43	1.41	1.40	1.39	1.38	1.37	1.36	1.35	1.34	1.33
18	1.42	1.40	1.39	1.38	1.37	1.36	1.35	1.34	1.33	1.32
19	1.41	1.40	1.38	1.37	1.36	1.35	1.34	1.33	1.32	1.30
20	1.40	1.39	1.37	1.36	1.35	1.34	1.33	1.32	1.31	1.29
21	1.39	1.38	1.37	1.35	1.34	1.33	1.32	1.31	1.30	1.28
22	1.39	1.37	1.36	1.34	1.33	1.32	1.31	1.30	1.29	1.28
23	1.38	1.37	1.35	1.34	1.33	1.32	1.31	1.30	1.28	1.27
24	1.38	1.36	1.35	1.33	1.32	1.31	1.30	1.29	1.28	1.26
25	1.37	1.36	1.34	1.33	1.32	1.31	1.29	1.28	1.27	1.25
26	1.37	1.35	1.34	1.32	1.31	1.30	1.29	1.28	1.26	1.25
27	1.36	1.35	1.33	1.32	1.31	1.30	1.28	1.27	1.26	1.24
28	1.36	1.34	1.33	1.31	1.30	1.29	1.28	1.27	1.25	1.24
29	1.35	1.34	1.32	1.31	1.30	1.29	1.27	1.26	1.25	1.23
30	1.35	1.34	1.32	1.30	1.29	1.28	1.27	1.26	1.24	1.23
40	1.33	1.31	1.30	1.28	1.26	1.25	1.24	1.22	1.21	1.19
60	1.30	1.29	1.27	1.25	1.24	1.22	1.21	1.19	1.17	1.15
120	1.28	1.26	1.24	1.22	1.21	1.19	1.18	1.16	1.13	1.10
∞	1.25	1.24	1.22	1.19	1.10	1.16	1.14	1.12	1.08	1.00

Table A-8

GRAPHS FOR BINOMIAL CONFIDENCE INTERVAL*
CONFIDENCE LEVEL 95%

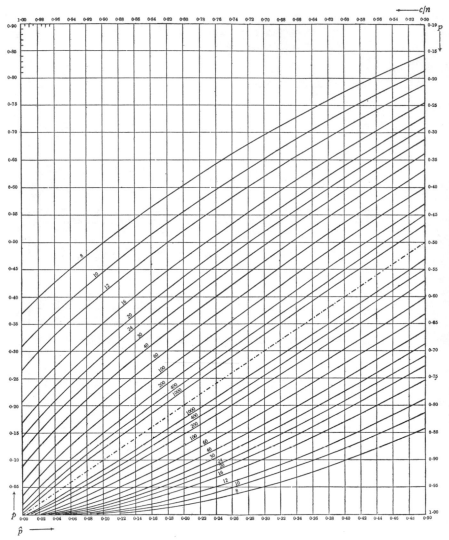

* The data of this table are extracted with kind permission from *Biometrika Tables for Statisticians*, 3rd Ed., Vol. I, Table 41, 1966. London: Bentley House.

GRAPHS FOR BINOMIAL CONFIDENCE INTERVAL
CONFIDENCE LEVEL 99%

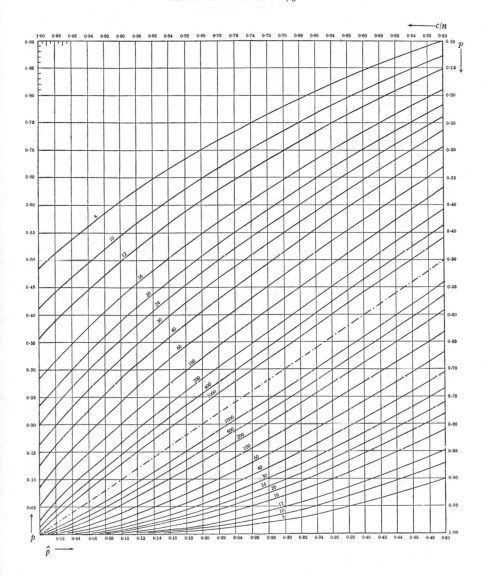

Table A-9

CONFIDENCE LIMITS FOR μ, THE MEAN OF A POISSON DISTRIBUTION*

Confidence Level

	0.99		0.95		0.90		
x	Lower	Upper	Lower	Upper	Lower	Upper	x
0	0.00000	5.30	0.0000	4.61	0.0000	3.00	0
1	0.00501	7.43	0.0253	5.57	0.0513	4.74	1
2	0.103	9.27	0.242	7.22	0.355	6.30	2
3	0.338	10.98	0.619	8.77	0.818	7.75	3
4	0.672	12.59	1.09	10.24	1.37	9.15	4
5	1.08	14.15	1.62	11.67	1.97	10.51	5
6	1.54	15.66	2.20	13.06	2.61	11.84	6
7	2.04	17.13	2.81	14.42	3.29	13.15	7
8	2.57	18.58	3.45	15.76	3.98	14.43	8
9	3.13	20.000	4.12	17.08	4.70	15.17	9
10	3.72	21.40	4.80	18.39	5.43	16.96	10
11	4.32	22.78	5.49	19.68	6.17	18.21	11
12	4.94	24.14	6.20	20.96	6.92	19.44	12
13	5.58	25.50	6.92	22.23	7.69	20.67	13
14	6.23	26.84	7.65	23.49	8.46	21.89	14
15	6.89	28.16	8.40	24.74	9.25	23.10	15
16	7.57	29.48	9.15	25.98	10.04	24.30	16
17	8.25	30.79	9.90	27.22	10.83	25.50	17
18	8.94	32.09	10.67	28.45	11.63	26.69	18
19	9.64	33.38	11.44	29.67	12.44	27.88	19
20	10.35	34.67	12.22	30.89	13.25	29.06	20
21	11.07	35.95	13.00	32.10	14.07	30.24	21
22	11.79	37.22	13.79	33.31	14.89	31.42	22
23	12.52	38.48	14.58	34.51	15.72	32.59	23
24	13.25	39.74	15.38	35.71	16.55	33.75	24
25	14.00	41.00	16.18	36.90	17.38	34.92	25
26	14.74	42.25	16.98	38.10	18.22	36.08	26
27	15.49	43.50	17.79	39.28	19.06	37.23	27
28	16.24	44.74	18.61	40.47	19.90	38.39	28
29	17.00	45.98	19.42	41.65	20.75	39.54	29
30	17.77	47.21	20.24	42.83	21.59	40.69	30
35	21.64	53.32	24.38	48.68	25.87	46.40	35
40	25.59	59.36	28.58	54.47	30.20	52.07	40
45	29.60	65.34	32.82	60.21	34.56	57.69	45
50	33.66	71.27	37.11	65.92	38.96	63.29	50

* The data of this table are extracted with kind permission from *Biometrika Tables for Statisticians*, 3rd Ed., Vol. I, Table 40, 1966. London: Bentley House.

Table A-10

PERCENTILES OF THE SKEWNESS TEST STATISTIC a_3 *

Size of sample n	$(a_3)_{0.95}$	$(a_3)_{0.99}$	Size of sample n	$(a_3)_{0.95}$	$(a_3)_{0.99}$
25	0.711	1.061	200	0.280	0.403
30	0.662	0.986	250	0.251	0.360
35	0.621	0.923	300	0.230	0.329
40	0.587	0.870	350	0.213	0.305
45	0.558	0.825	400	0.200	0.285
50	0.534	0.787	450	0.188	0.269
			500	0.179	0.255
60	0.492	0.723	550	0.171	0.243
70	0.459	0.673	600	0.163	0.233
80	0.432	0.631	650	0.157	0.224
90	0.409	0.596	700	0.151	0.215
100	0.389	0.567	750	0.146	0.208
			800	0.142	0.202
125	0.350	0.508	850	0.138	0.196
150	0.321	0.464	900	0.134	0.190
175	0.298	0.430	950	0.130	0.185
200	0.280	0.403	1000	0.127	0.180

For a full explanation see Sec. 8-5-1.

* The data of this table are extracted with kind permission from *Biometrika Tables for Statisticians*, 3rd Ed., Vol. I, Table 34, part B, 1966. London: Bentley House.

Table A-11

PERCENTILES OF GEARY'S KURTOSIS TEST STATISTIC g^*

Size of sample n	$g_{0.01}$	$g_{0.05}$	$g_{0.10}$	$g_{0.90}$	$g_{0.95}$	$g_{0.99}$
11	0.6675	0.7153	0.7400	0.8899	0.9073	0.9359
16	0.6829	0.7236	0.7452	0.8733	0.8884	0.9137
21	0.6950	0.7304	0.7495	0.8631	0.8768	0.9001
26	0.7040	0.7360	0.7530	0.8570	0.8686	0.8901
31	0.7110	0.7404	0.7559	0.8511	0.8625	0.8827
36	0.7167	0.7440	0.7583	0.8468	0.8578	0.8769
41	0.7216	0.7470	0.7604	0.8436	0.8540	0.8722
46	0.7256	0.7496	0.7621	0.8409	0.8508	0.8682
51	0.7291	0.7518	0.7636	0.8385	0.8481	0.8648
61	0.7347	0.7554	0.7662	0.8349	0.8434	0.8592
71	0.7393	0.7583	0.7683	0.8321	0.8403	0.8549
81	0.7430	0.7607	0.7700	0.8298	0.8376	0.8515
91	0.7460	0.7626	0.7714	0.8279	0.8353	0.8484
101	0.7487	0.7644	0.7726	0.8264	0.8344	0.8460
201	0.7629	0.7738	0.7796	0.8178	0.8220	0.8322
301	0.7693	0.7781	0.7828	0.8140	0.8183	0.8260
401	0.7731	0.7807	0.7847	0.8118	0.8155	0.8223
501	0.7757	0.7825	0.7861	0.8103	0.8136	0.8198
601	0.7776	0.7838	0.7873	0.8092	0.8123	0.8179
701	0.7791	0.7848	0.7878	0.8084	0.8112	0.8164
801	0.7803	0.7857	0.7885	0.8077	0.8103	0.8152
901	0.7814	0.7864	0.7890	0.8071	0.8096	0.8142
1001	0.7822	0.7869	0.7894	0.8066	0.8090	0.8134

For a full explanation see Sec. 8-5-2.
* The data of this table are extracted with kind permission from *Biometrika Tables for Statisticians*, 3rd Ed., Vol. I, Table 34, part A, 1966. London: Bentley House.

Table A-12

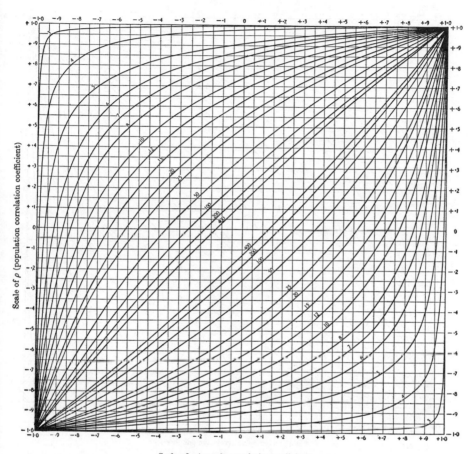

Scale of r (sample correlation coefficient)

* The data of this table are extracted with kind permission from *Biometrika Tables for Statisticians*, 3rd Ed., Vol. I, Table 15, 1966. London: Bentley House.

Table A-12 *(cont.)*

CHARTS FOR COMPUTING CONFIDENCE LIMITS FOR THE
POPULATION CORRELATION COEFFICIENT ρ
CONFIDENCE LEVEL 99%*

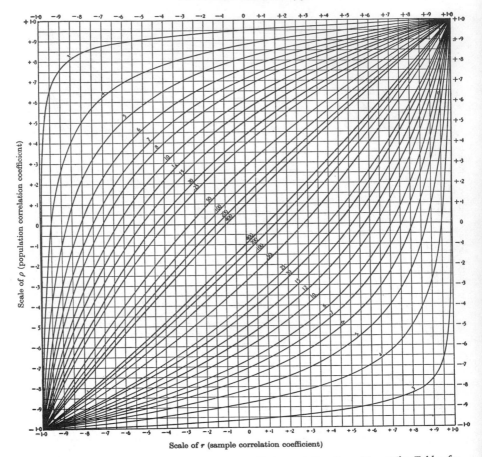

Scale of *r* (sample correlation coefficient)

* The data of this table are extracted with kind permission from *Biometrika Tables for Statisticians*, 3rd Ed., Vol. I, Table 15, 1966. London: Bentley House.

Table A-13

WILCOXON RANK SUM TEST*

2 α = 0.10

N_2 \ N_1	4 (T_l–T_r)	5 (T_l–T_r)	6 (T_l–T_r)	7 (T_l–T_r)	8 (T_l–T_r)	9 (T_l–T_r)	10 (T_l–T_r)	11 (T_l–T_r)	12 (T_l–T_r)	13 (T_l–T_r)	14 (T_l–T_r)
4	11–25	17–33	24–42	32–52	41–63	51–75	52–88	74–102	87–117	101–133	116–150
5	12–28	19–36	26–46	34–57	44–68	54–81	66–94	78–109	91–125	106–141	121–159
6	13–31	20–40	28–50	36–62	46–74	57–87	69–101	82–116	95–133	110–150	126–168
7	14–34	21–44	29–55	39–66	49–79	60–93	72–108	85–124	99–141	115–158	131–177
8	15–37	23–47	31–59	41–71	51–85	63–99	75–115	89–131	104–148	119–167	136–186
9	16–40	24–51	33–63	43–76	54–90	66–105	79–121	93–138	108–156	124–178	141–195
10	17–43	26–54	35–67	45–81	56–96	69–111	82–128	97–145	112–164	128–184	146–204
11	18–46	27–58	37–71	47–86	59–101	72–117	86–134	100–153	116–172	133–192	151–213
12	19–49	28–62	38–76	49–91	62–106	75–123	89–141	104–160	120–180	138–200	156–222
13	20–52	30–65	40–80	52–95	64–112	78–129	92–148	108–167	125–187	142–209	161–231
14	21–55	31–69	42–84	54–100	67–117	81–135	96–154	112–174	129–195	147–217	166–240
15	22–58	33–72	44–88	56–105	69–123	84–141	99–161	116–181	133–203	152–225	171–249
16	24–60	34–76	46–92	58–110	72–128	87–147	103–167	120–188	138–210	156–234	176–258
17	25–63	35–80	47–97	61–114	75–133	90–153	106–174	123–196	142–218	161–242	181–267
18	26–66	37–83	49–101	63–119	77–139	93–159	110–180	127–203	146–226	165–251	186–276
19	27–69	38–87	51–105	65–124	80–144	96–165	113–187	131–210	150–234	170–259	191–285
20	28–72	40–90	53–109	67–129	83–149	99–171	117–193	135–217	154–242	175–267	196–294
21	29–75	41–94	55–113	69–134	85–155	102–177	120–200	139–224	159–249	180–275	202–302
22	30–78	43–97	57–117	72–138	88–160	105–183	123–207	142–232	163–257	184–284	207–311
23	31–81	44–101	58–122	74–143	90–166	108–189	127–213	146–239	167–265	189–292	212–320
24	32–84	45–105	60–126	76–148	93–171	111–195	130–220	150–246	171–273	194–300	217–329
25	33–87	47–108	62–130	78–153	96–176	114–201	133–227	154–253	176–280	199–308	222–338
26	34–90	48–112	64–134	81–157	98–182	117–207	137–233	158–260	180–288	203–317	227–347
27	35–93	50–115	66–138	83–162	101–187	120–213	140–240	162–267	184–296	208–325	233–355
28	36–96	51–119	67–143	85–167	103–193	123–219	144–246	166–274	189–303	213–333	238–364
29	37–99	53–122	69–147	87–172	106–198	126–225	147–253	170–281	193–311	218–341	243–373
30	38–102	54–126	71–151	89–177	109–203	129–231	151–259	174–288	197–319	222–350	248–382
31	39–105	55–130	73–155	92–181	111–209	132–237	154–266	178–295	202–326	227–358	253–391
32	40–108	57–133	75–159	94–186	114–214	135–243	158–272	181–303	206–334	232–366	259–399
33	41–111	58–137	77–163	96–191	117–219	138–249	161–279	185–310	210–342	237–374	264–408
34	42–114	60–140	78–168	98–196	119–225	141–255	165–285	189–317	215–349	241–383	269–417
35	43–117	61–144	80–172	100–201	122–230	144–261	168–292	193–324	219–357	246–391	274–426
36	44–120	62–148	82–176	102–206	124–236	148–266	172–298	197–331	223–365	251–399	279–435
37	45–123	64–151	84–180	105–210	127–241	151–272	175–305	201–338	228–372	256–407	285–443
38	46–126	65–155	85–185	107–215	130–246	154–278	179–311	205–345	232–380	260–416	290–452
39	47–129	67–158	87–189	109–220	132–252	157–284	182–318	209–352	236–388	265–424	295–461
40	48–132	68–162	89–193	111–225	135–257	160–290	186–324	213–359	241–395	270–432	300–470
41	49–135	69–166	91–197	114–229	138–262	163–296	189–331	217–366	245–403	275–440	305–479
42	50–138	71–169	93–201	116–234	140–268	166–302	193–337	221–373	249–411	279–449	311–487
43	51–141	72–173	95–205	118–239	143–273	169–308	196–344	224–381	254–418	284–457	316–496
44	52–144	74–176	96–210	120–244	146–278	172–314	200–350	228–388	258–426	289–465	321–505
45	53–147	75–180	98–214	123–248	148–284	175–320	203–357	232–395	262–434	294–473	326–514
46	55–149	77–183	100–218	125–253	151–289	178–326	207–363	236–402	267–441	299–481	331–523
47	56–152	78–187	102–222	127–258	154–294	181–332	210–370	240–409	271–449	303–490	337–531
48	57–155	79–191	104–226	129–263	156–300	184–338	214–376	244–416	275–457	308–498	342–540
49	58–158	81–194	106–230	132–267	159–305	187–344	217–383	248–423	280–464	313–506	347–549
50	59–161	82–198	107–235	134–272	162–310	190–350	221–387	252–430	284–472	318–514	352–558

N_2 \ N_1	15 (T_l–T_r)	16 (T_l–T_r)	17 (T_l–T_r)	18 (T_l–T_r)	19 (T_l–T_r)	20 (T_l–T_r)	21 (T_l–T_r)	22 (T_l–T_r)	23 (T_l–T_r)	24 (T_l–T_r)	25 (T_l–T_r)
4	132–168	150–186	168–206	187–227	207–249	228–272	250–296	273–321	297–347	322–374	348–402
5	138–177	155–197	173–218	193–239	213–262	235–285	257–310	281–335	305–362	330–390	357–418
6	143–187	161–207	179–229	199–251	220–274	242–298	265–323	289–349	313–377	339–405	366–434
7	148–197	166–218	186–239	206–262	227–286	249–311	272–337	297–363	322–391	348–420	375–450
8	153–207	172–228	192–250	212–274	234–298	257–323	280–350	305–377	330–406	357–435	385–465
9	159–216	178–238	198–261	219–285	241–310	264–336	288–363	313–391	339–420	366–450	394–481
10	164–226	184–248	204–272	226–297	248–322	272–348	296–376	321–405	348–434	375–465	403–497
11	170–235	190–258	210–283	232–308	255–334	279–361	304–389	329–419	356–449	384–480	413–512
12	175–245	196–268	217–293	239–319	262–346	286–374	312–402	338–432	365–463	393–495	423–527
13	181–254	201–279	223–304	245–331	269–358	294–386	320–415	346–446	374–477	403–509	433–542
14	186–264	207–289	229–315	252–342	276–370	301–399	328–428	355–459	383–491	412–524	442–558
15	191–274	213–299	235–326	259–353	284–381	309–411	336–441	363–473	392–505	422–538	452–573
16	197–283	219–309	242–336	266–364	291–393	317–423	344–454	372–486	401–519	431–553	462–588
17	202–293	225–319	248–347	273–375	298–405	325–435	352–467	380–500	410–533	440–568	472–603
18	208–302	231–329	255–357	280–386	305–417	332–448	360–480	389–513	419–547	450–582	482–618
19	214–311	237–339	261–368	286–398	313–428	340–460	368–493	398–526	428–561	459–597	492–633
20	219–321	243–349	268–378	293–409	320–440	348–472	376–506	406–540	437–575	469–611	501–649
21	225–330	249–359	274–389	300–420	327–452	355–485	385–518	415–553	446–589	478–626	511–664
22	230–340	255–369	280–400	307–431	335–463	363–497	393–531	423–567	455–603	488–640	521–679
23	236–349	261–379	287–410	314–442	342–475	371–509	401–544	432–580	464–617	497–655	531–694
24	242–358	267–389	293–421	321–453	349–487	379–521	409–557	441–593	473–631	507–669	541–709
25	247–368	273–399	300–431	328–464	357–498	386–534	417–570	449–607	482–645	516–684	551–724
26	253–377	279–409	306–442	335–475	364–510	394–546	426–582	458–620	491–659	526–698	561–739
27	258–387	285–419	313–452	342–486	371–522	402–558	434–595	467–633	500–673	535–713	571–754
28	264–396	291–429	319–463	348–498	379–533	410–570	442–608	475–647	510–686	545–727	581–769
29	270–405	297–439	326–473	355–509	386–545	418–582	450–621	484–660	519–700	554–742	591–784
30	275–415	303–449	332–484	362–520	393–557	425–595	459–633	493–673	528–714	564–756	601–799
31	281–424	309–459	339–494	369–531	401–568	433–607	467–646	501–687	537–728	574–770	611–814
32	286–434	315–469	345–505	376–542	408–580	441–619	475–659	510–700	546–742	583–785	621–829
33	292–443	321–479	352–515	383–553	415–592	449–631	483–672	519–713	555–756	593–799	631–844
34	298–452	327–489	358–526	390–564	423–603	457–643	492–684	527–727	564–770	602–814	641–859
35	303–462	333–499	365–536	397–575	430–615	464–656	500–697	536–740	574–783	612–828	651–874
36	309–471	340–508	371–547	404–586	438–626	472–668	508–710	545–753	583–797	621–843	661–889
37	315–480	346–518	378–557	411–597	445–638	480–680	516–723	554–766	592–811	631–857	671–904
38	320–490	352–528	384–568	418–608	452–650	488–692	525–735	562–780	601–825	641–871	681–919
39	326–499	358–538	391–578	425–619	460–661	496–704	533–748	571–793	610–839	650–886	691–934
40	331–509	364–548	397–589	432–630	467–673	504–716	541–761	580–806	619–853	660–900	702–948
41	337–518	370–558	404–599	439–641	474–685	511–729	549–774	588–820	628–867	669–915	712–963
42	343–527	376–568	410–610	446–652	482–696	519–741	558–786	597–833	638–880	679–929	722–978
43	348–537	382–578	417–620	452–664	489–708	527–753	566–799	606–846	647–894	689–943	732–993
44	354–546	388–588	423–631	459–675	497–719	535–765	574–812	615–859	656–908	698–958	742–1008
45	360–555	394–598	430–641	466–686	504–731	543–777	583–824	623–873	665–922	708–972	752–1023
46	365–565	400–608	436–652	473–697	511–743	551–789	591–837	632–886	674–936	718–986	762–1038
47	371–574	406–618	443–662	480–708	519–754	558–802	599–850	641–899	683–950	727–1001	772–1053
48	377–583	412–628	449–673	487–719	526–766	566–814	607–863	649–913	693–963	737–1015	782–1068
49	382–593	418–638	456–683	494–730	534–777	574–826	616–875	658–926	702–977	746–1030	792–1083
50	388–602	425–647	462–694	501–741	541–789	582–838	624–888	667–939	711–991	756–1044	802–1098

WILCOXON RANK SUM TEST*

2 α = 0.05

Table A-13 (*cont.*)

WILCOXON RANK SUM TEST*

$2\alpha = 0.02$

N_2 \ N_1 (T_l–T_r)	4	5	6	7	8	9	10	11	12	13	14	15	16	17	18	19	20	21	22	23	24	25
4	—	15–35	22–44	29–55	38–66	48–78	58–92	70–106	83–121	96–138	111–155	127–173	143–193	161–213	180–234	199–257	220–280	242–304	264–330	288–356	313–383	338–412
5	10–30	16–39	23–49	31–60	40–72	50–85	61–99	73–114	86–130	100–147	115–165	131–184	148–204	166–225	185–247	205–270	226–294	248–319	271–345	295–372	320–400	346–429
6	11–33	17–43	24–54	32–66	42–78	52–92	63–107	75–123	89–139	103–157	118–176	135–195	152–216	171–237	190–260	210–284	232–308	254–334	277–361	302–388	327–417	354–446
7	11–37	18–47	25–59	34–71	43–85	54–99	66–114	78–131	92–148	107–166	122–186	139–206	157–227	176–249	195–273	216–297	238–322	261–348	284–376	309–404	335–433	361–464
8	12–40	19–51	27–63	35–77	45–91	56–106	68–122	81–139	95–157	111–175	127–195	144–216	162–238	181–251	201–285	222–310	244–336	267–363	291–391	316–420	342–450	370–480
9	13–43	20–55	28–68	37–82	47–97	59–112	71–129	84–147	99–165	114–185	131–205	148–227	167–249	186–273	207–297	228–323	250–350	274–377	298–406	324–435	350–466	378–497
10	13–47	21–59	29–73	39–87	49–103	61–119	74–136	88–154	102–174	118–194	135–215	153–237	172–260	191–285	212–310	234–336	257–363	281–391	306–420	331–451	358–482	386–514
11	14–50	22–63	30–78	40–93	51–109	63–126	77–143	91–162	106–182	122–203	139–225	157–248	177–271	197–296	218–322	240–349	263–377	288–405	313–435	339–466	366–498	394–531
12	15–53	23–67	32–82	42–98	53–115	66–132	79–151	94–170	109–191	126–212	143–235	162–258	182–282	202–308	224–334	246–362	270–390	295–419	320–450	347–481	374–514	403–547
13	15–57	24–71	33–87	44–103	56–120	68–139	82–158	97–178	113–199	130–221	148–244	167–268	187–293	208–319	230–346	253–374	276–404	301–434	327–465	354–497	382–530	411–564
14	16–60	25–75	34–92	45–109	58–126	71–145	85–165	100–186	116–208	134–230	152–254	171–279	192–304	213–331	235–359	259–387	283–417	308–448	335–479	362–512	391–545	420–580
15	17–63	26–79	36–96	47–114	60–132	73–152	88–172	103–194	120–216	138–239	156–264	176–289	197–315	218–343	241–371	265–400	290–430	315–462	342–494	370–527	399–561	429–596
16	17–67	27–83	37–101	49–119	62–138	76–158	91–179	107–201	124–224	142–248	161–273	181–299	202–326	224–354	247–383	271–413	296–444	323–475	350–508	378–542	407–577	437–613
17	18–70	28–87	39–105	51–124	64–144	78–165	93–187	110–209	127–233	146–257	165–283	185–310	207–337	229–366	253–395	278–425	303–457	330–489	357–523	386–557	416–592	446–629
18	19–73	29–91	40–110	52–130	66–150	81–171	96–194	113–217	131–241	150–266	169–293	190–320	212–348	235–377	259–407	284–438	310–470	337–503	365–537	394–572	424–608	455–645
19	19–77	30–95	41–115	54–135	68–156	83–178	99–201	116–225	134–250	154–275	174–302	195–330	217–359	241–388	265–419	290–451	317–483	344–517	373–551	402–587	432–624	464–661
20	20–80	31–99	43–119	56–140	70–162	85–185	102–208	119–233	138–258	157–285	178–312	200–340	222–370	246–400	271–431	297–463	323–497	351–531	380–566	410–602	441–639	473–677
21	21–83	32–103	44–124	58–145	72–168	88–191	105–215	123–240	142–266	161–294	182–322	204–351	228–380	252–411	277–443	303–476	330–510	359–544	388–580	418–617	449–655	482–693
22	21–87	33–107	45–129	59–151	74–174	90–198	108–222	126–248	145–275	165–303	187–331	209–361	233–391	257–423	283–455	310–488	337–523	366–558	395–595	426–632	458–670	490–710
23	22–90	34–111	47–133	61–156	76–180	93–204	110–230	129–256	149–283	169–312	191–341	214–371	238–402	263–434	289–467	316–501	344–536	373–572	403–609	434–647	466–686	499–726
24	23–93	35–115	48–138	63–161	78–186	95–211	113–237	132–264	152–292	173–321	196–350	219–381	243–413	269–445	295–479	322–514	351–549	380–586	411–623	442–662	475–701	508–742
25	23–97	36–119	50–142	64–167	81–191	98–217	116–244	135–272	156–300	177–330	200–360	224–391	248–424	274–457	301–491	329–526	358–562	388–599	418–638	450–677	483–717	517–758
26	24–100	37–123	51–147	66–172	83–197	100–224	119–251	138–280	159–309	181–339	204–370	229–401	254–434	280–468	307–503	335–539	365–575	395–613	426–652	458–692	492–732	526–774
27	25–103	38–127	52–152	68–177	85–203	103–230	121–259	142–287	163–317	185–348	209–379	233–412	259–445	285–480	313–515	342–551	371–589	402–627	434–666	466–707	500–748	535–790
28	26–106	39–131	54–156	70–182	87–209	105–237	124–266	145–295	167–325	189–357	213–389	238–422	264–456	291–491	319–527	348–564	378–602	409–641	441–681	475–721	509–763	544–806
29	26–110	40–135	55–161	71–188	89–215	108–243	127–273	148–303	170–334	193–366	218–398	243–432	269–467	297–502	325–539	355–576	385–615	417–654	449–695	483–736	517–779	553–822
30	27–113	41–139	56–166	73–193	91–221	110–250	130–280	151–311	174–342	198–374	222–408	248–442	275–477	302–514	331–551	361–589	392–628	424–668	457–709	491–751	526–794	562–838
31	28–116	42–143	58–170	75–198	93–227	112–257	133–287	155–318	178–350	202–383	227–417	253–452	280–488	308–525	337–563	368–601	399–641	431–682	465–723	499–766	534–810	571–854
32	28–120	43–147	59–175	77–203	95–233	115–263	136–294	158–326	181–359	206–392	231–427	258–462	285–499	314–535	343–575	374–614	406–654	439–695	472–738	507–781	543–825	580–870
33	29–123	44–151	61–179	78–209	97–239	117–270	139–301	161–334	185–367	210–401	235–437	262–473	290–510	319–543	349–587	381–626	413–667	446–709	480–752	515–796	552–840	589–886
34	30–126	45–155	62–184	79–215	99–245	120–276	141–309	164–342	189–375	214–410	240–446	267–483	296–520	325–559	356–598	387–639	420–680	453–723	488–766	523–811	560–856	598–902
35	30–130	46–159	63–189	81–220	101–251	122–283	144–316	168–349	192–384	218–419	244–456	272–493	301–531	331–570	362–610	394–651	427–693	461–736	496–780	532–825	569–871	607–918
36	31–133	47–163	65–193	83–225	103–257	125–289	147–323	171–357	196–392	222–428	249–465	277–503	306–542	336–582	368–622	400–664	433–707	468–750	503–795	540–840	577–887	616–934
37	32–136	48–167	66–198	84–231	105–263	127–296	150–330	174–365	199–401	226–437	253–475	282–513	311–553	342–593	374–634	407–676	440–720	475–764	511–809	548–855	586–902	625–950
38	32–140	49–171	67–203	86–236	107–269	129–303	153–337	177–373	203–409	230–446	258–484	287–523	317–563	348–604	380–646	413–689	447–733	483–777	519–823	556–870	595–917	634–966
39	33–143	50–175	69–207	88–241	109–275	132–309	156–344	181–380	207–417	234–455	262–494	292–533	322–574	353–616	386–658	420–701	454–746	490–791	527–837	564–885	603–933	643–982
40	34–146	51–179	70–212	90–246	111–281	134–316	159–351	184–388	210–426	238–464	267–503	296–544	327–585	359–627	392–670	426–714	461–759	497–805	534–852	573–899	612–948	652–998
41	34–150	52–183	72–216	91–252	113–287	137–322	161–359	187–395	214–434	242–473	271–513	301–554	333–595	365–638	398–682	433–726	468–772	505–818	542–866	581–914	620–964	661–1014
42	35–153	53–187	73–221	93–257	116–292	139–329	164–366	190–404	218–442	246–482	276–522	306–564	338–606	371–649	404–694	439–739	475–785	512–832	550–880	589–929	629–979	670–1030
43	35–157	54–191	74–226	95–262	118–298	142–335	167–373	194–411	221–451	250–491	280–532	311–574	343–617	376–661	410–706	446–751	482–798	519–846	558–894	597–944	638–994	679–1046
44	36–160	55–195	76–230	97–267	120–304	144–342	170–380	197–419	225–459	254–500	284–542	316–584	348–628	382–672	416–718	452–764	489–811	527–859	565–909	605–959	646–1010	688–1062
45	37–163	56–199	77–235	98–273	122–310	147–348	173–387	200–427	229–467	258–509	289–551	321–594	354–638	388–683	423–729	459–776	496–824	534–873	573–923	614–973	655–1025	697–1078
46	37–167	57–203	78–240	100–278	124–316	149–355	176–394	203–435	232–476	262–518	293–561	326–604	359–649	393–695	429–741	465–789	503–837	541–887	581–937	622–988	663–1041	706–1094
47	38–170	58–207	80–244	102–283	126–322	152–361	179–401	207–442	236–484	266–527	298–570	330–615	364–660	399–706	435–753	472–801	510–850	549–900	589–951	630–1003	672–1056	715–1110
48	39–173	59–211	81–249	103–289	128–328	154–368	181–409	210–450	240–492	270–536	302–580	335–625	369–671	405–717	441–765	478–814	517–863	556–914	597–965	638–1018	681–1071	724–1126
49	39–177	60–215	82–254	105–294	130–334	157–374	184–416	213–458	243–501	274–545	307–589	340–635	375–681	410–729	447–777	485–826	524–876	563–928	604–980	646–1033	689–1087	733–1142
50	40–180	61–219	84–258	107–299	132–340	159–381	187–423	216–466	247–509	279–553	311–599	345–645	380–692	416–740	453–789	491–839	531–889	571–941	612–994	655–1047	698–1102	743–1157

Table A-13 (*cont.*)

WILCOXON RANK SUM TEST*

$2\alpha = 0.01$

Each cell gives T_l–T_r.

N_2 \ N_1	4	5	6	7	8	9	10	11	12	13	14
4	—	—	21–45	28–56	37–67	46–80	57–93	68–108	81–123	94–140	109–157
5	—	15–40	22–50	29–62	38–74	48–87	59–101	71–116	84–132	98–149	112–168
6	10–34	16–44	23–55	31–67	40–80	50–94	61–109	73–125	87–141	101–159	116–178
7	10–38	16–49	24–60	32–73	42–86	52–101	64–116	76–133	90–150	104–169	120–188
8	11–41	17–53	25–65	34–78	43–93	54–108	66–124	79–141	93–159	108–178	123–199
9	11–45	18–57	26–70	35–84	45–99	56–115	68–132	82–149	96–168	111–188	127–209
10	12–48	19–61	27–75	37–89	47–105	58–122	71–139	84–158	99–177	115–197	131–219
11	12–52	20–65	28–80	38–95	49–111	61–128	73–147	87–166	102–186	118–207	135–229
12	13–55	21–69	30–84	40–100	51–117	63–135	76–154	90–174	105–195	122–216	139–239
13	13–59	22–73	31–89	41–106	53–123	65–142	79–161	93–182	109–203	125–226	143–249
14	14–62	22–78	32–94	43–111	54–130	67–149	81–169	96–190	112–212	129–235	147–259
15	15–65	23–82	33–99	44–117	56–136	69–156	84–176	99–198	115–221	133–244	151–269
16	15–69	24–86	34–104	46–122	58–142	72–162	86–184	102–206	119–229	136–254	155–279
17	16–72	25–90	36–108	47–128	60–148	74–169	89–191	105–214	122–238	140–263	160–288
18	16–76	26–94	37–113	49–133	62–154	76–176	92–198	108–222	125–247	144–272	164–298
19	17–79	27–98	38–118	50–139	64–160	78–183	94–206	111–230	129–255	148–281	168–308
20	18–82	28–102	39–123	52–144	66–166	81–189	97–213	114–238	132–264	152–290	172–318
21	18–86	29–106	40–128	53–150	68–172	83–196	99–221	117–246	136–272	155–300	176–328
22	19–89	29–111	42–132	55–155	70–178	85–203	102–228	120–254	139–281	159–309	180–338
23	19–93	30–115	43–137	57–160	71–185	88–209	105–235	123–262	142–290	163–318	184–348
24	20–96	31–119	44–142	58–166	73–191	90–216	107–243	126–270	146–298	167–327	188–358
25	20–100	32–123	45–147	60–171	75–197	92–223	110–250	129–278	149–307	170–337	193–367
26	21–103	33–127	46–152	61–177	77–203	94–230	113–257	132–286	152–316	174–346	197–377
27	22–106	34–131	48–156	63–182	79–209	97–236	115–265	135–294	156–324	178–355	201–387
28	22–110	35–135	49–161	64–188	81–215	99–243	118–272	138–302	159–333	182–364	205–397
29	23–113	36–139	50–166	66–193	83–221	101–250	120–280	141–310	163–341	185–374	209–407
30	23–117	37–143	51–171	68–198	85–227	103–257	123–287	144–318	166–350	189–383	213–417
31	24–120	37–148	53–175	69–204	87–233	106–263	126–294	147–326	169–359	193–392	217–427
32	24–124	38–152	54–180	71–209	89–239	108–270	128–302	150–334	173–367	197–401	222–436
33	25–127	39–156	55–185	72–215	90–246	110–277	131–309	153–342	176–376	200–411	226–446
34	26–130	40–160	56–190	73–221	92–252	112–284	134–316	156–350	180–384	204–420	230–456
35	26–134	41–164	57–195	75–226	94–258	114–291	136–324	159–358	183–393	208–429	234–466
36	27–137	42–168	58–200	76–232	96–264	117–297	139–331	162–366	186–402	212–438	238–476
37	28–140	43–172	60–204	78–237	98–270	119–304	141–339	165–374	190–410	216–447	242–486
38	28–144	44–176	61–209	79–243	100–276	121–311	144–346	168–382	193–419	219–457	247–495
39	29–147	45–180	62–214	81–248	102–282	123–318	147–353	171–390	197–427	223–466	251–505
40	29–151	46–184	63–219	82–254	103–289	126–324	149–361	174–398	200–436	227–475	255–515
41	30–154	46–189	65–223	84–259	105–295	128–331	152–368	177–406	203–445	231–484	259–525
42	31–157	47–193	66–228	85–265	107–301	130–338	155–375	180–414	207–453	234–494	263–535
43	31–161	48–197	67–233	87–270	109–307	133–344	157–383	183–422	210–462	238–503	268–544
44	32–164	49–201	68–238	88–276	111–313	135–351	160–390	186–430	214–470	242–512	272–554
45	32–168	50–205	69–243	90–281	113–319	137–358	162–398	189–438	217–479	246–521	276–564
46	33–171	51–209	71–247	91–287	115–325	139–365	165–405	192–446	220–488	250–530	280–574
47	34–174	52–213	72–252	93–292	117–331	142–371	168–412	195–454	224–496	253–540	284–584
48	34–178	53–217	73–257	95–297	118–338	144–378	170–420	198–462	227–505	257–549	289–593
49	35–181	54–221	74–262	96–303	120–344	146–385	173–427	201–470	231–513	261–558	293–603
50	36–184	55–225	76–266	98–308	122–350	148–392	176–434	204–478	234–522	265–567	297–613

N_2 \ N_1	15	16	17	18	19	20	21	22	23	24	25
4	125–175	141–195	159–215	177–237	197–259	218–282	239–307	262–332	285–359	310–386	335–415
5	128–187	145–207	163–228	182–250	202–273	223–297	245–322	267–349	291–376	316–404	342–433
6	132–198	149–219	168–240	187–263	207–287	228–312	250–338	274–364	298–392	323–421	349–451
7	136–209	154–230	172–253	192–276	212–301	234–326	256–353	280–380	305–408	330–438	357–468
8	140–220	158–242	177–265	197–289	218–314	240–340	263–367	287–395	311–425	337–455	364–486
9	144–231	163–253	182–277	202–302	223–328	246–354	269–382	293–411	319–440	345–471	372–503
10	149–241	167–265	187–289	208–314	229–341	252–368	275–397	300–426	326–456	352–488	380–520
11	153–252	172–276	192–301	213–327	235–354	258–382	282–411	307–441	333–472	360–504	388–537
12	157–263	177–287	197–313	218–340	241–367	264–396	289–425	314–456	340–488	368–520	396–554
13	162–273	181–299	202–325	224–352	247–380	271–409	295–440	321–471	348–503	376–536	404–571
14	166–284	186–310	208–336	230–364	253–393	277–423	302–454	328–486	355–519	383–553	413–587
15	171–294	191–321	213–348	235–377	259–406	283–437	309–468	335–501	363–534	391–569	421–604
16	175–305	196–332	218–360	241–389	265–419	290–450	315–483	342–516	370–550	399–585	429–621
17	180–315	201–343	223–372	246–402	271–432	296–464	322–497	350–530	378–565	407–601	437–638
18	184–326	206–354	228–384	252–414	277–445	302–478	329–511	357–545	385–581	415–617	446–654
19	189–336	211–365	234–395	258–426	283–458	309–491	336–525	364–560	393–596	423–633	454–671
20	193–347	216–376	239–407	263–439	289–471	315–505	343–539	371–575	401–611	431–649	463–687
21	198–357	220–388	244–419	269–451	295–484	322–518	350–553	378–590	408–627	439–665	471–704
22	202–368	225–399	250–430	275–463	301–497	328–532	356–568	386–604	416–642	447–681	480–720
23	207–378	230–410	255–442	280–476	307–510	335–545	363–582	393–619	424–657	455–697	488–737
24	211–389	235–421	260–454	286–488	313–523	341–559	370–596	400–634	431–673	464–712	497–753
25	216–399	240–432	265–466	292–500	319–536	348–572	377–610	408–648	439–688	472–728	505–770
26	220–410	245–443	271–477	298–512	325–549	354–586	384–624	415–663	447–703	480–744	514–786
27	225–420	250–454	276–489	303–525	332–561	361–599	391–638	422–678	455–718	488–760	522–803
28	229–431	255–465	281–501	309–537	338–574	367–613	398–652	430–692	462–734	496–776	531–819
29	234–441	260–476	287–512	315–549	344–587	374–626	405–666	437–707	470–749	504–792	540–835
30	239–451	265–487	292–524	321–561	350–600	380–640	412–680	444–722	478–764	513–807	548–852
31	243–462	270–498	298–535	326–574	356–613	387–653	419–694	452–736	486–779	521–823	557–868
32	248–472	275–509	303–547	332–586	362–626	394–666	426–708	459–751	494–794	529–839	565–885
33	252–483	280–520	308–559	338–598	368–639	400–680	433–722	467–765	501–810	537–855	574–901
34	257–493	285–531	314–570	344–610	375–651	407–693	440–736	474–780	509–825	545–871	583–917
35	261–504	290–542	319–582	349–623	381–664	413–707	447–750	481–795	517–840	554–886	591–934
36	266–514	295–553	324–594	355–635	387–677	420–720	454–764	489–809	525–855	562–902	600–950
37	270–525	300–564	330–605	361–647	393–690	426–734	461–778	496–824	533–870	570–918	608–967
38	275–535	305–575	335–617	367–659	399–703	433–747	468–792	504–838	540–886	578–934	617–983
39	280–545	310–586	340–629	372–672	406–715	440–760	475–806	511–853	548–901	587–949	626–999
40	284–556	314–598	346–640	378–684	412–728	446–774	482–820	518–868	556–916	595–965	634–1016
41	289–566	319–609	351–652	384–696	418–741	453–787	489–834	526–882	564–931	603–981	643–1032
42	293–577	324–620	357–663	390–708	424–754	459–801	496–848	533–897	572–946	611–997	652–1048
43	298–587	329–631	362–675	396–720	430–767	466–814	503–862	541–911	580–961	620–1012	660–1065
44	303–597	334–642	367–687	401–733	437–779	473–827	510–876	548–926	587–977	628–1028	669–1081
45	307–608	339–653	373–698	407–745	443–792	479–841	517–890	556–940	595–992	636–1044	678–1097
46	312–618	344–664	378–710	413–757	449–805	486–854	524–904	563–955	603–1007	644–1060	687–1113
47	316–629	349–675	384–721	419–769	455–818	493–867	531–918	570–970	611–1022	653–1075	695–1130
48	321–639	354–686	389–733	425–781	461–831	499–881	538–932	578–984	619–1037	661–1091	704–1146
49	325–650	359–697	394–745	430–794	468–843	506–894	545–946	585–999	627–1052	669–1107	713–1162
50	330–660	364–708	400–756	436–806	474–856	512–908	552–960	593–1013	635–1067	677–1123	721–1179

* The data of this table are extracted with kind permission from *Documenta Geigy Scientific Tables*, 6th Ed., pp. 124–127, Geigy Pharmaceuticals, Division of Geigy Chemical Corporation, Ardsley, N.Y.; calculations kindly supplied by Prof. C. White, Department of Public Health, Yale University, New Haven, Conn. and published in part in "The Use of Ranks in a Test of Significance for Comparing Two Treatments," *Biometrics* 8, 33–41, 1952; and Wabeke, D. and Van Eeden, C., "Handleiding voor de toets van Wilcoxon," Report s 176 (M 65) Mathematisch Centrum, Statistische Afdeling, Amsterdam, 1955.

Table A-14

n	α(two-sided) 0.10 α(one-sided) 0.05	0.05 0.025	0.02 0.01	0.01 0.005
1				
2				
3				
4				
5	0, 5			
6	0, 6	0, 6		
7	0, 7	0, 7	0, 7	
8	1, 7	0, 8	0, 8	0, 8
9	1, 8	1, 8	0, 9	0, 9
10	1, 9	1, 9	0, 10	0, 10
11	2, 9	1, 10	1, 10	0, 11
12	2, 10	2, 10	1, 11	1, 11
13	3, 10	2, 11	1, 12	1, 12
14	3, 11	2, 12	2, 12	1, 13
15	3, 12	3, 12	2, 13	2, 13
16	4, 12	3, 13	2, 14	2, 14
17	4, 13	4, 13	3, 14	2, 15
18	5, 13	4, 14	3, 15	3, 15
19	5, 14	4, 15	4, 15	3, 16
20	5, 15	5, 15	4, 16	3, 17
21	6, 15	5, 16	4, 17	4, 17
22	6, 16	5, 17	5, 17	4, 18
23	7, 16	6, 17	5, 18	4, 19
24	7, 17	6, 18	5, 19	5, 19
25	7, 18	7, 18	6, 19	5, 20
26	8, 18	7, 19	6, 20	6, 20
27	8, 19	7, 20	7, 20	6, 21
28	9, 19	8, 20	7, 21	6, 22
29	9, 20	8, 21	7, 22	7, 22
30	10, 20	9, 21	8, 22	7, 23
31	10, 21	9, 22	8, 23	7, 24
32	10, 22	9, 23	8, 24	8, 24
33	11, 22	10, 23	9, 24	8, 25
34	11, 23	10, 24	9, 25	9, 25
35	12, 23	11, 24	10, 25	9, 26
36	12, 24	11, 25	10, 26	9, 27
37	13, 24	12, 25	10, 27	10, 27
38	13, 25	12, 26	11, 27	10, 28
39	13, 26	12, 57	11, 28	11, 28
40	14, 26	13, 27	12, 28	11, 29
41	14, 27	13, 28	12, 29	11, 30
42	15, 27	14, 28	13, 29	12, 30
43	15, 28	14, 29	13, 30	12, 31
44	16, 28	15, 29	13, 31	13, 31
45	16, 29	15, 30	14, 31	13, 32
46	16, 30	15, 31	14, 32	13, 33
47	17, 30	16, 31	15, 32	14, 33
48	17, 31	16, 32	15, 33	14, 34
49	18, 31	17, 32	15, 34	15, 34
50	18. 32	17, 33	16, 34	15, 35
51	19, 32	18, 33	16, 35	15, 36
52	19, 33	18, 34	17, 35	16, 36
53	20, 33	18, 35	17, 36	16, 37
54	20, 34	19, 35	18, 36	17, 37
55	20, 35	19, 36	18, 37	17, 38
56	21, 35	20, 36	18, 38	17, 39
57	21, 36	20, 37	19, 38	18, 39
58	22, 36	21, 37	19, 39	18, 40
59	22, 37	21, 38	20, 39	19, 40
60	23, 37	21, 39	20, 40	19, 41

CRITICAL VALUES OF n_+ FOR THE SIGN TEST

n	α(two-sided) 0.10 α(one-sided) 0.05	0.05 0.025	0.02 0.01	0.01 0.005
61	23, 38	22, 39	20, 41	20, 41
62	24, 38	22, 40	21, 41	20, 42
63	24, 39	23, 40	21, 42	20, 43
64	24, 40	23, 41	22, 42	21, 43
65	25, 40	24, 41	22, 43	21, 44
66	25, 41	24, 42	23, 43	22, 44
67	26, 41	25, 42	23, 44	22, 45
68	26, 42	25, 43	23, 45	22, 46
69	27, 42	25, 44	24, 45	23, 46
70	27, 43	26, 44	24, 46	23, 47
71	28, 43	26, 45	25, 46	24, 47
72	28, 44	27, 45	25, 47	24, 48
73	28, 45	27, 46	26, 47	25, 48
74	29, 45	28, 46	26, 48	25, 49
75	29, 46	28, 47	26, 49	25, 50
76	30, 46	28, 48	27, 49	26, 50
77	30, 47	29, 48	27, 50	26, 51
78	31, 47	29, 49	28, 50	27, 51
79	31, 48	30, 49	28, 51	27, 52
80	32, 48	30, 50	29, 51	28, 52
81	32, 49	31, 50	29, 52	28, 53
82	33, 49	31, 51	30, 52	28, 54
83	33, 50	32, 51	30, 53	29, 54
84	33, 51	32, 52	30, 54	29, 55
85	34, 51	32, 53	31, 54	30, 55
86	34, 52	33, 53	31, 55	30, 56
87	35, 52	33, 54	32, 55	31, 56
88	35, 53	34, 54	32, 56	31, 57
89	36, 53	34, 55	33, 56	31, 58
90	36, 54	35, 55	33, 57	32, 58
91	37, 54	35, 56	33, 58	32, 59
92	37, 55	36, 56	34, 58	33, 59
93	38, 55	36, 57	34, 59	33, 60
94	38, 56	37, 57	35, 59	34, 60
95	38, 57	37, 58	35, 60	34, 61
96	39, 57	37, 59	36, 60	34, 62
97	39, 58	38, 59	36, 61	35, 62
98	40, 58	38, 60	37, 61	35, 63
99	40, 59	39, 60	37, 62	36, 63
100	41, 59	39, 61	37, 63	36, 64

Table A-15

n = number of pairs

n	Critical Values			
	$\alpha \leq 0.10$	$\alpha \leq 0.05$	$\alpha \leq 0.02$	$\alpha \leq 0.01$
1				
2				
3				
4				
5	0, 15			
6	2, 19	0, 21		
7	3, 25	2, 26	0, 28	
8	5, 31	3, 33	1, 35	0, 36
9	8, 37	5, 40	3, 42	1, 44
10	10, 45	8, 47	5, 50	3, 52
11	13, 53	10, 56	7, 59	5, 61
12	17, 61	13, 65	9, 69	7, 71
13	21, 70	17, 74	12, 79	9, 82
14	25, 80	21, 84	15, 90	12, 93
15	30, 90	25, 95	19, 101	15, 105
16	35, 101	29, 107	23, 113	19, 117
17	41, 112	34, 119	28, 125	23, 130
18	47, 124	40, 131	32, 139	27, 144
19	53, 137	46, 144	37, 153	33, 158
20	60, 150	52, 158	43, 167	37, 173
21	67, 164	58, 173	49, 182	42, 189
22	75, 178	66, 187	55, 198	48, 205
23	83, 193	73, 203	62, 214	54, 222
24	91, 209	81, 210	69, 231	61, 239
25	100, 225	89, 236	76, 249	68, 257

* The data of this table are extracted with kind permission from *Documenta Geigy Scientific Tables*, 6th Ed., p. 128, Geigy Pharmaceuticals, Division of Geigy Chemical Corporation, Ardsley, N.Y. as prepared by Tukey, J., *Memorandum Report* 17, Statistical Research Group, Princeton University, 1949.

Table A-16

DISTRIBUTION-FREE CONFIDENCE LIMITS FOR THE MEDIAN OF A CONTINUOUS DISTRIBUTION*

n	0.95	0.99	n	0.95	0.99	n	0.95	0.99	n	0.95	0.99
6	1–6		31	10–22	8–24	56	21–36	18–39	81	32–50	29–53
7	1–7		32	10–23	9–24	57	21–37	19–39	82	32–51	29–54
8	1–8	1–8	33	11–23	9–25	58	22–37	19–40	83	33–51	30–54
9	2–8	1–9	34	11–24	10–25	59	22–38	20–40	84	33–52	30–55
10	2–9	1–10	35	12–24	10–26	60	22–39	20–41	85	33–53	31–55
11	2–10	1–11	36	12–25	10–27	61	23–39	21–41	86	34–53	31–56
12	3–10	2–11	37	13–25	11–27	62	23–40	21–42	87	34–54	32–56
13	3–11	2–12	38	13–26	11–28	63	24–40	21–43	88	35–54	32–57
14	3–12	2–13	39	13–27	12–28	64	24–41	22–43	89	35–55	32–58
15	4–12	3–13	40	14–27	12–29	65	25–41	22–44	90	36–55	32–58
16	4–13	3–14	41	14–28	12–30	66	25–42	23–44	91	36–56	33–59
17	5–13	3–15	42	15–28	13–30	67	26–42	23–45	92	37–56	34–59
18	5–14	4–15	43	15–29	13–31	68	26–43	23–46	93	37–57	34–60
19	5–15	4–16	44	16–29	14–31	69	26–44	24–46	94	38–57	35–60
20	6–15	4–17	45	16–30	14–32	70	27–44	24–47	95	38–58	35–61
21	6–16	5–17	46	16–31	14–33	71	27–45	25–47	96	38–59	35–62
22	6–17	5–18	47	17–31	15–33	72	28–45	25–48	97	39–59	36–62
23	7–17	5–19	48	17–32	15–34	73	28–46	26–48	98	39–60	36–63
24	7–18	6–19	49	18–32	16–34	74	29–46	26–49	99	40–60	37–63
25	8–18	6–20	50	18–33	16–35	75	29–47	26–50	100	40–61	37–64
26	8–19	7–20	51	19–33	16–36	76	29–48	27–50			
27	8–20	7–21	52	19–34	17–36	77	30–48	27–51			
28	9–20	7–22	53	19–35	17–37	78	30–49	28–51			
29	9–21	8–22	54	20–35	18–37	79	31–49	28–52			
30	10–21	8–23	55	20–36	18–38	80	31–50	29–52			

* The data of this table are extracted with modifications with kind permission from *Documenta Geigy Scientific Tables*, 6th Ed., p. 105, Geigy Pharmaceuticals, Division of Geigy Chemical Corporation, Ardsley, N.Y.

Table A-17

RUNS ABOVE AND BELOW THE MEDIAN*

n'	Critical Values $\alpha = 0.10$	$\alpha = 0.05$	$\alpha = 0.02$	$\alpha = 0.01$
1				
2				
3				
4	2, 8			
5	3, 9	2, 10	2, 10	
6	3, 11	3, 11	2, 12	2, 12
7	4, 12	3, 13	3, 13	3, 13
8	5, 13	4, 14	4, 14	3, 15
9	6, 14	5, 15	4, 16	4, 16
10	6, 16	6, 16	5, 17	5, 17
11	7, 17	7, 17	6, 18	5, 19
12	8, 18	7, 19	7, 19	6, 20
13	9, 19	8, 20	7, 21	7, 21
14	10, 20	9, 21	8, 22	7, 23
15	11, 21	10, 22	9, 23	8, 24
16	11, 23	11, 23	10, 24	9, 25
17	12, 24	11, 25	10, 26	10, 26
18	13, 25	12, 26	11, 27	11, 27
19	14, 26	13, 27	12, 28	11, 29
20	15, 27	14, 28	13, 29	12, 30
21	16, 28	15, 29	14, 30	13, 31
22	17, 29	16, 30	14, 32	14, 32
23	17, 31	16, 32	15, 33	14, 34
24	18, 32	17, 33	16, 34	15, 35
25	19, 33	18, 34	17, 35	16, 36
26	20, 24	19, 35	18, 36	17, 37
27	21, 35	20, 36	19, 37	18, 38
28	22, 36	21, 37	19, 39	18, 40
29	23, 37	22, 38	20, 40	19, 41
30	24, 38	22, 40	21, 21	20, 42
31	25, 39	23, 41	22, 42	21, 43
32	25, 41	24, 42	23, 43	22, 44
33	26, 42	25, 43	24, 44	23, 45
34	27, 43	26, 44	24, 46	23, 47
35	28, 44	27, 45	25, 47	24, 48
36	29, 45	28, 46	26, 48	25, 49
37	30, 46	29, 47	27, 49	26, 50
38	31, 47	30, 48	28, 50	27, 51
39	32, 48	30, 50	29, 51	28, 52
40	33, 49	31, 51	30, 52	29, 53
41	34, 50	32, 52	31, 53	29, 55
42	35, 51	33, 53	31, 54	30, 56
43	35, 53	34, 54	32, 56	31, 57
44	36, 54	35, 55	33, 57	32, 58
45	37, 55	36, 56	34, 58	33, 59
46	38, 56	37, 57	35, 59	34, 60
47	39, 57	38, 58	36, 60	35, 61
48	40, 58	38, 60	37, 61	36, 63
49	41, 59	39, 61	38, 62	36, 64
50	42, 60	40, 62	38, 64	37, 65

n' = number of observations above or below median.
* The data of this table are extracted with kind permission from *Documenta Geigy Scientific Tables*, 6th Ed., p. 129, Geigy Pharmaceuticals, Division of Geigy Chemical Corporation, Ardsley, N.Y., as presented by Swed, F. S., and Eisenhart, C., "Tables for Testing Randomness of Grouping in a Sequence of Alternatives," *Annals of Mathematical Statistics*, 14, 66–87, 1943.

RUNS ABOVE AND BELOW THE MEDIAN

	Critical Values			
n'	$\alpha = 0.10$	$\alpha = 0.05$	$\alpha = 0.02$	$\alpha = 0.01$
51	43, 61	41, 63	39, 65	38, 66
52	44, 62	42, 64	40, 66	39, 67
53	45, 63	43, 65	41, 67	40, 68
54	45, 65	44, 66	42, 68	41, 69
55	46, 66	45, 67	43, 69	42, 70
56	47, 67	46, 68	44, 70	42, 72
57	48, 68	47, 69	45, 71	43, 73
58	49, 69	47, 71	46, 72	44, 74
59	50, 70	48, 72	46, 74	45, 75
60	51, 71	49, 73	47, 75	46, 76
61	52, 72	50, 74	48, 76	47, 77
62	53, 73	51, 75	49, 77	48, 78
63	54, 74	52, 76	50, 78	49, 79
64	55, 75	53, 77	51, 79	49, 81
65	56, 76	54, 78	52, 80	50, 82
66	57, 77	55, 79	53, 81	51, 83
67	58, 78	56, 80	54, 82	52, 84
68	58, 80	57, 81	54, 84	53, 85
69	59, 81	58, 82	55, 85	54, 86
70	60, 82	58, 84	56, 86	55, 87
71	61, 83	59, 85	57, 87	56, 88
72	62, 84	60, 86	58, 88	57, 89
73	63, 85	61, 87	59, 89	57, 91
74	64, 86	62, 88	60, 90	58, 92
75	65, 87	63, 89	61, 91	59, 93
76	66, 88	64, 90	62, 92	60, 94
77	67, 89	65, 91	63, 93	61, 95
78	68, 90	66, 92	64, 94	62, 96
79	69, 91	67, 93	64, 96	63, 97
80	70, 92	68, 94	65, 97	64, 98
81	71, 93	69, 95	66, 98	65, 99
82	71, 95	69, 97	67, 99	66, 100
83	72, 96	70, 98	68, 100	66, 102
84	73, 97	71, 99	69, 101	67, 103
85	74, 98	72, 100	70, 102	68, 104
86	75, 99	73, 101	71, 103	69, 105
87	76, 100	74, 102	72, 104	70, 106
88	77, 101	75, 103	73, 105	71, 107
89	78, 102	76, 104	74, 106	72, 108
90	79, 103	77, 105	74, 108	73, 109
91	80, 104	78, 106	75, 109	74, 110
92	81, 105	79, 107	76, 110	75, 111
93	82, 106	80, 108	77, 111	75, 113
94	83, 107	81, 109	78, 112	76, 114
95	84, 108	82, 110	79, 113	77, 115
96	85, 109	82, 112	80, 114	78, 116
97	86, 110	83, 113	81, 115	79, 117
98	87, 111	84, 114	82, 116	80, 118
99	87, 113	85, 115	83, 117	81, 119
100	88, 114	86, 116	84, 118	82, 120

Table A-18

FACTORS, k, FOR TWO-SIDED TOLERANCE LIMITS FOR NORMAL DISTRIBUTIONS*

n \ Π	\multicolumn{5}{c}{$1 - \alpha = 0.75$}	\multicolumn{5}{c}{$1 - \alpha = 0.90$}
	0.75	0.90	0.95	0.99	0.999	0.75	0.90	0.95	0.99	0.999
2	4.498	6.301	7.414	9.531	11.920	11.407	15.978	18.800	24.167	30.227
3	2.501	3.538	4.187	5.431	6.844	4.132	5.847	6.919	8.974	11.309
4	2.035	2.892	3.431	4.471	5.657	2.932	4.166	4.943	6.440	8.149
5	1.825	2.599	3.088	4.033	5.117	2.454	3.494	4.152	5.423	6.879
6	1.704	2.429	2.889	3.779	4.802	2.196	3.131	3.723	4.870	6.188
7	1.624	2.318	2.757	3.611	4.593	2.034	2.902	3.452	4.521	5.750
8	1.568	2.238	2.663	3.491	4.444	1.921	2.743	3.264	4.278	5.446
9	1.525	2.178	2.593	3.400	4.330	1.839	2.626	3.125	4.098	5.220
10	1.492	2.131	2.537	3.328	4.241	1.775	2.535	3.018	3.959	5.046
11	1.465	2.093	2.493	3.271	4.169	1.724	2.463	2.933	3.849	4.906
12	1.443	2.062	2.456	3.223	4.110	1.683	2.404	2.863	3.758	4.792
13	1.425	2.036	2.424	3.183	4.059	1.648	2.355	2.805	3.682	4.697
14	1.409	2.013	2.398	3.148	4.016	1.619	2.314	2.756	3.618	4.615
15	1.395	1.994	2.375	3.118	3.979	1.594	2.278	2.713	3.562	4.545
16	1.383	1.977	2.355	3.092	3.946	1.572	2.246	2.676	3.514	4.484
17	1.372	1.962	2.337	3.069	3.917	1.552	2.219	2.643	3.471	4.430
18	1.363	1.948	2.321	3.048	3.891	1.535	2.194	2.614	3.433	4.382
19	1.355	1.936	2.307	3.030	3.867	1.520	2.172	2.588	3.399	4.339
20	1.347	1.925	2.294	3.013	3.846	1.506	2.152	2.564	3.368	4.300
21	1.340	1.915	2.282	2.998	3.827	1.493	2.135	2.543	3.340	4.264
22	1.334	1.906	2.271	2.984	3.809	1.482	2.118	2.524	3.315	4.232
23	1.328	1.898	2.261	2.971	3.793	1.471	2.103	2.506	3.292	4.203
24	1.322	1.891	2.252	2.959	3.778	1.462	2.089	2.489	3.270	4.176
25	1.317	1.883	2.244	2.948	3.764	1.453	2.077	2.474	3.251	4.151
26	1.313	1.877	2.236	2.938	3.751	1.444	2.065	2.460	3.232	4.127
27	1.309	1.871	2.229	2.929	3.740	1.437	2.054	2.447	3.215	4.106
30	1.297	1.855	2.210	2.904	3.708	1.417	2.025	2.413	3.170	4.049
35	1.283	1.834	2.185	2.871	3.667	1.390	1.988	2.368	3.112	3.974
40	1.271	1.818	2.166	2.846	3.635	1.370	1.959	2.334	3.066	3.917
45	1.262	1.805	2.150	2.826	3.609	1.354	1.935	2.306	3.030	3.871
50	1.255	1.794	2.138	2.809	3.588	1.340	1.916	2.284	3.001	3.833
55	1.249	1.785	2.127	2.795	3.571	1.329	1.901	2.265	2.976	3.801
60	1.243	1.778	2.118	2.784	3.556	1.320	1.887	2.248	2.955	3.774
65	1.239	1.771	2.110	2.773	3.543	1.312	1.875	2.235	2.937	3.751
70	1.235	1.765	2.104	2.764	3.531	1.304	1.865	2.222	2.920	3.730
75	1.231	1.760	2.098	2.757	3.521	1.298	1.856	2.211	2.906	3.712
80	1.228	1.756	2.092	2.749	3.512	1.292	1.848	2.202	2.894	3.696
85	1.225	1.752	2.087	2.743	3.504	1.287	1.841	2.193	2.882	3.682
90	1.223	1.748	2.083	2.737	3.497	1.283	1.834	2.185	2.872	3.669
95	1.220	1.745	2.079	2.732	3.490	1.278	1.828	2.178	2.863	3.657
100	1.218	1.742	2.075	2.727	3.484	1.275	1.822	2.172	2.854	3.646
110	1.214	1.736	2.069	2.719	3.473	1.268	1.813	2.160	2.839	3.626
120	1.211	1.732	2.063	2.712	3.464	1.262	1.804	2.150	2.826	3.610
130	1.208	1.728	2.059	2.705	3.456	1.257	1.797	2.141	2.814	3.595
140	1.206	1.724	2.054	2.700	3.449	1.252	1.791	2.134	2.804	3.582
150	1.204	1.721	2.051	2.695	3.443	1.248	1.785	2.127	2.795	3.571
160	1.202	1.718	2.047	2.691	3.437	1.245	1.780	2.121	2.787	3.561
170	1.200	1.716	2.044	2.687	3.432	1.242	1.775	2.116	2.780	3.552
180	1.198	1.713	2.042	2.683	3.427	1.239	1.771	2.111	2.774	3.543
190	1.197	1.711	2.039	2.680	3.423	1.236	1.767	2.106	2.768	3.536
200	1.195	1.709	2.037	2.677	3.419	1.234	1.764	2.102	2.762	3.529
250	1.190	1.702	2.028	2.665	3.404	1.224	1.750	2.085	2.740	3.501
300	1.186	1.696	2.021	2.656	3.393	1.217	1.740	2.073	2.725	3.481
400	1.181	1.688	2.012	2.644	3.378	1.207	1.726	2.057	2.703	3.453
500	1.177	1.683	2.006	2.636	3.368	1.201	1.717	2.046	2.689	3.434
600	1.175	1.680	2.002	2.631	3.360	1.196	1.710	2.038	2.678	3.421
700	1.173	1.677	1.998	2.626	3.355	1.192	1.705	2.032	2.670	3.411
800	1.171	1.675	1.996	2.623	3.350	1.189	1.701	2.027	2.663	3.402
900	1.170	1.673	1.993	2.620	3.347	1.187	1.697	2.023	2.658	3.396
1000	1.169	1.671	1.992	2.617	3.344	1.185	1.695	2.019	2.654	3.390
∞	1.150	1.645	1.960	2.576	3.291	1.150	1.645	1.960	2.576	3.291

* The values in this table are abstracted from Eisenhart, C., Hastay, M. W., and Wallis, W. A., *Techniques of Statistical Analysis*, Table 2.1, pp. 102–107, 1947. New York: McGraw-Hill Book Company.

FACTORS, k, FOR TWO-SIDED TOLERANCE LIMITS FOR NORMAL DISTRIBUTIONS

n \ Π	$1 - \alpha = 0.95$					$1 - \alpha = 0.99$				
	0.75	0.90	0.95	0.99	0.999	0.75	0.90	0.95	0.99	0.999
2	22.858	32.019	37.674	48.430	60.573	114.363	160.193	188.491	242.300	303.054
3	5.922	8.380	9.916	12.861	16.208	13.378	18.930	22.401	29.055	36.616
4	3.779	5.369	6.370	8.299	10.502	6.614	9.398	11.150	14.527	18.383
5	3.002	4.275	5.079	6.634	8.415	4.643	6.612	7.855	10.260	13.015
6	2.604	3.712	4.414	5.775	7.337	3.743	5.337	6.345	8.301	10.548
7	2.361	3.369	4.007	5.248	6.676	3.233	4.613	5.488	7.187	9.142
8	2.197	3.136	3.732	4.891	6.226	2.905	4.147	4.936	6.468	8.234
9	2.078	2.967	3.532	4.631	5.899	2.677	3.822	4.550	5.966	7.600
10	1.987	2.839	3.379	4.433	5.649	2.508	3.582	4.265	5.594	7.129
11	1.916	2.737	3.259	4.277	5.452	2.378	3.397	4.045	5.308	6.766
12	1.858	2.655	3.162	4.150	5.291	2.274	3.250	3.870	5.079	6.477
13	1.810	2.587	3.081	4.044	5.158	2.190	3.130	3.727	4.893	6.240
14	1.770	2.529	3.012	3.955	5.045	2.120	3.029	3.608	4.737	6.043
15	1.735	2.480	2.954	3.878	4.949	2.060	2.954	3.507	4.605	5.876
16	1.705	2.437	2.903	3.812	4.865	2.009	2.872	3.421	4.492	5.732
17	1.679	2.400	2.858	3.754	4.791	1.965	2.808	3.345	4.393	5.607
18	1.655	2.366	2.819	3.702	4.725	1.926	2.753	3.279	4.307	5.497
19	1.635	2.337	2.784	3.656	4.667	1.891	2.703	3.221	4.230	5.399
20	1.616	2.310	2.752	3.615	4.614	1.860	2.659	3.168	4.161	5.312
21	1.599	2.286	2.723	3.577	4.567	1.833	2.620	3.121	4.100	5.234
22	1.584	2.264	2.697	3.543	4.523	1.808	2.584	3.078	4.044	5.163
23	1.570	2.244	2.673	3.512	4.484	1.785	2.551	3.040	3.993	5.098
24	1.557	2.225	2.651	3.483	4.447	1.764	2.522	3.004	3.947	5.039
25	1.545	2.208	2.631	3.457	4.413	1.745	2.494	2.972	3.904	4.985
26	1.534	2.193	2.612	3.432	4.382	1.727	2.469	2.941	3.865	4.935
27	1.523	2.178	2.595	3.409	4.353	1.711	2.446	2.914	3.828	4.888
30	1.497	2.140	2.549	3.350	4.278	1.668	2.385	2.841	3.733	4.768
35	1.462	2.090	2.490	3.272	4.179	1.613	2.306	2.748	3.611	4.611
40	1.435	2.052	2.445	3.213	4.104	1.571	2.247	2.677	3.518	4.493
45	1.414	2.021	2.408	3.165	4.042	1.539	2.200	2.621	3.444	4.399
50	1.396	1.996	2.379	3.126	3.993	1.512	2.162	2.576	3.385	4.323
55	1.382	1.976	2.354	3.094	3.951	1.490	2.130	2.538	3.335	4.260
60	1.369	1.958	2.333	3.066	3.916	1.471	2.103	2.506	3.293	4.206
65	1.359	1.943	2.315	3.042	3.886	1.455	2.080	2.478	3.257	4.160
70	1.349	1.929	2.299	3.021	3.859	1.440	2.060	2.454	3.225	4.120
75	1.341	1.917	2.285	3.002	3.835	1.428	2.042	2.433	3.197	4.084
80	1.334	1.907	2.272	2.986	3.814	1.417	2.026	2.414	3.173	4.053
85	1.327	1.897	2.261	2.971	3.795	1.407	2.012	2.397	3.150	4.024
90	1.321	1.889	2.251	2.958	3.778	1.398	1.999	2.382	3.130	3.999
95	1.315	1.881	2.241	2.945	3.763	1.390	1.987	2.368	3.112	3.976
100	1.311	1.874	2.233	2.934	3.748	1.383	1.977	2.355	3.096	3.954
110	1.302	1.861	2.218	2.915	3.723	1.369	1.958	2.333	3.066	3.917
120	1.294	1.850	2.205	2.898	3.702	1.358	1.942	2.314	3.041	3.885
130	1.288	1.841	2.194	2.883	3.683	1.349	1.928	2.298	3.019	3.857
140	1.282	1.833	2.184	2.870	3.666	1.340	1.916	2.283	3.000	3.833
150	1.277	1.825	2.175	2.859	3.652	1.332	1.905	2.270	2.983	3.811
160	1.272	1.819	2.167	2.848	3.638	1.326	1.896	2.259	2.968	3.792
170	1.268	1.813	2.160	2.839	3.527	1.320	1.887	2.248	2.955	3.774
180	1.264	1.808	2.154	2.831	3.616	1.314	1.879	2.239	2.942	3.759
190	1.261	1.803	2.148	2.823	3.606	1.309	1.872	2.230	2.931	3.744
200	1.258	1.798	2.143	2.816	3.597	1.304	1.865	2.222	2.921	3.731
250	1.245	1.780	2.121	2.788	3.561	1.286	1.839	2.191	2.880	3.678
300	1.236	1.767	2.106	2.767	3.535	1.273	1.820	2.169	2.850	3.641
400	1.223	1.749	2.084	2.739	3.499	1.255	1.794	2.138	2.809	3.589
500	1.215	1.737	2.070	2.721	3.475	1.243	1.777	2.117	2.783	3.555
600	1.209	1.729	2.060	2.707	3.458	1.234	1.764	2.102	2.763	3.530
700	1.204	1.722	2.052	2.697	3.445	1.227	1.755	2.091	2.748	3.511
800	1.201	1.717	2.046	2.688	3.434	1.222	1.747	2.082	2.736	3.495
900	1.198	1.712	2.040	2.682	3.426	1.218	1.741	2.075	2.726	3.483
1000	1.195	1.709	2.036	2.676	3.418	1.214	1.736	2.068	2.718	3.472
∞	1.150	1.645	1.960	2.576	3.291	1.150	1.645	1.960	2.576	3.291

Table A-19

n	1 − α = 0.75 Π = 0.75	0.90	0.95	0.99	1 − α = 0.90 Π = 0.75	0.90	0.95	0.99
50	5, 5	2, 1			5, 4	1, 1		
55	6, 6	2, 2	1, 1		5, 5	2, 1		
60	7, 6	2, 2	1, 1		6, 5	2, 1		
65	7, 7	3, 2	1, 1		6, 6	2, 2		
70	8, 7	3, 2	1, 1		7, 6	2, 2		
75	8, 8	3, 3	1, 1		7, 7	2, 2		
80	9, 8	3, 3	2, 1		8, 7	3, 2	1, 1	
85	10, 9	4, 3	2, 1		8, 8	3, 2	1, 1	
90	10, 10	4, 3	2, 1		9, 8	3, 2	1, 1	
95	11, 10	4, 3	2, 1		9, 9	3, 3	1, 1	
100	11, 11	4, 4	2, 1		10, 10	3, 3	1, 1	
110	12, 12	5, 4	2, 2		11, 11	4, 3	2, 1	
120	14, 13	5, 5	2, 2		12, 12	4, 4	2, 1	
130	15, 14	6, 5	3, 2		13, 13	5, 4	2, 1	
140	16, 15	6, 6	3, 2		14, 14	5, 5	2, 2	
150	17, 17	6, 6	3, 3		16, 15	5, 5	2, 2	
170	20, 19	7, 7	4, 3		18, 17	6, 6	3, 2	
200	23, 23	9, 8	4, 4		21, 21	8, 7	3, 3	
300	35, 35	13, 13	6, 6	1, 1	33, 32	12, 11	5, 5	
400	47, 47	18, 18	9, 8	2, 1	45, 44	16, 16	8, 7	1, 1
500	59, 59	23, 22	11, 11	2, 1	57, 56	21, 20	10, 9	1, 1
600	72, 71	28, 27	13, 13	2, 2	68, 68	26, 25	12, 11	2, 1
700	84, 83	33, 32	16, 15	3, 2	80, 80	30, 30	14, 14	2, 2
800	96, 96	37, 37	18, 18	3, 3	92, 92	35, 34	16, 16	3, 2
900	108, 108	42, 42	21, 20	4, 3	104, 104	40, 39	19, 18	3, 2
1000	121, 120	47, 47	23, 22	4, 4	117, 116	44, 44	21, 20	3, 3

n	1 − α = 0.95 Π = 0.75	0.90	0.95	0.99	1 − α = 0.99 Π = 0.75	0.90	0.95	0.99
50	4, 4	1, 1			3, 3			
55	5, 4	1, 1			4, 3			
60	5, 5	1, 1			4, 4			
65	6, 5	2, 1			5, 4	1, 1		
70	6, 6	2, 1			5, 5	1, 1		
75	7, 6	2, 1			5, 5	1, 1		
80	7, 7	2, 2			6, 5	1, 1		
85	8, 7	?, 2			6, 6	2, 1		
90	8, 8	3, 2			7, 6	2, 1		
95	9, 8	3, 2	1, 1		7, 7	2, 1		
100	9, 9	3, 2	1, 1		8, 7	2, 2		
110	10, 10	3, 3	1, 1		9, 8	2, 2		
120	11, 11	4, 3	1, 1		10, 9	3, 2		
130	13, 12	4, 4	2, 1		11, 10	3, 3	1, 1	
140	14, 13	4, 4	2, 1		12, 11	3, 3	1, 1	
150	15, 14	5, 4	2, 1		13, 13	4, 3	1, 1	
170	17, 16	6, 5	2, 2		15, 15	5, 4	2, 1	
200	20, 20	7, 6	3, 2		18, 18	6, 5	2, 2	
300	32, 31	11, 11	5, 4		29, 29	10, 9	4, 3	
400	43, 43	15, 15	7, 6		40, 40	14, 13	6, 5	
500	55, 54	20, 19	9, 8	1, 1	52, 51	18, 17	7, 7	
600	67, 66	24, 24	11, 10	1, 1	63, 63	22, 22	9, 9	
700	78, 78	29, 28	13, 13	2, 1	75, 74	26, 26	11, 11	1, 1
800	90, 90	33, 33	15, 15	2, 2	86, 86	31, 30	13, 13	1, 1
900	102, 102	38, 37	18, 17	2, 2	98, 97	35, 35	15, 15	2, 1
1000	114, 114	43, 42	20, 19	3, 2	110, 109	40, 39	18, 17	2, 1

* The data of this table are abstracted with kind permission from Somerville, Paul N., "Tables for Obtaining Non-Parametric Tolerance Limits," *Annals of Mathematical Statistics*, 29, pp. 599–601, 1958. Π is the minimum proportion of the population to be covered and 1 − α is the level of confidence.

Index